THE THEORY OF PHILOSOPHICAL CONSCIENCISM
Practice Foundations of Nkrumaism in Social Systemicity

Kofi Kissi Dompere

Adonis & Abbey Publishers Ltd
St James House
13 Kensington Square,
London, W8 5HD
United Kingdom

Website: http://www.adonis-abbey.com
E-mail Address: editor@adonis-abbey.com

Nigeria:
Suites C4 & C5 J-Plus Plaza
Asokoro, Abuja, Nigeria
Tel: +234 (0) 7058078841/08052035034

British Library Cataloguing-in-Publication Data
A catalogue record for this book is available from the British Library

ISBN: 9781909112667 (Paper Back)
       9781909112735 (Hard Back)

THE THEORY OF PHILOSOPHICAL CONSCIENCISM
Practice Foundations of Nkrumaism in Social Systemicity

Kofi Kissi Dompere

ADONIS & ABBEY
PUBLISHERS LTD

Table of Contents

CHAPTER THREE
CHARACTERISTIC-BASED ANALYTICS, A COUNTRY'S IDENTITY, SOCIAL ACTIONS AND PHILOSOPHICAL CONSCIENCISM IN SOCIAL SYSTEMICITY

CHAPTER FOUR
PHILOSOPHICAL CONSCIENCISM AND GAMES IN THE CATEGORIAL CONVERSIONS OF SOCIAL POLARITIES WITHIN SOCIAL SYSTEMICITY

DEDICATIONS

To all scholars, researchers, supporters and authors in the reference list whose efforts and hard works have made the work in this monograph possible, less tasking and more enjoyable.

This monograph has benefited from the works of Drs. Kwame Nkrumah, Cheik Anta Diop, and W. E. Abraham, Carter G. Woodson, W. E.B. DuBoise, Theophile Obenga, Schwaller de Lubicz, Molefi Asante, and all those who have worked and are working in the Africentric paradigm of thought seeking a unified approach to the understanding of human progress and its interactive conditions with nature by their ways of thought and motivation. This book is dedicated to them and more.

ACKNOWLEDGEMENTS

The general theory of decision-choice system is here viewed in its role to effect categorial conversions through the operative force of intentionality which, affects all areas of human action and the practice to bring about social transformations. The understanding of the path of socio-natural transformations is very important to the success of progressive socio-natural transformations under intentionality. This intentionality is both subjective and objective requiring different forms of analytics and multidisciplinary approaches where each approach is an illumination of an aspect of the transformation problem. New approaches are required. Here, I simply express my appreciation and thanks to all those practicing the principle of doubt by refusing to be imprisoned in the walls of ideological and scientific credulity that depletes energies from creative imagination due to a choice to remain in the comfort zone of intellectual ignorance and analytical arrogance. Under these conditions, I express my thanks to all working under the principle of doubt. This monograph has benefited from the contributions of those who I do agree and do not agree as seen from the references. I express my gratitude to all the authors in the list. I express my thanks to Professor Akwasi Osei whose encouragement provided a continual energy to bring the work to a finish. My great appreciation also goes to Mr. Francis Kwarteng for his substantial interest which provided me with inspirations to combine philosophy and mathematics to solve the problem of Philosophical Consciencism and its linkage to that of categorial conversion. I also thank Ms. Jasmin Blackman for all her proofreading as well as her interest in understanding the core of the analytical process. All errors in this monograph rest with me.

PREAMBLE

The new Egyptological ideology, born at the opportune moment, reinforced the theoretical bases of imperialist ideology. That is why it easily drowned out the voice of science, by throwing the veil of falsification over historical truth. This ideology was spread with the help of considerable publicity and taught the world over, because it alone had the material and financial means for its own propagation.

Thus imperialism, like the prehistoric hunter, first killed the being spiritually and culturally. Before trying to eliminate it physically. The negation of the history and intellectual accomplishments of Black Africans was cultural, mental murder, which preceded and paved the way for their genocide here and there in the world.... [R 1.86, pp. 1-2]

This is also the place to say that no thought, and particularly no philosophy, can develop outside of its historical terrain. Our young philosophers must understand this, and rapidly equip themselves with the necessary intellectual means in order to reconnect with the home of philosophy in Africa, instead of getting bogged down in the wrong battles of ethnophilosophy. By renewing ties with Egypt [Back] we soon discover an historical perspective of five thousand years that makes possible the diachronic study, on our own land, of all the scientific disciplines that we are trying to integrate into modern African thought [R1.86, p.4].

Only the implanting of such scientific discipline (African Egyptology) in Black Africa would, one day, lead to the grasping of the richness and the novelty of the cultural conscience that we want to awaken, its quality, its depth, its creative power [R1.86, p.6].

Today each group of people, armed with its rediscovered or reinforced cultural identity, has arrived at the threshold of the postindustrial era. An atavistic, but vigilant, African optimism inclines us to wish that all nations would join hands in order to build a planetary civilization instead of sinking down to barbarism [R1.86, p. 7].

What is called for as a first step is a body of connected thought which will determine the general nature of our action in unifying the society which we have inherited, this unification to take account , all times, of the elevated ideals underlying the traditional African society. Social revolution must therefore have, standing firmly behind it, an intellectual revolution, a revolution in which our thinking and philosophy are directed towards the redemption of our society. Our philosophy must find its weapons in the environment and living conditions of African people. It is from those conditions that the intellectual content of our philosophy must be created [R1.203, p. 78].

When we say that the traditional African view of the world is one of extraordinary harmony, then except for the word 'African' every single word in the sentence is both right and wrong. For in the first place the traditional world view is still alive today; secondly it is a question not of a world view in the European sense, since things that are contemplated, experienced and lived are not separable in it; thirdly it can be called extraordinary only in the European sense, while for the African it is entirely commonplace; and fourth, the expression 'harmony' is entirely inadequate, since it does not indicate what parts are being harmonized in what whole. And if we say 'everything' is harmonized, that tells us less than ever [Janheinz Jahn, Muntu, quoted in R1.8, p.11].

Mythology has had an almost limitless descent. It was in a savage or crudely primitive state in the most ancient Egypt, but the Egyptians who continued to repeat the myths did not remain savages. The same mythical mode of representing nature that was probably extant in Africa 100,000 years ago survives to-day among races who are no longer the producers of the Myths and Marchen than they are of language itself [R1.179, vol. I, p.5].

By indivisible duality (usually contracted to "individuality", it is obviously meant that the two modes of being are complementary halves. ... (The Dualization of Absolute Being resides simultaneously in Subjective and Objective Realms within the qualitative-quantitative dispositions from the diagram). This duality manifests itself in all areas, and on all levels of being as a major organizing force. In order to understand God, ourselves, the world and life we must be able to identify, understand, and live in harmony with the dualizing shaping forces of life.

In the Subjective Realm the duality manifests itself, on one hand, as Consciousness, Will (two polarities of the same reality), and on the other, as energy/matter. The former is referred to in the Kabalistical tradition as Ain, and Amen in the Kamitian. The latter is Soph, and Nu/Nut in the Kabalistical, and the Kamitian traditions respectively [R1.8, Vol. I, pp. 77-78].

The Tree of Life shows that the creative progress of the world is based on a plan in which all the things in the world are modifications of one and the same material substance and Being. Although they are different in their needs, mode of existence, and appearance, they are all parts of One Whole. The equilibrium between this oneness at the top and difference on the bottom, must be maintained [R1.8a, Vol. I, p. 81].

PREFACE

Brief and powerless is Man's life; on him and all his race the slow, sure doom falls pitiless and dark. Blind to good and evil, reckless of destruction, omnipotent matter rolls on its relentless way; for Man, condemned to-day to lose his dearest, to-morrow himself to pass through the gates of darkness, it remains only to cherish, ere yet the blow falls, the lofty thoughts that ennoble his little day; disdaining the coward terrors of the slave of Fate, to worship at the shrine that his own hands have built; undismayed by the empire of chance, to preserve a mind free from the wanton tyranny that rules his outward life; proudly defiant of the irresistible forces that tolerate, for a moment, his knowledge and his condemnation, to sustain alone, a weary but unyielding Atlas, the world that his own ideals have fashioned despite the trampling march of unconscious power [R12.24b, p. 87].

I The Basic Structure of the Monograph

It has been argued under the theory of categorial conversion that the socio-natural elements are on continual self-transformations in an infinite order. In this process of self-transformation, every ontological element resides in temporary qualitative equilibrium where the element is continually going through structural changes of its internal self-arrangements and identity due to its *internal decision-choice activities* in the time continuum. The supporting principle of temporary qualitative equilibrium of categorial elements is that every categorial element exists under the conditions of *actual-potential polarity* in relational continuum and unity, where there is an interchangeable messaging process between the actual and the potential. The potential is logically composed of infinite elements that can be actualized depending on the experiential information structure that defines the possibility and probability spaces.

The problem of the general and technical understanding of the never-ending dynamics of qualitative dispositions in relation to the structure of quantitative dispositions is what is called the *transformation problem*. The explanatory and prescriptive epistemic systems of the transformation constitute the *theory of transformation* of socio-natural elements and categories. The theory of transformation is about automatic self-transformations of ontological elements. It presents itself as two interdependent theories in relational continuum and unity. The two theories are the *theory of categorial conversion* and the *theory of Philosophical Consciencism*. The theory of categorial conversion presents the *necessary conditions* for socio-natural transformations. These conditions constitute the necessary conditions of convertibility of categorial elements. They establish the kind of decision-choice instruments that must be created for internal command and control of the transformation process. The conceptual existence of categories for categorial dynamics must be established on the basis of qualitative structural characteristic sets and sub-sets by the development of a sub-theory of *category formation*. Here, each ontological element or each epistemological element is placed in a category to create a family of families of categories. The existence of categories opens up an epistemic room for the study of categorial dynamics on the basis of internal decision-choice activities of categorial elements, where qualities are internally transformed.

To induce forces of qualitative motion and transformation from within there must be a process to generate the required internal energy. In this respect, each categorial element does not only reside as actual-potential polarity, but it is also postulated to exist as a duality under the principle of opposites with relational continuum and unity at all points of time. Within this epistemic framework, the ontological space is seen as a complex family of families of actual-potential polarities with residing dualities. The actual pole has its own residing duality and so also does the potential pole of the polarity. By establishing the conditions of an actual-potential polarity, a logical process is opened up in the epistemological space where the actual can be logically shown to fade into the potential by self-destruction, and the potential can also be shown to become the new actual by self-creation through the internal processes.

It is an imperative of the epistemic construct of the dynamic process to understand and establish the relational structure between the *polarities* and *dualities*. The polarities and dualities exit in an inseparably relational unities. The residing duality in each of the poles is made up of *negative dual* and *positive dual*. The negative and positive duals in the duality create internal contradictions and conflicts between the duals for the qualitative instability of the existence of the identity of categorial element. The negative dual is made up of the negative characteristic sub-set while the positive dual is made up of the positive characteristic sub-set. The existence of negative and positive duals to form a duality in the same categorial element creates not only contradictions and conflicts, but the conflicts and contradictions are transformed into a complex qualitatively dynamic game, where the command and control of the necessary conditions are the instruments that must be weaponized for the continual dynamic game actions under the experiential information structure to create the sufficient conditions for categorial conversion that is made up of internal and external information sub-structures. The payoff of the game is simply power and dominance to shape the direction of conversion as may be dictated by the sufficiency conditions. On the aggregate, the necessary conditions are established by the theory of categorial conversion as the *categorial moment, categorial transfer function* and *categorial transversality conditions.*

The categorial moment and categorial transfer function must be weaponized by either the positive dual to create power and dominance and establish the identity at its will, or the negative dual to create power and dominance and establish the identity of the categorial element. The payoff of power and dominance by either the positive dual or the negative dual is in relation to acceptance of the actual or the destruction of the actual and the actualization of a potential through the internal development of strategic command and control processes. The forces of the internal development must be shown to be the products of decision-choice activities acting on the categorial experiential information structure. It is important to follow the logical path of the theory of the transformation dynamics. This theory is isomorphic to the theory of info-dynamics in [R9.12]. The theory of info-dynamics is an extension of the general theory of information where old varieties are destroyed and new varieties are created [R9.11] in never-ending process. The epistemic geometry of the logic of self-transformation is provided in the main text of the monograph. This theory is general and applicable to the theory of economic development and the general theory of engineering. It also offer a framework in understanding life and death as self-transformation process. The core of self-transformation is the game between the positive and negative duals in relational continuum and unity for continual negation of negation or conversion of conversion under creative conflicts.

Given the necessary conditions of self-transformation, sufficient conditions for convertibility must be established to show how the internal power combined with the experiential information structure to create a categorial moment and transfer function to bring about categorial conversion by satisfying the categorial transversality conditions. The knowledge of the sufficiency of self-transformation in theory and practice is about critically examining the *controllability conditions* of the internal self-transformation process and how these controllability conditions are connected to the necessary conditions of the convertibility of the categorial elements. The theory that explains or prescribes the conditions of controllability of convertibility of categorial elements is the *theory of Philosophical Consciencism*. The concept of Philosophical Consciencism was advanced and reflected on by Nkrumah in his conceptual system [R1.202]. The current monograph and the one in [R1.90b] are used to develop the required theoretical framework for the understanding of general transformation in theory and application. The theory of Philosophical Consciencism is about organization of internal decision-choice activities and management of command and control to create actions and bring about a change in categorial varieties. The input of the decision-choice activities is knowledge which is produced by the internal mechanism that combines internal ideology with internal capacity for processing elemental experiential information structure.

The theory of Philosophical Consciencism lays down the conceptual map to guide the management of command and control of competitive strategic decision-choice actions in the dynamic game of negation of negation in socio-natural transformations through the residing dualities of the *actual pole* of the actual-potential polarity. The duality is composed of the negative and positive duals under contradictions and conflicts seeking for relative power and dominance. In correspondence with the duality under relational continuum and unity within the conceptual framework of the principle of opposites, the negative dual is equipped with *negative Philosophical Consciencism* while the positive dual is equipped with *positive Philosophical Consciencism*. The negative Philosophical Consciencism provide the map for the development of negative strategic decision-choice actions for the negation-of-negation game process in the actual-potential polarity. In fact. it guides the creation of the necessary conditions as controlled instruments for weaponization that will be under command and control of the negative dual to acquire strategic power and dominance to either maintain the conditions of the actual or destroy the conditions of the actual, create a *categorial power vacuum* and bring in a new actual. In other words, the negative Philosophical Consciencism is involved in creating conditions of instrumentation, instrumentality and controllability of the categorial conversion game through the negation-of negation of strategies and counter strategies.

Similarly, the positive Philosophical Consciencism provides the map for the development of positive strategic decision-choice actions for the negation-of-negation game process in the actual-potential polarity. It also guides the creation of the necessary conditions as controlled instruments for weaponization that will be under the command and control process of the positive dual to acquire strategic power and dominance to either maintain the conditions of the actual or destroy the conditions of the actual, create a categorial power vacuum and bring in a new actual. In other words, the positive Philosophical Consciencism like negative philosophical Consciencism is involved in creating conditions of *instrumentation, instrumentality* and *controllability* of the categorial-conversion game through the negation-of negation of strategies and counter strategies. The decision-choice strategies in designing command and control of the essential instruments of the categorial-conversion game is strategically opposite for the duals even though they reside in the same element. This relational structure of continuum and unity is beautifully captured by a symbolic representation called *funtummreku-denkyemmireku* (two-headed crocodile with one stomach). Here, the two heads struggle against one another to eat into the same stomach that supports them. This is another conceptual representation of duality with relational continuum and unity. Notice that the two heads are the negative and positive duals that constitute the duality in relational continuum and unity. The relational continuum and unity are represented by the stomach. The fact that they fight against each other also indicates that they are under contrary principles under contrary maps of Philosophical Consciencism for action and reaction. The theory of categorial conversion indicates the necessary socio-natural policy instruments that must be created for the command-control dynamics to bring about the transformation in the actual-potential polarity. The theory of Philosophical Consciencism indicates the goal-objective purpose, creates the necessary policy instruments as indicated in the theory of categorial conversion and guide the use of these policy instruments in the command-control dynamics to achieve the desired outcome. In a simple epistemic process, the theory of transformation is about the understanding of creation-destruction process of varieties in actual-potential socio-natural polarities and how this understanding will provide a decision-choice framework for the creation and destruction of social varieties to induce human progress. How general is the theory of Philosophical Consciencism?

II The Generality of Philosophical Consciencism in the General Mechanism of Transformation Dynamics

Let us try to offer a structural framework for the generality of the theory of Philosophical Consciencism as seen in relation to categorial conversion and socio-natural transformations. The generality of the theory of Philosophical Consciencism may be conceptually conceived in term of the principle of opposites with relational continuum and unity for all ontological elements. Just as the theory of categorial conversion deals with necessary conditions applicable to transformations of all socio-natural categorial elements, so also the theory of Philosophical Consciencism deals with sufficient conditions applicable to transformations of all socio-natural categorial elements. The generality is such that there is a conception of the existence of *natural decision-choice processes* which deal with the manufacturing of internal-controlled elements in nature, and the management of the command and control processes to bring about categorial conversions of ontological actual-potential polarities. The natural decision-choice processes involving the manufacturing of controlled tools to be weaponized for command and control in the strategic formation within the game of negation of negation or conversion and counter-conversion for strategic power and dominance is called *natural Philosophical Consciencism.* The generality of the theory will not be complete if it is not applicable to the transformations of social categorial elements within the family of social categories. The generality is extended to social categories in a way where there are *social decision-choice processes* that deal with the manufacturing of controlled elements in society, and the management of the command and control processes to bring about categorial conversions of epistemological actual-potential polarities. The social decision-choice processes involving the manufacturing of controlled elements to be weaponized for command and control in the strategic formation within the game of negation of negation or conversion and counter-conversion for strategic power and dominance is called *social Philosophical Consciencism.*

The combination of the natural and social types of Philosophical Consciencism constitutes the organic Philosophical Consciencism. The epistemic framework for the understanding of the working mechanism of Philosophical Consciencism is the *theory of Philosophical Consciencism.* Its basic premise is that every socio-natural transformation is decision-choice dependent conditional on internal preferences, the elemental experiential information structure and the logical mechanism for processing information into knowledge. It, therefore, deals with how the necessary conditions for categorial conversion of socio-natural polarities in the actual-potential space are given sufficiency for qualitative transformations by means of activities in the decision-choice space, where conflicts arise due to opposing preferences for the *categorial identity* by the duals of the duality. The game activities of the duals are to seek *categorial power* which is translated into *categorial dominance* to define the nature of the identity of the categorial element. The payoff of the game is thus power and dominance for decision-choice activities by the duals with opposing preferences for the actual which determine the identity of the categorial element. The whole transformation activities may be seen in terms of a pursue-avoidance control dynamic game with dualistic experiential information structures, where there is a never-ending process of learning and knowing by the duals in the game with irreversible outcomes and continual amplification of the experiential information structure. This continual amplification and updating of the experiential information structure leads to stock-flow disequilibria of information [R9.12]. The pursue-avoidance control dynamics is such that either the negative dual is pursuing to control the positive dual to acquire power and establish the identity of the element by dominance, or the positive dual is pursuing to control the negative dual to acquire power and establish the identity of the element by dominance.

By the principle of opposites, the categorial-conversion games are such that the decision-choice activities of the duals act as mutual constraints on the mutual search for power and dominance to establish the identity of the categorial element. The problems of the qualitative categorial conversions with relational continuum and unity cannot logically be posed with solutions in a logical dualism with excluded middle from the classical paradigm and the corresponding laws of thought and mathematics where the qualitative values are restricted to negative, positive and zero. One of the greatest ironies of the classical laws of thought in denying contradictions through the principle of excluded middle on the acceptance of the principle of non-contradiction is the failure to acknowledge that the whole theory of constrained optimization is based conceptually on the acceptance of the principle of opposites, where the objective set and the constraint set are in contradiction with one another in a restricted space of decision-choice action under duality with relational unity and continuum. These relational continuum and unity of constrained optimization may be seen as a relational continuum and unity when one considers costs-benefit decision-choice problems of categorial elements whose internal structure is defined by costs and benefits. The problem of existence of both costs and benefits in a relational continuum and unity for the same decision-choice element has been described as the *Asntrofi-anoma* problem, where the benefits cannot be selected without the costs and vice versa irrespective of the beneficiary and the cost bearer [R1.92] [R4.17][R.4.18]. It is also this relational continuum and unity of benefits and costs of the same element that provides the logical credibility of cost-benefit analysis and cost-benefit rationality in the decision-choice space. It is easy to see that in general that costs and benefits exist in contradictions and conflicts with relational continuum and not in excluded middle. In the theory of constrained optimization, the logical structure does not eliminate contradiction but assesses and defines it as a constraint in the system. It is this acceptance that allows the logical discussion of the problem of duality in socio-natural transformations. The problem of duality is substantially simplified when it is posed as a space-time problem where a solution may be abstracted approximately with classical mathematics and analyzed within the information space with use of the classical paradigm of thought.

III The Audacity of Motivation and a Change of the Paradigm of Thought

The theory of Philosophical Consciencism, like that of the theory of categorial conversion has motivations and reasons for its development in the construction-reduction duality in the epistemological space where the set of ontological conditions is taken as the identity. The set of motivations for the development of the theory of categorial conversion as a sub-theory of the general theory of transformation has been discussed in the relevant monograph [R1.90b]. Essentially, it is motivated by a need to search for required conditions of the theory of categorial conversion that will establish the general necessary conditions for socio-natural transformation. Given the success of abstracting the general necessary conditions, a question arises as to what are the sufficient conditions to complete the needed epistemic structure of the theory of socio-natural transformations that incorporates the internal transactions of any of categorial elements within a specific category. The need to

xxii

search for the general sufficiency conditions is the general underlying motivation for the development of the theory of Philosophical Consciencism as a sub-theory of the general theory of transformation. An added motivation is to put and show the relevance of Philosophical Consciencism in Nkrumah's conceptual system to the general theory of transformation of socio-natural elements. The sufficiency conditions as derived by the theory of Philosophical Consciencism bring into focus the decisive effects of the internal information-decision-choice interactive processes as generated by socio-natural duals of the dualities in causing intra-categorial and inter-categorial conversions transmitted to the poles of the socio-natural polarities. The internal decision-choice actions, conditional upon the experiential information structure and the internal mechanism of information processing, are the driving force of transformation.

The initial motivation to work on this monograph started from the period of my graduate studies in economics at Temple University in order to understand the process of social transformation, cultural difficulties, social problems and possible resolutions and solutions. The motivation stayed with me during my period of professorship at Howard University. The motivation was to search for factors that can help to explain the rise of different social formations as well as help to define the path of knowing in relation to social transformations. Here, it must be understood that knowing itself is a process under a continual self-transformation through a self-corrective mechanism at each round of claimed knowledge and within the knowledge-ignorance duality under contradictions and tensions with a relation continuum and unity, generally defined by the Africentric principle of opposites. The general theory of socio-natural transformation has special logical difficulties and epistemic problems in all attempts to connect the epistemological space to the ontological space in order to claim understanding and knowledge. The resolutions to the logical difficulties and the solutions to the epistemic problems are added elements which provided the motivation to the final conception and development of the monograph and its companion, to try to construct a general epistemic structure that will move general cognition to bring some light to the dark chambers of cognitive familiarity and pessimism, for a change of paradigm that maintains the same structure of reasoning in the logical walls of familiarity, where quality characteristics are difficult to handle. These motivations drove me to examine the explanatory and prescriptive theories as applied to economic development in terms of its epistemic structure viewed in the framework of transformation that must deal with the problems of handling the simultaneous existence of dynamics of quantitative and qualitative dispositions [R15.14] [R15.38]. The motivations and the problems encountered propelled me to work on paradigms of thought where I discussed the classical paradigm and the fuzzy paradigm of thought and their applications to sciences in general including mathematics and decision-choice rationality [R3.7] [R3.8] [R3.9] [R3.10] [R3.13]. It may be useful to keep in mind that all activities including development of language, knowledge construction, knowing, teaching, learning and other cognitive actions over the epistemic space are decision-choice dependent. The effectiveness of the decision-choice structure depends on Philosophical Consciencism that provides the logical framework for individual and collective actions.

Armed with an audacity of motivation and philosophical-scientific curiosity, I tried to examine the scientific and logical contributions of Marx [R15.34] [R7.23] [R15.37], Engels, [R15.19] Schumpeter [R15.45] [R15.46]], Lenin [R12.63][R15.24][R15.25] and Rostov [R15.45] [R15.46]] to see what they offer from their works and understanding of dialectics, contradictions and transformations in the quality-quantity-time space and how their analytical systems oppose or lend support to Nkrumah's analytics. For comparative analysis among Marx, Schumpeter and Rostov in relations to theories of development see the discussions in [R15.14]. It is noted that elemental transformations do not come by themselves neither does the passage of time alone bring about changes in the quality-quantity space. Any transformation is the result of battles of opposite to resolve contradictions that create internal conflicts in the categorial elements. The resolutions are the results of internal decision-choice activities and the mechanism of how these decision-choice activities are translated into actions by the opposites which are the duals of the duality. The theory of internal transformation must be seen in terms of game actions of the opposites. In this respect, the works mentioned above are examined as to how they may be related to strategic decision-choice actions and games from within the internal struggle of individual self-changes of categorial identity. Disappointedly, the works offered no meaningful logical connection to establish such a game theoretic framework or decision-choice framework.

These works placed emphasis on the use of the instruments of the necessary conditions as if these instruments can bring about categorial conversion. The utilization of the decision-choice system in the creation of the instruments of necessary conditions of transformations of varieties is neglected. Similarly, the utilization of the decision-choice system for the management of command-control actions to create the sufficient conditions for internal self-transformation is also neglected. The theory of categorial conversion is an explanatory theory of socio-natural forces which points to the necessary conditions of transformations of categorial varieties. The theory of Philosophical Consciencism is prescriptive theory of how to create the internal sufficient conditions from the necessary conditions to bring about the self-transformations of varieties. The theory of categorial conversion is rational foundations of all transformations of categorial varieties, and the theory of Philosophical Consciencism is the practice foundations of all transformations of categorial varieties. Nkrumah's analytics and practice placed emphasis on the joint operations of categorial conversion and Philosophical Consciencism with the greater stress on Philosophical Consciencism particularly as applied to social transformation.

The preoccupation with ideology, social-class problems and conditions of the battle field of politics, capitalism, imperialism and communism prevented the aforementioned great minds to develop a self-contained theory for a general mechanism of socio-natural transformation. They failed to create a logical structure that must connect the mechanism of change to a set of factors such as information, classification of universal elements, language, decision-choice activities, action space, games, ideology and power of dominance such as colonialism, religious subjugation and racism where transformations find expressions and elemental identities in qualitative differences. This epistemic failure is amplified by the lack of initial conditions, intermediate identities, final destinations and conditions to indicate when the transformation has occurred with a change in categorial identity of the element.

Here, the major epistemic principle is that any transformation is a qualitative phenomenon that must be related to the effects of changes of quantitative disposition. It is this lack of abstracting a possible self-contained logical structure from the above mentioned works that provided an added motivation to examine the critical contemplations of the theoretical foundations for a possible development of a general epistemic framework that may be used for the development of a general theory of socio-natural transformation that fits into the methodological construction-reduction duality. The critical contemplation points to an epistemic structure that must link the epistemological space to the ontological space, where quality and quantity of ontological elements exist in a relationally inseparable unit. The audacity of motivation resides in an epistemic attempt to establish a logical linkage between categorial conversion and Philosophical Consciencism in order to maintain the relational continuum and unity in the quantity-quality duality. This cognitive motivational audacity is a driving force to define a necessary framework for the development of *qualitative mathematics* where qualitative equation of motion may be abstracted for the construct of a theory of qualitative dynamics of varieties.

It is this critical contemplation that brought into focus the motivation to reread Nkrumah's works not in terms of politics and ideology but in terms of their scientific contents and compare his approach to the approaches of Marx,[R15.34] [R15.35] [RI5.36] [R15.37], Engels [R7.23][R15.19][R15.20],Schumpeter,[R15.45][R15.46][RI5.47][R1 5.48][R15.49],Lenin[R15.24] [R15.25], Mao [R12.63] and others. The frameworks of Marx, Engels, Schumpeter, Linin, Mao and others are helpful in illuminating different sides of the complex process of transformation. The illuminations are good for general understanding but they lack the satisfaction of the scientific requirement of establishing the initial conditions, the conditions of intermediate identities and terminal points for the dynamic process of transformations of the ontological elements. Nkrumah's conceptual framework seems to offer a possible alternative path to construct the complex theory of transformation of socio-natural elements. Viewed as internal self-transformation in terms of decision-choice activities Nkrumah set up to develop two sets of interdependent analytical concepts which offer an epistemic framework for the development of a set of necessary and a set of sufficient conditions for socio-natural transformations in quality-quantity duality where behind every qualitative disposition is a quantitative disposition within the dynamics of all actual-potential polarities. His introduction of analytical concepts of category, categorial conversion, primary category, derived category, quantitative disposition, qualitative disposition and his logical power to link these concepts to methodological constructionism, reductionism and nominalism provide both philosophical and scientific frame to develop axioms for the decision-choice complexity of transformations.

These concepts provide a good starting point into the development of a general theory of transformation of socio-natural elements from within. The concept of *category* provides a notion of similarity and difference in which elements as varieties may belong. The concept of *categorial conversion* indicate a process where one categorial element changes form and enters into another category. The concepts of *qualitative and quantitative dispositions* introduce the *African conceptual duality* which exists in relational continuum and unity such that quantity is quality and quality is quantity in some complicated appearances in the universal system of things. The quantitative and qualitative dispositions provide us with a way to form categories and examine categorial differences and similarities at a state and in time when transformation occurs within the elements in the quantity-quality duality. The concept of *primary category* initializes the time-process of any transformation while that of the *derived category* indicates the intermediary categories of the temporary final transformations where such categorial elements exist in temporary qualitative equilibrium. The concept of *nominalism* provides a linkage between ontological elements and linguistic vocabulary on the basis of which experiential information structure is developed, a paradigm is created, and explanatory and prescriptive structures are formed for validity through the concept of reductionism in verifying the events in the primary-derived duality in relational continuum and unity. The audacity of motivation under the principle of epistemic doubt allowed a development of important linkage between categorial conversion and Africentric principle of opposite composed of polarity and duality in relational continuum and unity.

It is importantly useful to notice Nkrumah's use of African traditional conceptual system involving the principle of opposites where the Africentric concept of duality existing in a relational continuum and unity is logically employed at every opportune time. Here transformation comes about through the struggle of opposites to acquire power and dominance to establish categorial identity. To illustrate the power conflicts in the struggle of opposites under the concept of duality and to show the permanency of contradictions in all things, Nkrumah introduced the concepts of *negative and positive actions* as an aggregate strategy of transformation available to the duals of the duality in the game of negation of negation for categorial dominance. The African conceptual system of the principle of opposites within the structural continuum and inseparability allow the theory of socio-natural transformation to be linked with game theory where every categorial element is in temporary equilibrium and experiential information structure is continually in stock-flow disequilibrium. These concepts and Nkrumah's discussions allow one to construct the theory of categorial conversion with philosophical and mathematical rigor within the domain of science from which the *necessary conditions* of transformation of socio-natural categorial elements are established. The understanding of the theory of transformation is of general interest to everyone in both academy and the general public as such that the mathematics and philosophical elegance have been reduced to a minimum and bare essentials. In doing so the internal logical consistency and self-containment as well as the epistemic beauty have not been compromised. The logical connectivity of the theory of categorial conversion and the African principle of opposite where conflicts between the duals are brought to the fore.

The necessary conditions for the theory of transformation as established by the theory of categorial conversion under the Africentric principle opposites indicate the instruments that must be internally created then controlled through the management of a command-control process to bring about either a negative or positive action. In other words, the theory that created the necessary conditions must be supported by a theory that will indicate and establish the *sufficiency conditions* to complete the general theory of transformation of socio-natural elements. This theory to construct the sufficient condition of transformation must also be derived on the basis of Africentric principle of opposite. To develop a theory that will allow the sufficiency conditions to be derived, Nkrumah observed that the sufficiency conditions must be derived from the internal decision-choice activities conditional upon the elemental experiential information structure of categorial elements within actual-potential polarities and the residing dualities. The required theory must be related to the decision-choice behaviors of the duals in their strategic relation to acquire power and dominance. In this respect, the theory must account for the strategies and counter strategies of the duals in a game setting where each dual works to produce the necessary instruments and design the management of the command-control process. The theory must be designed to solve strategic decision-choice relational problems between the duals, and resolve categorial crises in the duality. This was recognized by Nkrumah. It was through this recognition, that Nkrumah reasoning within the

African traditions of the principle of opposites, introduced negative and positive action under the control of Philosophical Consciencism as a guide to deal with the internal decision-choice activities, where an epistemic link must be developed and the theory must be developed to relate the negative action to the negative dual and the positive action to the positive dual.

The manner in which Nkrumah's Philosophical Consciencism constituted a framework of thought to guide the internal action fields and decision-choice processes in the duality must be shown by a development of the *theory of Philosophical Consciencism* for the internal decision-choice processes to manufacture the required necessary instruments, and then weaponize them for a strategic optimal control game to acquire power and dominance by either the negative or the positive dual. The development of the required theory is the audacity of motivation and curiosity in the work of this monograph. The theory of Philosophical Consciencism is developed in this monograph to give an epistemic support to Nkrumah's concept of Consciencism as a general foundational system that must guide categorical conversion in all decision-choice spaces. In this respect, the theory of Philosophical Consciencism links collective conscience to individual and collective decision-choice actions and then to transformations under the principle of intentionality of conscience where such conscience resides in the experiential information structure. With the theory Philosophical Consciencism, the sufficient conditions will be indicated to provide the understanding of how elemental policies may be designed for natural and social interventions through social decision-choice actions by mimicking the natural internal decision-choice actions.

The social decision-choice actions are reflected in social formations and engineering of all forms. In this respect, the theory of transformation is seen in terms of the *zero-sum game* between the internal negative and positive duals of all the categories of the socio-natural dualities and then mapped into all the categories of the socio-natural actual-potential polarities where there is never-ending organic control processes under the principle of opposites in relational continuum and unity. The conceptual and application meaning of the zero-sum game finds expression in the negation-of-negation or conversion-of-conversion process where the positive gain of one pole is exactly the negative gain of the other pole in power and dominance to establish the categorical identities of varieties. In this framework, the decision-choice agents have little control over perpetual transformation as it relates to the disappearance of the old and the emergence of the new. Their only control is to influence the potential element that may be actualized. An important question arises as to how are the natural decision-choice processes generated in the duals of the natural duality, and how are the decision-choice results translated onto the action space and then to the poles of the natural actual-potential polarity. Furthermore, do the natural processes and the social processes in the use of experiential information structure follow the same path? In this respect, another important motivation for the development of this monograph is to examine the nature of the socio-natural internal decision making processes under incomplete, vague and qualitative information, and how subjective and objective phenomena play in the outcomes of the dualistic zero-sum game as expressed in the payoff of power and dominance to establish categorial identities at each transformation point. The categorial identity expresses the nature of the categorial power and dominance of the duals in the duality and the dualities in the actual pole of the actual-potential polarity.

IV The Goals, Objectives and Intents of the Monograph

The central objective of this monograph is the development of a unified and self-contained theory of Philosophical Consciencism from its conceptual foundation of philosophy and mathematics in order to establish the sufficient conditions of the theory of transformation. The structure of the theory must provide a linkage to the theory of categorial conversion in relational continuum and unity. Such relational continuum and unity must be found in how the theory develops analytical processes to establish sufficient conditions to support the necessary conditions that are established by the theory of categorial conversion. In this respect, the transformation problem is solved by the general theory of transformation which is divided into an interdependent two sub-theories of the theory of categorial conversion and the theory of Philosophical Consciencism. The theory of categorial conversion establishes the *necessary instruments and conditions* for change. The theory of Philosophical Consciencism establishes how the instruments may be manufactured and used to create sufficient conditions for transformations of categorial elements. By retaining category formation and the principle of opposites which is made up of polarity and duality as have been discussed in the theory of categorial conversion [R15.15][R15.16], an epistemic path is opened to link the Africentric concepts of polarity, duality and principles of categorial conflicts to the dynamics of qualitative and quantitative dispositions as the driving forces of categorial conversions . Similarly, by a judicious utilization of the fuzzy paradigm of thought, composed of its logic and mathematics, the sufficiency of the categorial-conversion process is given a definition in the decision-choice space conditional upon fuzzy experiential information structure to create the theory of Philosophical Consciencism in the optimal control of categorial-conversions in a fuzzy game space.

The essential organic goal and objective of the theory of Philosophical Consciencism is thus to explain how the internal decision-choice processes combined with experiential information structure brings about socio-natural transformations in the ontological space. The other objective is to present a decision-choice framework that will constitute a thought system to guide social actions in the dynamics of politico-economic formations where transformations are defined in the quantity-quality duality and mapped onto the subjective-objective space. Additionally, it is to provide a methodological framework of constructionism-reductionism duality to guide the development of either an explanatory or prescriptive theory of economic development and social transformations. The two monographs entitled the *theory of categorial conversion* and *the theory of Philosophical Consciencism* are also intended to bring into the general understanding the commonness of the African epistemic view of paradigm of thought, where everything is viewed in terms of relational continuum and unity as well as every elements is internally dualized in the negative-positive space to generate energy, power and force for the internal dynamics of actual-potential polarities of categorial elements where old varieties are destroyed and new varieties are created under the substitution-transformation principle. It is hoped that these two monographs, in addition to the one on Polyrhythmicity with polyrhythmics and the laws of polyrhythms are unique in the African art forms [R1.92], and will be sufficient to establish the African epistemic unity in a manner that will point to the support of the African cultural unity that will undoubtedly promote the political unification of Africa and her people.

Methodologically, the two monographs establish scientifically epistemic framework for the development of theories of economic development, politico-legal development and all forms of theories of engineering sciences in the construction-reduction duality, where old varieties are destroyed and new varieties are created in the technological space of information-knowledge dynamics. The use of this methodological structure for the enveloping of dynamics of human knowledge is discussed in [3.7] [R3.8] [R3.10] [R3.13] [R8.44]. The grand objective of the theory of categorial conversion is to produce an explanatory-prescriptive process to establish the necessary conditions for socio-natural transformation. The grand objective of the theory of Philosophical Consciencism is to produce an *explanatory-prescriptive process* to establish the sufficient conditions for *socio-natural transformation* on the basis decision-control actions. The unity of *the theory of categorial conversion* and *the theory of philosophical Consciencism* forms the *general theory of transformation* in systemicity. The necessary conditions are ontological and the sufficient conditions are both ontological and epistemological where the emphasis is on decision-choice process in effecting transformation outcomes. The African who have taken time to understand the two monographs in a relational continuum and unity will have a self-rejuvenation in the epistemic space to acquire new tools produced by the interrelated structure of the principle of opposite in the general understanding of internally induced socio-natural transformation. The non-African who have also taken time to understand the two monographs in a relational continuum and unity will also have a self-rejuvenation in the epistemic space to acquire new logical tools of thought produced from the interrelated structures of the principle of opposites in the general understanding of internally induced socio-natural transformations. Both the African and non-African will have an appreciation for the African foundations of conflict theory, game theory and fuzzy logic under the principle of opposites in the creation and destruction of socio-natural categorial varieties as internal processes of multiplicity of rhythms of matter and energy, the forms of enveloping of which define the paths of socio-natural processes [R1.8] [R1.36] [R1.57] [R1.92][R1.218] [R1.243] [R1.247].

They will also presented with the understanding that the study of socio-natural transformations is the study of problem-solution enveloping through a family of decision-choice systems which are governed by categories of Philosophical Consciencism. It is this problem-solution enveloping that gives meaning to both progress and life. Progress and life without the problem-solution duality are deprived of form and meaning. It is this problem-solution duality, seen in terms of

Africentric principle of opposites with relational continuum and unity in the sphere of existence, which gives meaning to science, general knowledge, and their development and progress. The Africentric conceptual tradition of the principle of opposites in relational continuum and unity presents a notion of change through the internal behavior of the problem-solution process under never-ending conditions of conflicts in universal existence. The organic problem-solution process is permanent and is composed of a family of families of problem-solving processes that are individualized at the level of persons, at the level of nations, demographical environment and at the level climatological and geomorphological environments. Each of the families of the problem-solution processes is defined by the corresponding experiential information structure. The family of experiential information structures defines the family of problem-solution processes which then creates individuation in the transformation game in actual-potential polarity. It is this individualization of problem-solution processes that allows for the statement that the meaning of life depends on the individual and the meaning of national progress also depends on individual nation regarding the constitution of a good society. It is this individualization at the levels of society and nations in relation to experiential information structure and specific constitution of the concept of a *good society* that shapes the contents and uses of Philosophical Consciencism to manufacture the sufficient conditions for destruction of the old and the creation of new in the substitution-transformation dynamics. It is within this epistemic structure, that the concept of violence-nonviolence, diplomacy-non-diplomacy and negotiation-non-negotiation dualities, under the principle of opposites in affecting the control dynamics of transformation game at all human levels, find scientific understanding in defining decision-choice outcomes.

Each of the individual solution-problem process is mapped into the violence-nonviolence duality to abstract a solution toolbox assessed on the basis of a particular Philosophical Consciencism. In this epistemic system dynamics, the greatest contribution of the African-centered principle of opposites is not the affirmation of the permanency of the organic problem-solving process and the family of families of problem-solution individuations, but how the understanding will inform us to use nonviolence and negotiation process in the game of transformation that projects creation-destruction polarities of socio-natural actual-potential varieties. The creation-destruction process is the general foundation in the study of information in static and dynamic spaces to produce theories of *info-statics* and *info-dynamics*. Under these epistemic conditions, the theory of info-statics is constructed to study the time-point characteristics and communications of these characteristic of objects. Similarly, the theory of info-dynamics is constructed to study the results of the problem-solution processes that generate continual enveloping of transformation dynamics of actual-potential polarities. These are the strengths that both the theories of categorial conversion and Philosophical Consciencism present for epistemic understanding at the level of socio-natural processes and at the level of progress of human civilization.

V Organization of the Monograph (from Categorial Conversion)

The main body of the monograph is organized into seven chapters. These chapters are introduced with a preamble and a prologue. The preamble provides some reflections on paradigms, laws of thought and African traditions as they relate to thinking and the structure of creative and controllability conditions in the decision-choice activities conditional upon experiential information structure. The prologue presents some key concepts within the African traditions, and the general and specific epistemic frame of reasoning in the development of the theory. It provides an introduction to the monograph, in terms of exist and entry points into the general logical framework in the constructionism-reductionism methodological duality with relational continuum and unity for the development of the theory of Philosophical Consciencism as decision-choice framework for strategic transformation. The contents of the preamble and prologue introduce the principle of opposites that provide us both exist and entry points to the development of polarity and duality as logical instruments to deal with the dynamics of qualitative and quantitative dispositions in relational continuum and unity, as well as categorial dynamics as an optimal control phenomenon within the African conceptual system.

The objective, here, is to connect the theoretical developments of both categorial conversion and Philosophical Consciencism to the fundamentals of African cognitive unity in understanding the mechanism of general transformational dynamics as it has been introduced in [R1.92] and in support of the African cultural unity as it has been presented in [R1.84] and linked to thought foundations of African socio-political unity as discussed in [R1.91], [R1.202]. The driving force of this methodological approach is simply the notion that the African conceptual traditions, the supporting paradigm of thought, logical, philosophical and mathematical contributions will remain in an epistemic obscurity and cannot be fully developed until the members of African intelligentsia become awakened to connect it with the conceptual foundation of the *principle of opposites* that forms the core of African thought system as it was in Kamet (Ancient Egypt and Nile Valley Civilization) and still reflected in the logics of Africentric thinking in the West, East, South and parts of the North Africa [R1.218a] [R1.218b] [R1.243][R1.248] [R1.251a]such as Dogon . In this monograph on the theory of Philosophical Consciencism and the follow up monograph of the theory of categorial conversion, the interest is not simply on Nkrumah, but on the development of an African-centered foundations of thinking that supported his decision-choice framework for decolonization and development involving reconstruction and management of African social formation from within Africa. The theory of Philosophical Consciencism is a study in decision-choice processes in the general mechanism of transformations of actual-potential polarities in socio-natural systemicity.

Chapter One deals with the development of a general framework of Nkrumaism and relates it to the problem of the development of abstract ideas and the practice of ideas in social settings. It brings into focus the discussions on relational structure of information requirements, internal decision-choice processes, philosophy, ideology, freedom, liberation and human action as derived from within on the basis of decision-choice activities to control and resolve conflicts of preferences in the individual-collective duality and in reference to imperialism, neocolonialism and emancipation under a general and specific framework of Philosophical Consciencism. Chapter Two deals with the relational structure of categorial conversion, culture and Philosophical Consciencism, and the policy dynamics in Nkrumaism. Here, a specific meaning of culture is defined by its DNA, related to social polarities such as the colonialism-decolonization polarity and the nature of social progress with special reference to Africa. Contending ideologies and opposing philosophies are discussed where the task of Nkrumah on the road to Africa's decolonization, independence and emancipation is analytically presented in relation to the development of a theory that justifies sufficient conditions for internal self-motion. Chapter Three deals with characteristic-based analytics that allows countries' identities to be specified for the type of Philosophical Consciencism that will be needed for the development of a specific set of social actions tailored for a specific entity. Negative and positive types of Philosophical Consciencism as well as natural and social types of Philosophical Consciencism are introduced and discussed in relation to positive and negative actions that unite the duals in the game of dominance and definition of identity of the duality. The chapter further discusses an initial introduction of linkages among polarity, duality, positive action and negative action and then to dynamic zero-sum games in the duality. The intellectual tasks in creating thoughts to guide Africa's decolonization and emancipation from within and under the principle of self-motion that relate to the experiential information structure to the whole zero-sum control dynamic game of dominance are discussed.

Chapters Four is devoted to an extensive and intensive development of Philosophical Consciencism and dynamic games in the categorial conversion of social polarities where the people are linked to negative and positive actions, social conscience and consciousness and a collective personality with special reference to African people. The methods of creating the necessary instruments, the management of command and control are discussed. These include the discussion of violent and non-violent strategies and tactics. The chapter is concluded with the effect of information requirements on the use of the Philosophical Consciencism. Chapter 5 deals with leadership and institution under the use of Philosophical Consciencism in categorial conversion of social polarities with special reference to the African personality. The concept and the role of cost-benefit analysis and rationality are discussed in relation to the game of the duals in the duality and polarity.

Chapter 6 expands on Chapter 5, where Philosophical Consciencism and socio-political decision strategies are discussed for an optimal categorial-conversion dynamic game between the duals in relation to the transformation behavior in the actual-potential polarity. Here the chapter deals with social bases for policy development under Philosophical Consciencism, where the concept of social progress is explicated. In reference to African conditions, a set of different types of Philosophical Consciencism are created and distinguished from African-centered Philosophical Consciencism on the basis of the African experiential information structure. The elements of the set of the different types are distinguished by the internal experiential information structures that are specific to each entity. Additionally, the chapter is used to discuss the connecting relational structure of political rationality, economic rationality and legal rationality as advanced by Nkrumah under African-centered Philosophical Consciencism to deal with African conditions. The monograph is concluded with Chapter 7 which deals with the complex issues of contents and curriculum of African studies which are relevant for the construction of African-centered Philosophical Consciencism and the design of African education to support Africa's complete emancipation, freedom and justice. Here, the discussions involve the intentionalities of knowledge production and the relevance of undistorted experiential information structure. A distinction is placed between Western imperial intentionality and African intentionality and their relationships to the studies of the art and the science of thinking in research, teaching and learning, and how they relate to African consciousness and conscience in the battle of power and dominance. The Western imperial intentionality of knowledge production relates to the imperial dominance of colonialism and the neo-imperial dominance of neo-colonialism and resource exploitation. In contrast, the African intentionality of knowledge production relates to emancipation, freedom and justice from imperialism and neo-imperialism as seen from the framework of the theory of Philosophical Consciencism in the global decision space and international socio-political game over the resource space.

PROLOGUE

THE MORPHOLOGY OF THE THEORY OF PHILOSOPHICAL CONSCIENCISM: THE PRACTICE FOUNDATIONS OF NKRUMAISM IN SOCIAL SYSTEMICITY

I Reflections on Relational Structure of Categorial Conversion, Philosophical Consciencism and Socio-Natural Transformations

In the theory of categorial conversion, a number of philosophical and mathematical issues are discussed. These philosophical and mathematical issues are related to problems and solutions in the dynamics of qualitative and quantitative dispositions of the categorial existence of universal elements. In the bare essentials, the problems and solutions are aggregately reduced to the transformation problem of the qualitative disposition of ontological elements and its solution as applied to elements in socio-natural categories. The transformation problem and its solution for necessary conditions are the epistemic problems of the construction of a general mechanism of qualitative dynamics. This is the problem which is taken up in the theory of categorial conversion to explain and abstract the necessary conditions for socio-natural transformations as may be observed in the dynamics of actual-potential polarities. The challenge of the theory of the general mechanism of transformation is first to derive the conditions of existence of categories in the relevant spaces by a *theory of category formation* which is the identification and establishment of categorial varieties, and secondly, to derive the necessary conditions of categorial convertibility by a *theory of categorial conversion*. Given the existence of categories of universal varieties, the necessary conditions of categorial convertibility are shown by the theory of categorial conversion to consist of *categorial moment, categorial transfer function, categorial crisis zone, categorial distance, categorial difference and categorial transversality conditions*. The theory of socio-natural transformation, therefore, is composed of a *theory of category formation* that explains the formation of categories which presents the general foundations of linguistic forms under the principle of nominalism, and a *theory of categorial conversion* that explains the necessary conditions of categorial conversions, and where conditions of categorial conversion initialize self-motion and self-transformation in qualitative and quantitative dispositions under relational continuum and unity of varieties.

The conceptual foundation in formulating the categorial-conversion problem contains the Africentric principles of opposites in the most general form of the universal system of both ontology and epistemology. The principles of opposites are composed of conceptual tools of polarity and duality defined by negative and positive characteristic sub-sets within the same element. The structure of the duality is such that negative characteristic sub-set constitutes the negative dual and the positive characteristic sub-set constitutes the positive dual. The concept of the general characteristic set allows the partitioning of the ontological space into ontological categories. The concept of the relative negative-positive characteristic sub-sets in the same element provides definitions for the categorial identities of socio-natural varieties. The categorial identities allow one to know if the qualitative disposition of an element or a category has been self-transformed and moved into another category with a defined identity different from the previous identity. The process where one element in a category and changes identity in order to enter into another category is called *inter-categorial conversion*. Inter-categorial conversion relates to changes in qualitative disposition of qualitative varieties. Given any category as established by its qualitative disposition, there is an intra-categorial conversion. The intra-categorial conversion relates to changes of the quantitative disposition of quantitative variety.

The concept of polarity allows the identification and establishment of actual and potential poles, differences and similarities in relational continuum and unity in addition to the organic distribution of the family of families of actual-potential polarities in both the ontological and epistemological spaces. The concepts of the negative and positive characteristic sub-sets allow us to establish the existence of internal contradictions and tensions in relational continuum and unity within categories and elements, thus creating dualities in relational continuum and unity with internal contradictions and conflicts without excluded middle. The concept of the relative characteristic sub-set allows the identification of negative and positive dualities. The concepts of negative and positive dualities allow the identification and characterization of the respective poles of the actual-potential polarity as negative or positive. The concepts of negative and positive characteristic sub-sets, not only reveal the internal structures of the duals in the duality, but define the nature of contradictions and conflicts which produce the energy and forces of opposite of qualitative motions within the categorial element as well as the dominating pole.

The categorial identities and internal behavior are the result of interplay of matter, energy and information. The concept of matter, broadly defined, allows one to establish the content of socio-natural elements. The concept of energy, also broadly defined, allows one to establish force and energy that relate to categorial conversions. The concept of information, also broadly defined, allows the establishment of energy-material differences, similarities and commonness of elements to create socio-natural varieties required for the development theories of info-statics and info-dynamics and link them to the theory of socio-natural transformations. All of these concepts are revealed and established by the information connector between the ontological space and epistemological space, where a paradigm of thought operating on the information structure allows the establishment of similarities and differences between the ontological information structure and the epistemological information structure to produce results from the knowledge-development process, where these results are used as inputs into the socio-natural decision-choice systems .

The ontological information structure presents ontological categories with ontological qualitative and quantitative dispositions on the basis of which ontological varieties are established. Similarly, the epistemological information structure presents epistemological categories with epistemological qualitative and quantitative dispositions. The ontological information structure is exact and hence the ontological qualitative disposition is also exact in terms of identities and inherent varieties. The ontological qualitative and quantitative dispositions of elements and categories introduce into the epistemological space vagueness, subjectivity, inexactness and complexity in terms of linguistic and conceptual meaning of information attributes and cognitive abilities to correctly establish the epistemological information and process it to derive conditional knowledge as an input for the decision-choice controllers of the production of categorial convertibility conditions. The construct of the epistemological information structure is the work of cognitive agents. This construct is affected by quantitative limitations in terms of volume of information, vagueness in terms of clarity of information, subjectivity in terms of interpretation of information and disinformation in terms of trust of the sources of the information regarding misinformation, disinformation and propaganda.

The logical tools for dealing with the presence of quantitative information limitationality and limitativeness are probability and probabilistic reasoning. The logical tools in dealing with the presence of qualitative disposition, vagueness, inexactness, complexity and subjective phenomena are fuzzy logic, fuzzy mathematics, possibilistic reasoning through the analytical processes of fuzzification and defuzzification [R3.1] [R3.14] [R3.7] [R3.8] [R3.9] [R3.13] [R3.27] [R4][R5.3][R6]. The fuzzification process is used to formulate the convertibility problems under the principles of opposites with relational continuum and unity involving contradictions and conflicts. The defuzzification process is used to solve the convertibility problems to obtain the conditions of categorial convertibility that define the conditions of transformation of the actual-potential polarity. If the information structure is both vague and incomplete then the categorial-conversion problems and their solutions will be treated with the fuzzy-stochastic tools of reasoning in the fuzzy-stochastic decision-choice control space. The successful creation of convertibility conditions to bring about categorial conversion of actual-potential polarity is the work of *reasoning conscience* that is an internal part of qualitative disposition of the categorial identity. This reasoning conscience which provides a map of categorial-conversion game strategies for resolving the crises in categorial identity is what Nkrumah called *Philosophical Consciencism*. Since categorial identity is unique, there are as many forms of categorial Philosophical Consciencism as there are categories and categorial elements. If the information structure is exact and complete and the principle of opposites is seen in terms of dualism with excluded middle then the appropriate logical tools are the classical logic and corresponding mathematics which will be extended to include probabilistic reasoning if the information is incomplete but exact. For further discussions involving conditions of exactness and inexactness in reasoning see [R3.7] [R3.9][R3.10][R3.13][R3.15][R3.20].

II Conceptual Problems and Analytical Pillars of Philosophical Consciencism

To specify the conceptual problems of Philosophical Consciencism and the analytical pillars of its theory, it is useful to have a reasonable understanding of its relational structure to categorial conversion. Let us keep in mind that the theory of categorial conversion presents the conditions required for qualitative change in the actual-potential relational structure. These conditions of change are the categorial-convertibility conditions. The conditions mainly define states of being in the sense that they define *temporary qualitative equilibria* for categorial elements as they exist in the actual-potential polarities.

The temporary qualitative categorial equilibrium of any element exists under intense conflicts and contradictions under the principle of opposites in relational continuum and unity. The disturbances of the categorial equilibrium for any categorial element are induced by decision-choice actions from within that provide directional instructions to the behavior of the negative and positive characteristic sub-set of the same element. The instructions are generated by a central processing unit (CPU) whether conscious or unconscious that receives and processes information and distributes the results to the positive and negative characteristic sub-sets for positive and negative actions at the general level of actual-potential polarities. The instructions of the central processing unit are drawn from the conscience that sets the intentionalities relating to any qualitative equilibrium of any actual-potential polarity of change and maintenance of the actual. A question may be asked as to whether the property of the central processing unit exists for all universal elements including cognitive and non-cognitive agents. Are these central processing units endowed with *conscience*, if they exist, for receiving, processing information signals and distributing instructions to relevant opposing duals given their common environments for initiating qualitative self-motion of the element?

The conceptual framework that is being advanced here is that every element in the universal system is endowed with a central processing unit (CPU) which is composed of a negative central processing sub-unit and a positive central processing sub-unit in relational continuum and unity. In other words, the elemental or categorical central processing unit (CPU) exists as a duality of negative dual and positive dual in an uneasy elemental or categorical crisis from within. It is this internal crisis that places each element on temporary qualitative equilibrium and provides the possibility and force of transformation. Are these central processing units of elements or categories endowed with conscience relative to the common environment of the element that is shared by the duals?

The answer projected by the basic principles of categorial conversion is that each universal element or category is not only endowed with a central processing unit for the information contained in the duality but the central processing unit is endowed with conscience. The elemental conscience is composed of negative and positive sub-conscience in relational continuum and unity to define the identity of the element or category. The negative central processing sub-unit with the negative sub-conscience receives information from the common environment relative to the joint activities of the duals, processes it as a decision-choice input, and creates and sends instructions to the negative characteristic sub-set (the negative dual) for negative action on behalf of the negative dual. Similarly, the positive central processing sub-unit with the positive sub-conscience receives information from the common environment relative to the joint activities of the duals, processes it as a decision-choice input, and creates and sends instructions to the positive characteristic sub-set (the positive dual) for positive action on behalf of the positive dual. The negative and positive actions are strategies and counter-strategies in the game of negation of negation towards categorial conversion of socio-natural actual-potential polarities. As presented, the analytical pillars are such that every universal element belongs to a category as well as resides in an actual-potential polarity which is endowed with both negative and positive central processing sub-units, with negative and positive elemental or categorial conscience and capacity of decision-choice activities, as well as capacity for action in the struggle against one another for the categorial conversion games in all actual-potential polarities given the universal system, where the instruments of gamming are the system of dualities with each duality structured in relational continuum and unity. This problem if conflict in duality with relational continuum and unity between the duals is the *funtummereku-denkyemmereku* problem in the Africentric principle of opposites. This problem in Africentric principle of opposites is captured by the problem of two-headed crocodile with one stomach (*funtummereku-denkyemmereku*), where the two heads have a bloody fight against each other in order to feed the stomach that sustains them and, where the heads are in a relational continuum in existence and the stomach connects them in a relational unity. The problem-solution structure defines a complex philosophical concept of unity in diversity and shared destiny for continual transformation of the contents of actual-potential polarity [R15.15][R15.17].

III The Principle of Opposites, Conscience and Conscious Existence

The central pillars of the principle of opposites as seen in the African conceptual system are polarity, duality, and negative and positive characteristic sub-sets which exist in relational continuum and unity under contradictions and conflicts. The concept of polarity is used to specify the conflicts and contradictions between the actual and potential as opponents with relational continuum and unity in actual-potential polarity. The concept of actual-potential polarity under the principle of opposites is a powerful linguistic tool in dealing with categorial transformational or developmental or engineering phenomena. The concept of development in socio-natural space is meaningless without reference to actual-potential relational structure. Here the distinction between development and growth must be asserted where development is associated with the dynamics of qualitative disposition in quality-time space, and growth is associated with the dynamics of quantitative disposition in quantity-time space where quality and quantity are defined as duality with relational continuum and unity. The dynamics of quantitative disposition requires the constancy of qualitative disposition and not the other way around. The explication of the concept demands a clear understanding of dynamics of opposites as an internal structure of all universal elements. Each pole of the polarity has a residing duality that provides the pole's identity in terms of either negative or positive pole in accordance with whether the residing duality is a negative or positive duality, where the negative pole has a residing negative duality and the positive pole has a residing positive duality.

Depending on the situation, the actual pole may be associated with either the negative or positive duality while the potential may be associated with either the positive or negative duality respectively. The negative duality has the dominance of the negative dual as established by the negative characteristic sub-set relative to its internal positive characteristic sub-set, while the positive duality has the dominance of the positive dual as established by the positive characteristic sub-set relative to its internal negative characteristic sub-set. The activities of the opposing characteristic sub-sets are to preserve or create dominance of the individual self as a dual relative to its opposite. It is these activities of the negative and positive characteristic sub-set that create contradictions and conflicts which then set in motion the battles of negation and initialize that transformative action of any of the dualities, which act on the actual pole to either destroy or affirm the conditions of its existence. The activities of the characteristic sub-sets imply the existence of *categorical convertibility conditions* and *categorical controllability conditions* in socio-natural transformations of universal elements. The implication here is that there must be two categories of theories that must be related to socio-natural transformations.

IV Conscience, Consciousness, Convertibility and Transformative Actions

The theory of categorial conversion provides the necessary and sufficient convertibility conditions for qualitative transformation in socio-natural actual-potential polarities. The *convertibility conditions,* therefore, constitute a *set of programs* that must be acted on if conversions are to occur in either negative, neutral or positive qualitative direction. They do not provide the set of decision-choice actions for implementing the conditions. They merely state what factors must be controlled in order to implement the program of convertibility conditions to bring about conversion of the actual pole and actualization of an element in the potential space. It is this process of bringing the actual into the potential space and actualization of a potential element that the control action is required from the *controllability conditions* to define the optimal action for an optimal process of conversion of the pole and the realization of a new actual. The controllability conditions, therefore, constitute an *action set* or the basis of actions by the internal processing unit without which conversion is impossible.

It is at this point of search for controllability conditions that it is claimed, here, that the central processing unit of every element is endowed with conscience in various degree of complexity relative to the corresponding degree of consciousness. The existence of conscience due to the internal organization of the negative and positive characteristic sub-sets and the degree of conscience are determined by the degree of efficient organization of the relative negative-positive characteristic set. The same internal organizations of the negative-positive characteristic sub-sets endow the elements with the possibility of consciousness and how the conscience and consciousness are related and integrated for internal dynamic behavior of elements and categories. The understanding of the roles played by conscience and consciousness in controllability of the categorial conversions requires a discussion and explication of these two concepts. Transformative and convertibility actions are decision-choice determined, while the decision-choice process depends on the intentionalities which are shaped by consciousness from experiential social information structure from which Philosophical Conscience is manufactured.

V. Explication of Concepts of Conscience, Consciousness and Cotrollability in Categorial Conversions of Actual-Potential Polarities under Philosophical Consciencism

What does it mean to say that the internal system of elements or categories has conscience and consciousness and what is the relative scientific meaning? The question arises as to how do conscience and consciousness relate to the internal convertibility, observability, controllability and categorial conversion of actual-potential polarities. Consciousness has a simpler meaning relative to conscience which has many contextual meanings. The relevant meanings to controllability and categorial conversion will be examined to give clarity to the theory of Philosophical Consciencism as presenting a framework for automatic internal controllability of elemental transformations as viewed in relation to natural and social categorial conversions. The theory of Philosophical Consciencism will also provide a justification in the study of self-exited systems, self-correcting systems, self-creating systems and self-transforming systems and their relationship to the understanding of natural automatic control systems as creating foundations for engineering artificially automatic control systems. Let us keep in mind that the theory of categorial conversion derives the convertibility conditions which constitute the necessary program set for internal self-transformation on the basis of internal decision-choice process.

In categorial conversions, the system is specified in terms of polarities which are defined by a set of dualities, where the dualities are defined by a family of negative and positive characteristic sub-sets under the principle of oppositeness with relational continuum and unity. Each categorial element exists in contradictions and conflicts in terms of the funtummereku-denkyemmereku problem generating continual *internal crisis* due to the competitive acts of the negative and positive duals. This internal crisis is a crisis in *categorial identity* that defines the dualistic *categorial conscience* of the elements or the category, where the categorial conscience is composed of *negative sub-conscience* and *positive sub-conscience* in relational continuum and unity that, together, define the conscience of the elements in terms of the maintenance or change of categorial identity. The competitive behaviors of the negative sub-conscience and positive sub-conscience to create a dominance of either the negative dual or the positive dual define the crisis in categorial conscience and the corresponding identity.

Conscience is, here, claimed to be an internal qualitative characteristic of all universal elements and categories in varying degrees of complexity. In this respect, what is conscience and how does it relate to internal controllability in the categorial conversion process of actual-potential polarities? Conscience concerns the internalization and individualization of categorial circumstances and conditions of the categorial stability leading to a decision-choice action of either the rejection or acceptance of the existence of the temporary qualitative equilibrium state. Conscience, in this situation, is an internal reasoning and assessment faculty held by the individual dual to arrive at a negative or positive action to accept the existing internal categorial conditions or to reject and change the existing internal conditions of an element or category in order to destroy the actual and bring in a new actual from the potential space. It is a decision-choice quality whose activities involve constructive destruction of the actual in transformations of socio-natural actual-potential polarities. The conscience relates to the decision-choice quality on the basis of which control action is exercised. The internal conditions define the environment of categorial existence which is reflected in the outer appearance (the categorial superstructure). The acceptance of the existing condition is in fact, an acceptance of the actual pole in the actual-potential polarity. Every silence is an indirect acceptance of the conditions of the actual pole on the basis of cost-benefit rationality. The rejection of the conditions of the actual is a necessary decision-choice action for change and actualization of a desired potential element under a given information structure. Here, an important logical and application complication arises. Rejection is necessary but not sufficient for a change and an actualization of a potential. This provides a framework to explain why similar categorial elements with the same experiential information structure behave differently towards transformation.

Under the principle of opposites, everything exists in dualities and polarities, as such that the organic categorial conscience is composed of *negative sub-conscience* and *positive sub-conscience* where each exists as simultaneously *ordinary sub-conscience* and *critical sub-conscience* in relational continuum and unity. The ordinary sub-conscience involves a situation where an individual dual rejects the conditions of the existing actual from a benefit-cost implication as unfavorable or unjust but endures the overwhelming cost without taking up the responsibility of action for change. The critical sub-conscience, on the other hand, is associated with the capacity to have a weighing principle of responsibility of action distribution between the actual pole and potential pole in the actual-potential polarity. The action distribution is derived from the environmental circumstance and experiential information structure as well as their effects on the dynamic behaviors in the quantity-quality space. Both the positive and negative central processing sub-units are endowed with capacities to develop the necessary and sufficient conscience required to maintain or change the conditions of the actual pole where the residing duality is under acute contradiction and bitter internal struggle of the duals for dominance and negation of dominance.

A transformation of the existing actual and the actualization of a desired potential are not possible, and cannot be realized, until the positive central processing unit has acquired, by developing within itself, the necessary and sufficient sub-conscience. In this transformational process, conscience is viewed as a quality of the internal central processing sub-unit to resolve the qualitative crisis produced by the opposing forces for self-preservation under mutual negation of the poles of the polarity. The qualitative crisis is a crisis in identity and categorial belonging. The resolution of the crisis may be a decision-choice action in favor of a transformation of the existing actual into the potential space where a new actual from the potential space will be brought into existence with new contradictions, conflicts and struggles at a new temporary qualitative equilibrium for a given experiential information structure, where the new actual may be contrary to preference as judged by cognitive agents. Under the same experiential information structure, the decision-choice action in the resolution of the categorial crisis may also lead to the maintenance of the existing actual under continual temporary qualitative equilibrium with increasing contradictions, conflicts and intense struggles for *categorial dominance*. Conscience is, simply, an elemental or categorial drive that promotes the positive and negative characteristic sub-sets (duals) to produce negative and positive actions for competitive struggle for transformation or maintenance of an existing actual.

The organic categorial conscience of every element exists in deferential degrees of intensity under categorial crisis in the process of the negation game involving the resolution of its internal contradictions and conflicts produced by the negative and positive duals whose relative characteristic set gives the categorial and elemental identity. Here, the requirement of conscience is the capacity to *sense* the negative and the positive characteristic sub-sets from the point of view of their relative existence and mutual destruction in their common environment to provide the drives to transform a pole in the actual-potential polarity, or to affirm the existence of a particular pole in a defined actual-potential polarity. The important conceptual idea is that every universal element is perceived to exist as a duality in relational continuum and unity under contradictions and conflicts from which categorical and elemental conscience tend to internally emerge from the crisis within a division of the characteristic sets into negative and positive duals as their contribution to the understanding of categorial or elemental destruction through mutual decision-choice negation to relationally alter the nature of the actual-potential polarity. Besides, every universal element existing as duality for its *destruction*, also exists as actual-potential polarity for its transformation and substitution where its destruction creates a vacuum to be filled with a substitution.

As presented, every category or element has an internal organic central processing unit that is endowed with an internal reasoning system based on some level of internal conscience that has varying degrees of quality and complexity depending on the internal organization to take part in the categorial-conversion game to develop strategies and counterstrategies for the qualitative dynamic game of negation of negation of the poles to create new actual-potential polarities for the continual creative-destruction process. The categorial-conversion game is an *optimal qualitative control game* where the payoff is categorial dominance by either the negative or positive dual. The *internal conscience* that will be needed to organize the internal reasoning and dynamic behavior will be referred to as *Philosophical Consciencism* which is positive and negative in duality with relational continuum and unity. The general concept of Philosophical Consciencism is derived from Nkrumah's particular conception as applied to Africa and in relation to the African experiential information structure [R1.203]. *Philosophical Consciencism* is a conceptual reasoning map in terms of the disposition of the distribution of internal forces and their directives which will enable the internal arrangements to

1

digest the internal contradictions in order to resolve the conflicts for the existing qualitative categorial equilibrium, or to transform it into a new qualitative categorial equilibrium in relation to its identity and experiential information structure. The thought process is such that when Philosophical Consciencism is viewed in relation to natural categorial conversion, then one speaks of *natural Philosophical Consciencism*, and similarly, when Philosophical Consciencism is viewed in relation to social categorial conversion, then one speaks of *social Philosophical Consciencism*. Natural Philosophical Consciencism and its uses in transformations in actual-potential polarities are ontological. However social Philosophical Consciencism is ontological and its uses in transformations are epistemological. It may be pointed out that the creation of self-excitement, self-organization, self-correction and self-controllability of mechanical robots is a process to equip the robots with *artificial Philosophical Consciencism* for internal self-directives. Socio-natural Philosophical Consciencism is viewed in relation to qualitative dispositions in time-relation phenomenon, while artificial Philosophical Consciencism is viewed in relation to quantitative disposition in time-space phenomenon. From the point of view of cost-benefit rationality, the *qualitative control game* of categorial conversion of actual-potential polarity and continual negation of negation may be a dynamic zero-sum game where either the negative dual wins to dominate to maintain the power relation and the actual pole, or the positive dual wins to dominate to change the power relation and actualize a potential. It is these possible game-theoretic outcomes that give meaning to progressive change or one-step forward and two-steps backward in a socio-natural spaces.

The concept of *capacity to sense* introduces the idea of *consciousness* of the internal environment and experiential information structure for the development of conscience. A general question then arises as to whether consciousness of the categorical elements is a necessary quality for the development of conscience relative to their internal environment and the external information structure as the internal conditions of the elements while other questions follow. What relation does the internal consciousness have with the external information structure? What relationships exist among the internal experiential information structure, the external experiential information structure, categorial consciousness and the internal Philosophical Conscience? Philosophical Consciencism provides the action map for dealing with controllability conditions of the internal conflicting processes of the negative and positive duals. It is the heartbeat of categorial conversion. It is linked to the internal decision-choice processes regarding the stability and equilibrium of categorial identity. It works with the internal and external information structures through the categorial consciousness of the elements. The categorial consciousness provides the map for processing the experiential information composed of the external and internal information structures to derive knowledge inputs for the central processing sub-units that operate at the guidance of the map of Philosophical Consciencism. Here, it must be noted that the *external experiential information structure* indicates the need to transform and the *internal experiential information structure* indicates the process to transform from within on the basis of the Philosophical Consciencism. It is this relational structure between internal and external experiential information structures that provides a justification to state that sustainable transformation cannot be externally induced because it is an internal process. One cannot drink a medicine for a sick person in order for the sick to get well or to be cured.

In this respect, it is useful to examine the essential qualitative characteristics of the *conscious state* and *unconscious state* in relation to categorial action and non-action in the behavior of the central processing sub-units. The unconscious state of the central processing unit is peaceful, passive, and free from contradiction and conflicts. The conscious state of the central processing unit is in relation to the general categorial environment and the information structure of the relative behaviors of the negative and positive duals which exist in an antagonistic mode. The conscious state of the central processing unit is non-peaceful, active and characterized by contradiction, conflicts, struggles, violence and crises which lead to creative destruction of the actual and emergence of the new actual as the resolution. The resolution of these internal antagonistic behaviors requires that the central processing sub-units must be aware and hence equipped with the *capacity of consciousness* in order to develop conscience for creating appropriate decision-choice internal controllers to affect the direction of qualitative transformation and categorial conversion of the actual-potential polarity.

As discussed, the theory of socio-natural Philosophical Consciencism is simply about the controllability conditions of decision-choice actions to resolve internal contradictions and conflicts of categorial elements that maintain the actual or transform the actual. In other words, the theory of Philosophical Consciencism is a *general control theory* of internal processes where the conditions of the optimal control may be abstracted from the use of internal and external experiential information structures from the logical map provided by the theory. The generality of the theory lies in its application to controllability of quality-time phenomena involving qualitative movements of transformations, quantity-time phenomena involving quantitative movements of time-space phenomena, and the simultaneity of quality-quantity-time phenomena involving interactive processes of dynamics of qualitative and quantitative dispositions. The controllability is undertaken by the negative and positive central processing sub-units in relational continuum and unity within the actual pole and not in the potential pole to create *categorial crisis* in the actual pole, destroy the conditions of its existence, create a *categorial power vacuum,* and allow a new *category power* to emerge in terms of actual-potential polarity with new duality which contains new contradictions, conflicts, new central processing sub-units under a new categorial environment and experiential information structure to resolve the crisis in the actual pole. The process of resolving the internal crisis is a two-element *control dynamic game* played between the negative and positive duals seeking for internal dominance of the duality. The game is about a change in relation of qualitative disposition. From the point of view of cognitive agents with the capacity of moral judgment, the actualized potential may or may not be preferred. Here, it may be observed that Philosophical Consciencism is a rational guide to resolve the funtummereku-denkyemmereku problem that defines a complex philosophical concept of unity in diversity and shared destiny in the Africentric principle of opposites for continual transformation of the contents of actual-potential polarity.

VI Another Reflection on Categorial Conversion and Philosophical Conciencism

It is useful, for increasing clarity, to take another simple reflection on control dynamics of qualitative disposition and its possible relation to dynamics of quantitative disposition. The simple reflection will present a relational structure of the theory of categorial conversion, the theory of Philosophical Consciencism and transformations of socio-natural elements. The study and understanding of transformations in nature offer an important framework to understand their applications to social transformations. The transformations of both natural elements and social elements may be either positive (good), neutral or negative (bad) as viewed by the use of subjective assessments of cognitive agents where conscience does not only have the capacity to judge moral conditions on the basis of cost-benefit rationality to transform the actual in the direction of a subjective preference for the element in the potential space but has capacity to accept, reject and influence action. Again, it must be emphasized that transformations of socio-natural elements and the construction of corresponding explanatory or prescriptive theories are not possible if the concepts of actual, potential and polarity are not claimed, defined and explicated in terms of scientific language. For transformations to occur, it must be axiomatic that every categorial element resides in an unstable equilibrium of actual-potential polarity under the Africentric principle of opposites where solutions to duality problems are guided by Philosophical Consciencism under the principle of *Asantrofi-anoma rationality* which in simplicity says that you cannot receive the benefits of a decision without the corresponding costs.

The general negative and positive progress of socio-natural elements in the universal space is understood in terms of qualitative dynamics in the construction-destruction duality with relational continuum to form never-ending creative-destruction dynamics in the universal system. The creative-destruction dynamics is the universal transformation where the old actual is destroyed and a new actual is created in substitution. The construction of the explanatory and prescriptive theories of the universal creative-destruction dynamics is the *transformation problem*. The transformation problem is divided into the *problem of necessary conditions* which is handled by the *theory of categorial conversion*; and the *problem of sufficient conditions* which is handled by the *theory of Philosophical Consciencism*. The *theory of transformation* is thus made up of the theory of categorial conversion and the theory of Philosophical Consciencism. The necessary conditions establish the *controllable (controlled) objects* that must be created if transformation is to occur. These controlled objects are equally available to both the negative dual and positive dual where the controllability depend on the strategic decision-choice actions on both duals.

The sufficient conditions establish the decision-choice medium for exercising productive activities in creating the controlled objects and programming of strategic decision-choice actions for the manipulations by the duals to establish maximum dominance in the quantity-quality-time space. Here, the theory of Philosophical Consciencism establishes a strategic decision-choice framework of maximization of dominance by each of the duals subject to the strategic decision-choice actions of the opposing dual. The degree of dominance becomes the objective set while the degree of strategic response becomes the constraint set defined in the fuzzy-stochastic space for each of the duals. For the theory of transformation to be self-contained, the concepts of actual and potential must be connected to the principle of opposites where the actual and potential exist in an antagonistic mode to form a polarity which is defined by contradictions and conflicts that exist in an interconnected mode of relational continuum and unity. Similarly, for the actual and potential to have self-transformation, the contradictions and conflicts with relational continuum and unity must be connected to the internal dualities of the actual and potential poles of categorial existence of universal elements. For the dualities to have the qualitative properties of contradictions and conflicts, they must be endowed with capacities to internally produce them.

Their endowments are identified as the presence of the characteristic set that exists as negative and positive characteristic sub-sets in an antagonistic and competitive mode for dominance, where the characteristic sub-sets define the duals which exist in relational continuum and unity with the relative negative characteristic set providing the categorial identity of the elements in the universal space. The negative and positive duals existing in an antagonistic mode for mutual negation of negation define a dynamic two-element zero-sum game where the payoff is dominance, the maintenance or the transformation of the actual. The results of the game become part of the experiential information structure as well as irreversible. The presence of negative and positive duals with contradictions, conflicts, central processing sub-units endowed with negative and positive Philosophical Consciencism introduces conflicts and contradictions into the continual transformation process of a never-ending dynamic control game phenomena as opposing strategic decision-choice activities that destroy the actual and usher in a new actual.

Here, the universe is seen as a family of categories (species) where each element is defined by a characteristic set of negative and positive duals. The characteristic set, its internal organization and arrangements provide the dualistic nature of the elements, categorial identities, and categorial belonging. The negative and positive characteristic sub-sets provide the categorial elements, the power of self-transformation, as well as the creation of the force to internally transform itself. For the analytical structure of internal transformations, every categorial element exists as a variety as well as an actual-potential polarity. Here, two important theoretical and application questions that form the analytical core of the explanatory and prescriptive processes tend to arise: 1) What are the necessary conditions of self-transformation and 2) what are the sufficient conditions for self-transformation? These two questions define the nature of the dynamic zero-sum qualitative transformation game where the payoff is a dominance to maintain the actual or a dominance to change the actual and bring in a new actual from the potential space. The answer to the first question is provided by the theory of categorial conversion. The answer to the second question is provided by the theory of Philosophical Consciencism. The two theories together provide a *unified theory of self-transformation* of socio-natural elements. They, thus constitute a *general mechanism of transformation* of socio-natural elements in the universal space.

VII Information Structure and Philosophical Consciencism

The theory of Philosophical Consciencism establishes a framework of a *strategic conceptual map* of decision-choice activities of the negative dual for the development of strategic *negative actions* from the *general action set* for the maximization of the degree of negative dominance subject to the reaction function of the positive dual. Similarly, it establishes a framework of a *strategic conceptual map* of decision-choice activities of the positive dual for the development of strategic *positive actions* from the *general action set* for the maximization of the positive dominance subject to the reaction function of the negative dual. The positive and negative actions constitute strategic interactions with relational continuum and unity under the principle of opposites. Here, the transformation game is more than an evolutionary game. The transformation process is also more than an adaptive process. It includes both evolutionary and adaptive phenomena. In the internal self-transformation zero-sum game, there is no permanent categorial equilibrium for the actual but there is temporary qualitative equilibrium under the *experiential information structure*. There is no *stable state*. However, one can speak of a *stable path* which is an enveloping of points of outcomes of categorial conversions as part of the evolving experiential information structure under the decision-choice actions of Philosophical Consciencism.

The strategically epistemic map is developed with an information structure that must be processed with conditions of Philosophical

Consciencism. Let us keep in mind that all transformations in the ontological space work with the relational structure of three basic organic elements which has been explained in the theory of categorial conversion. The three basic elements are *matter, energy and information*. The information is general to all and connects forms of matter and energy in any transformative process that relationally produces knowledge square of potential, possible, probable and actual [R3.7] [R3.10]. The transformation process of mater-energy elements continually generates never-ending distribution of forms of varieties with a continual expansion of stock of information. It is this information that becomes specific to objects in time-processes as experiential information structures which appear as social experiential information or natural experiential information depending on the nature of the object. Controllability cannot be exercised without information, the ability to process it into relevant knowledge, the know-how to create the specific controlled objects and the decision-choice capacity to use the knowledge to manipulate the controlled object. Here, the structure of information-knowledge-decision-choice processes is presented as implicit and explicit parts of the theory of Philosophical Consciencism in defining the antagonistic game activities of the negative and positive duals of the duality. The nature of the experiential information is critical to the outcome of the categorial conversion. The total experiential information structure at any time is made up of external and internal information structures. Given the decision-choice capacity of the duals for actions, the game space for which the theory of Philosophical Consciencism must function as a total decision-choice framework is completely defined by the total experiential information structure.

The nature of this game space for the application of Philosophical Consciencism is such that the properties of the experiential information structure are defective and deceptive information structures. The defective information structure is made up of incomplete information sub-structure and vague information sub-structure. The incomplete information sub-structure affects the quantity of information and hence gives rise to probabilistic reasoning, while the vague information structure affects the quality of information and hence gives rise to possibilistic reasoning. The deceptive information structure is also made up of disinformation sub-structure and misinformation sub-structure. Both the sub-structures affect the quantity and quality of information through the propaganda process for altering the game space, to invite confusion and subjectivity, and hence give rise to both possibilistic and probabilistic forms of reasoning. The traditional game space is completely replaced by a *fuzzy-stochastic game space* defined by defective and deceptive information structures [R2.23] [R13.8][R13.9], [R17] [R17.2][R17.3]. It is the nature of this fuzzy-stochastic game space why Philosophical Consciencism meets its greatest challenges in organization and mobilization of the control strategies of the categorial-conversion game in infinite space. It is also here, that mathematics and laws of thought find their critical challenges in answering a certain class of questions. These questions include but are not limited to:

1. Under what set of conditions does Philosophical Consciencism improve the behavior of either the negative or positive dual in the controlling activities of negation of negation in the dynamic zero-sum game of categorial conversion of actual-potential polarity?
2. What role does Philosophical Consciencism play in controlling the rate and direction of the resultant of the negative and positive forces in the contradictions and conflicts in the game space?
3. Can the theory of Philosophical Consciencism explain the diversity of time and behavior of the results of the duals in engineering the mutual negation of negation in the socio-natural actual-potential polarities?
4. What is strategic rationality embodied in the conceptual framework of Philosophical Consciencism in the categorial-conversion game of the actual-potential polarity and is it possible to uniquely determine the qualitative state of the system at any time?
5. Can the deliberative conceptual action implied by strategies that are derived by the logical process from Philosophical Consciencism meet the requirements of cost-benefit rationality?
6. Is it conceivable that the practical steps for strategic decision-choice action implicit in the Theory of Philosophical Consciencism for categorial conversion of actual-potential polarity create more confusion in the transformational cost-benefit space due to the actualization of undesirable potential elements?

7. Can we conceive of a socio-natural system that belongs to a family of classes of qualitatively stable rational systems in the ontological space? Alternatively, is it conceptually possible to conceive of a socio-natural system that can be rationally controlled by any of the duals in the sense of being crafted and directed to an outcome of negation of negation associated with the transformative dynamics of the actual-potential polarities according to calculated conscience, will and reason implied by Philosophical Consciencism?

8. In all these, what are the relational structures among Philosophical Consciencism, matter, energy and information, and how does Philosophical Consciencism deal with both defective and deceptive information structures where matter and energy are ontological, the information structure is both ontological and epistemological, and philosophical Consciencism is epistemological?

Reflections on these questions are essential in understanding the role of Philosophical Consciencism as a conceptual map in the development of strategic actions in the categorial-conversion zero-sum game in the actual-potential polarities.

VIII Organization, Mobilization and Philosophical Consciencism

The important role of the theory of Philosophical Consciencism must relate to *organizational intelligence* and *mobilization action* on the part of the negative and positive duals. Organization is required to mobilize the elements of the characteristic set and then create a mass support for either the destruction of the conditions of the actual and creation of a *categorial-power vacuum* for the actualization of a potential element, or the maintenance of the conditions of the actual to prevent a creation of a categorial-power vacuum and an actualization of any potential element. The theory of Philosophical Consciencism embraces the fundamental needs and experiential information structure to create an ideology in support of the mobilization of the elements of the characteristic set to bring about sufficient conditions to win the categorial zero-sum game at a maximum speed. The theory of Philosophical Consciencism provides a logical framework that would ensure the rational utilization of the internal resources of the duality in the game of categorial conversion.

The logical framework is used to create socio-natural groups and relevant institutional support where every characteristic belongs to a group as well as employs the power of organization.

The internal structure where every characteristic is isolated from others and exists by itself is ontologically unattainable. Every ontological element is an internal system of elements with various assigned roles for the identity with external and internal survivability of the element. Here, every element of the characteristic set is duty bound to serve a common interest which is either continual existence or continual transformation of identity creating history of varieties with continual updating of the organic information structure. The theory of Philosophical Consciencism, thus projects collectivism, with a relational continuum and unity as a result of the categorial experiential information structure. The use of the epistemic guidance to organize and mobilize brings fruitful results to the collective in the game of dominance as seen from the Africentric principle of opposites that establishes an organic concept of a problem-solution path as an ontological condition of continual creation and destruction of varieties.

The positive Philosophical Consciencism informs the positive dual in terms of strategies to efficiently form the relevant organization and implied set of institutions and mobilization of the characteristics to win and create dominance, manufacture categorial moments for intra-categorial conversion by destroying the conditions of the actual, and then manufacture a transfer function to bring about inter-categorial conversion that is powerful enough to satisfy the transversality conditions for the negation of the actual in the actual-potential polarity. The negative Philosophical Consciencism informs the negative dual in terms of strategies to efficiently form the relevant organization and implied set of institutions and mobilization of the characteristics to win and create dominance, manufacture categorial resistance against intra-categorial conversion; and the manufacture protective belts to contain the possible transfer function and weaken it from acquiring a sufficient power from bringing about inter-categorial conversion that will satisfy the transversality conditions for the negation of the actual in the actual-potential polarity.

The results of activities of organizing and mobilizing as well as the strategies executed by both the negative and positive duals are relationally connected to the experiential information structure by information expansion (quantity) and deepening (quality) on the basis of which new strategies are formed by each one of them from the instruction of the conceptual system of the relevant Philosophical Consciencism. The act of mutual updating of the experiential information structure by expansion and deepening becomes part of the game of the mutual negation and categorial conversion of the poles. The power of this act of mutual updating of the experiential information structure as part of the game is the development of information deception through manipulation and propaganda to create deceptive information. The development of deceptive information structure is available to each of the duals depending on the ideological and moral contents of the Philosophical Consciencism. This availability and possible use is amplified in Philosophical Consciencism in dealing with control strategies in the categorial game of social polarities. Basically, the deceptive information structure is an information strategy that may come from the possible-world space for misleading an opposing dual to adopt a sub-optimal strategy in the game of mutual negation. It becomes part of the arsenal of mixed strategies to abstract an advantage. This is especially the case in categorial conversions of social actual-potential polarities under cost-benefit rationality in categorial dominance where organization and mobilization affect the distribution of social costs and benefits and creates increasing tension. The development of the theory of Philosophical Consciencism must account for the possibility of the use of deceptive information structure as an effective strategy of the categorial-conversion game. It is under the problems of social categorial conversion with the use of deceptive information structure that the development of the theory of Philosophical Consciencism meets its analytical challenges in abstracting relevant decision-choice instrumentation of arsenals and the construct of mixed strategies for the resolution of conflicts and contradictions.

IX Policy Instrumentality, Controllability and Philosophical Consciencism

The power of the theory of Philosophical Consciencism is its defining structural framework for creating a set of general actions for implementation. This set of general actions constitutes the guidance of decision-choice controllability of the complex multi-variable categorial-conversion process of the actual-potential polarities. It also defines the set of mixed strategies for the game of negation of negation that is played by the duals of the duality as well as by the negative and positive dualities under the principle of opposites. The sequence of successes of the categorial controllability requires the creation of multi-dimensional instruments of control. The instruments of control are policies and corresponding institutions through which the needed resources are mobilized and policies are implemented within the organization. The categorial instrumentation and the decision-choice actions to determine relevant policies and corresponding institutions are under the conceptual and ideological guidance of the contents of Philosophical Consciencism.

The creation of the policies and institutions for the task of controllability of the categorial-conversion process is what is referred to as *categorial instrumentation*. The arming of policies and institutions to serve as the means or agency of conversion is called *categorial instrumentality* for the control of the categorial-conversion processes of actual-potential socio-natural polarities. For negative Philosophical Consciencism there will be corresponding categorial instrumentation and categorial instrumentality for strategic and tactical actions that will favor the activities of the negative dual for negative negation. Similarly, for positive Philosophical Consciencism there will be corresponding categorial instrumentation and categorial instrumentality for strategic and tactical actions that will favor the activities of the positive dual for positive negation.

The categorial instrumentation to bring the categorial instrumentality into effect is an indispensable part of the general development of the theoretical framework of the contents of any Philosophical Consciencism whose uniqueness is always determined by the experiential information structure composed of internal and external information structures. The framework includes the guidance for information collection, information storage, information processing into knowledge, the use of the knowledge to set vision and objective, and to craft strategies and tactics to manage the relevant categorial-conversion parameters of the qualitative controlled system to change or protect the actual in the actual-potential polarities. The characters and conceptual framework of the Philosophical Consciencism for creating conditions of instrumentality are different for social and natural categorial conversions, not in terms of the roles they play but in terms of the contents of guidance of the control process. The construct of the natural Philosophical Consciencism is an ontological process to create the relevant conditions for categorial instrumentation and instrumentality by the activities of intra-categorial conversion to manufacture the categorial moments and activities of inter-categorial conversion to manufacture the transfer functions that are sufficient enough for the satisfaction of the categorial transversality conditions to bring about the categorial conversion of an actual pole in natural actual-potential polarity. This is the natural process of categorial instrumentation.

The construct of social Philosophical Consciencism on the other hand, is an epistemological process that mimics the accumulated social knowledge of the ontological process to create the relevant conditions for categorial instrumentation and instrumentality by the activities of intra-categorial conversion, to manufacture the social categorial moments and activities of inter-categorial conversion to manufacture the social transfer functions that have sufficient force enough for the satisfaction of the social categorial transversality conditions to bring about the categorial conversion of an actual pole in social the actual-potential polarity. The theory of Philosophical Consciencism is to guide the processes of instrumentation, instrumentality and controllability for the categorial conversion of the socio-natural actuals in the actual-potential polarities. The social process of categorial instrumentation, categorial instrumentality and social controllability is understandable if one examines, for example, the independence-colonialism polarity in conditions of political-power struggles as governed by the principle of opposites with residing dualities whose duals are defined by negative and positive characteristic sub-sets.

In the conceptual framework of social Philosophical Consciencism, the instrumentality involves all policy constructs, broadly defined, and controllability involves all institutional constructs, where the construction of both policy and institutional instruments are under the decision-choice actions which are monitored and guided by social Philosophical Consciencism to reconcile conflicts in the cost-benefit space as it is related to contradictions in the collective decision-choice actions. The degree of success of instrumentality will depend on decision-choice agents who have understood the logical framework and ideological thrust of the theory of social Philosophical Consciencism and believe in the logical framework and the ideological thrust in order to utilize the contents of Philosophical Consciencism to create social vision, national interest and the supporting goals and objectives with appropriate institutions, to design categorial moments and transfer functions to bring about intra-categorial and inter-categorial conversion in order to overcome the barrier of categorial transversality conditions in the social space. For each experiential information structure and cultural DNA, the theory of Philosophical Consciencism is unique in its relational construct. It constitutes a program of thought and action that is in relational continuum and unity, the development of which must be undertaken from within to guide the categorial-conversion process.

X Common Elements of Categorial Conversion Useful for the Development and Use of Philosophical Consciencism

It is useful to conclude the prologue by enlisting some common static and dynamic properties about the categorial-conversion processes that will serve as foundation for the development and use of the theory of Philosophical Consciencism to bring about the sufficient conditions of transformation.

1. Categories are classes defined by negative and positive qualitative characteristic sub-sets, where the negative-positive relative characteristic sets define the elemental identities and categorial belongings.
2. Categorial conversion is about qualitative changes through the alteration of the relative characteristic set of the actual, and hence a redefinition of the elemental identity and the categorial belonging of the actual which is always one (that is a singleton set).
3. The categorial conversion is about the change of the actual by simultaneously destroying it and transforming it into the potential, and construction and actualization of a potential element through the substitution-transformation processes.
4. The potential elements to replace the actual are many and so also the directions to reach them (that is, there is a set of potential elements and a set of categorial-conversion paths). As such, the actualization of the potential element may be one out of many potential elements, where some may be undesirable in terms of real benefit. Similarly, the direction to reach any potential element may be one of many directions some of which are costly inefficient in real value.
5. Implied in any categorial-conversion process is decision-choice function. With the use of the knowledge of

experiential information structure and the evaluation of cost-benefit effects in the action space on the actual and the potentials, the decision-choice modules of each of the duals of the duality choose an alternative action that is favorable to it from the action space to promote its net-benefit in the game of power and dominance.

6. The decision-choice modules, the information collection and processing into knowledge, the cost-benefit evaluations, the selection of action from the action space, the instrumentation of required elements of control within the violence-nonviolence duality and the manner in which the actual is destroyed and a potential is actualized are all under the internal Philosophical Consciencism that specifies the guiding philosophy and ideology of the duals in negative and positive actions.

7. The objective of either the negative or the positive type of Philosophical Consciencism is to guide the decision-choice process of either the negative or positive dual to create the sufficient conditions through the enhancements of either the negative or positive categorial moment from the intra-categorial conversion process and categorial transfer function, to overcome the categorial transversality conditions through the inter-categorial conversion process to destroy the conditions that allow the actual to exist and create new conditions that will allow a potential element to be actualized, and hence bring about a transformation of the actual and the substitution of a new potential in a relational continuum and unity.

8. Any socio-natural transformation process is made up of two sub-processes of categorial conversion and Philosophical Consciencism. The combined sub-processes may not achieve the desired result of power and dominance by a dual since the transformation process is a zero-sum game.

9. At any time point in the game process, the outcome of the transformation process is irreversible, and it becomes added to the internal experiential information structure with a new category of actual-potential polarity defining a new game for power and dominance by the principle of opposites with continuum and unity. In this process, there is always primary

actual-potential polarity from which new actual-potential polarities are created as derivatives in a never ending process.

10. There are certain fundamental notions and concepts that are useful to understand and construct the theory of transformation as either an explanatory or prescriptive theory to the transformation problem of socio-natural ontological elements.

A) The theory of socio-natural transformation must be made up of three sub-theories of category formation, categorial conversion and Philosophical Consciencism.

B) The categories are formed by the negative and positive characteristics where the proportionality of the negative and positive characteristic sub-sets must be defined to establish elemental identity and categorial belonging.

C) Each element is viewed under the principle of opposites and defined as a duality with negative and positive duals in terms of statics and additionally defined as actual-potential polarity in terms of dynamics.

D) The type of duality and the nature of the duals must be specified where the relevant internal information structure must also be specified in addition to the type of information structure such as fuzzy-stochastic or stochastic-fuzzy information structure, where the information structure must include deception and propaganda in its representation.

E) The decision-choice processes as they relate to knowledge production, categorial moments, and transfer functions, valuation of states of the actual and the potential must be explicitly defined.

F) The initial conditions of transformation and the realized destination must be specified.

G) The decision-choice modules must be related to the game by the negative and positive duals of the duality, as well as indicate the necessary and sufficient conditions of convertibility and controllability of the transformations as they relate to actual-potential polarities.

H) In these conditions, the role of philosophy and the paradigm of thought with its logic and mathematics must be specified, and the manner in which they help to construct the theory of transformation and possible uses clearly indicated.

I) The two interdependently companion monographs, as they are presented, utilize at the level of philosophy, the Africentric principle of opposites which is made up of *dualities* and *polarities*. The polarities exist as a set of actual-potential polarities for transformation, and the dualities are the instruments to which the games of transformation are specified and played. The initial conditions are specified by the *primary category* of the actual-potential polarity and all other actual-potential polarities are derivatives for the primary. The information input is the internal experiential information structure composed of internal and external production. At the level of the paradigm, laws of thought and mathematics, the conditions of the fuzzy paradigm with its logic and mathematics are used to deal with opposites, duality and polarity with relational continuum and unity to specify the problem, derive the solution and discuss the results in the sense that negative and positive mutually create themselves under the principle negation of negation.

XI National Education, Curriculum, and Social Technology under Philosophical Consciencism

Finally, there must be a clear understanding of integrated relation among civic and non-civic national education, curriculum, social technology and experiential information structure under Philosophical Consciencism. In their conceptual structures and applications that may be demanded of national civics, curriculum, social technology and experiential information structure, Philosophical Consciencism relates to the character of the national education which defines the *diameter of the national consciousness* through a well-structured curriculum and its development. The diameter of the national consciousness helps to establish the *circumference of national conscience* and the philosophy that may be created to guide the national vision and the establishment of national goals and objective for a nation-building process to attain independence, freedom and justice for all. The character of national education sought will be affected by the character of *social vision* and the set of national goals and objectives as well as will influence the static nature and the dynamic process of the national curriculum development. The structure and character of the national education will affect the national information processing capacity and the decision-choice system required for the command and control of the management of the path of the national socio-economic development and politico-legal transformations that form the foundation of the nation-building effort.

The progress and direction of the national internal socio-economic development, politico-legal transformation, defense of sovereignty and a struggle against external threats will depend on the policies that are created through national decision-choice actions under the guidance of the national Philosophical Consciencism. The progress and direction of the national socio-economic development and politico-legal transformation will also depend on the perception of what constitutes a nation building, where the phenomenon of a nation building is encapsulated in the *national institutional configuration* and its development in intensive and extensive forms. The institutional configuration defines the structure of *social technology* on the basis of which *physical technology* and knowledge and other social know-hows are created from the experiential information structure and in relation to the national problem-solution path for social progress. The effectiveness of the social technology in the social transformation and the national problem-solution process depends on the national Philosophical Consciencism that provides guidance to the dynamics of national decision-choice system. This national decision-choice system also affects the character and development of the national Philosophical Consciencism and all elements that it guides. It is here that the idea, that great institutions are created by great minds and great mind and Philosophical Consciencism mutually create themselves, finds some epistemic rest in the sense that Philosophical Consciencism and its development are the derivatives of people who think as people of thought and acts as people of action. In other words, philosophical Consciencism unites thought and action under the phenomenon of continual internal transformation through its effects on social consciousness and the national decision system. The effectiveness of the theory of Philosophical Consciencism acquires a transformational intensity in the dynamics of socio-natural actual-potential polarity when it is united with the theory of categorial conversion to define the necessary and sufficient conditions through the interplay of education, curriculum, social technology and the dynamics of national decision-choice system, where practice is guided by thought and thought is translated in action for a continual problem-solution process in the sense that every solution to a problem brings in a new problem requiring a new solution. The path of this problem-solution process is the enveloping of elemental history of things.

CHAPTER ONE

THE GENERAL FRAMEWORK OF THE PRACTICE OF NKRUMAISM: The Theory of philosophical Consciencism in Social Systemicity

Nkrumaism may be defined by two interdependent logically mathematical and philosophical structures in logically relational unity which is composed of *Categorial Conversion* (CC) and *Philosophical Consciencism* (PC) as applied to social systemicity. Discussions on the unity of the two and their applications the African conditions are what Nkrumah called Consciencism. The two have powerful analytical roles in understanding the transformations of nature and society from a common logical frame in socio-natural systems. The socio-natural transformations are about destruction of existing varieties and the creation of new varieties through ontological and epistemological command and control of decision-choice actions. The common logical frame relates to *qualitative disposition, quantitative disposition* and *temporal disposition* (time). Here, time may be seen as neutral from the viewpoint of elements in quality-quantity-time dimensionality. The categorial conversion is discussed in a monograph which is entitled *The Theory of Categorial Conversion: The Rational Foundations of Nkrumaism* [R1.90b]. The rational foundations are composed of philosophy, mathematics and computable systems. The structure of categorial conversion in Nkrumaism is to provide a logical and philosophical argument in support of rational justification of internal self-motion of matter, and use this justification to motivate internal self-transformation of any social set-up by providing necessary conditions of convertibility of social varieties. It thus provides a general mechanism of change in the qualitative and quantitative dispositions of elements and categories where existing social varieties in the space of actual are intentionally destroyed, and new varieties from the potential space are created in transformation-substitute dynamics.

The philosophical and mathematical foundations merely laid down the theoretical framework as a system of thought that must guide the process of reconciling the conflicts between negative and positive actions in the society for socio-political transformations relating to qualitative and quantitative dispositions of social institutions and socio-economic development stages that present themselves as organizational varieties. The theory of Philosophical Consciencism presents a logical framework to mimic the categorial-conversion dynamics (transformation dynamics) of nature in a manner that will bring about controllability of socio-economic transformations by designing the sufficient condition through decision-choice actions to create sufficient conditions in support of the necessary condition indicated by the theory of Categorial Conversion. The transformation dynamics is such that the social system is seen to be composed of a triplet of economic, political and legal structures. The triplet forms an integrated unity of institutions that present a framework for the management of command-control actions of decision-choice system for designing and implementing transformations of the unity of economic, political and legal institutions, where each unity is viewed as institutional variety. At each time point there is one element of the actual social variety in the actual space for any country. However, there are infinite set of potential varieties that may be written as $\Box = \left(\varpi_i \mid i \in \mathbb{I}^\infty \right)$

where each ϖ_i is a variety that may be actualized from the potential space. The decision sovereignty of every institutional variety at each point of time may be specified to belong to a set of combinational degrees of domestic-foreign control structure. The identity of each social variety $\mathbb{V}$ at each time point is defined by a united combination of resource allocation, $\mathbf{A}$, output production $\mathbf{P}$ and output distribution $\mathbf{D}$ such that the identity of the social variety may be written

$$\mathbb{V} = \left(\mathbf{A} \otimes \mathbf{P} \otimes \mathbf{D} \right) = \left\{ \upsilon = \left(\mathbf{a}, \mathbf{p}, \mathbf{d} \right) \mid \mathbf{a} \in \mathbf{A}, \mathbf{p} \in \mathbf{P}, \mathbf{d} \in \mathbf{D}, \forall t \in \mathbb{T} \right\}$$

where $\mathbb{T}$ is a time set. The *allocation* is defined by the institutional mechanism. The output *production* is defined by a broad know-how within the institutional mechanism. The output *distribution* is defined by the institutional configuration and the value system that specifies the fundamental ethical postulate regarding the relational nature of individual-community duality at any time point. The logical system as set up for the development of the theory of socio-natural transformation may be represented in an epistemic geometry as in Figure 1.0.1.

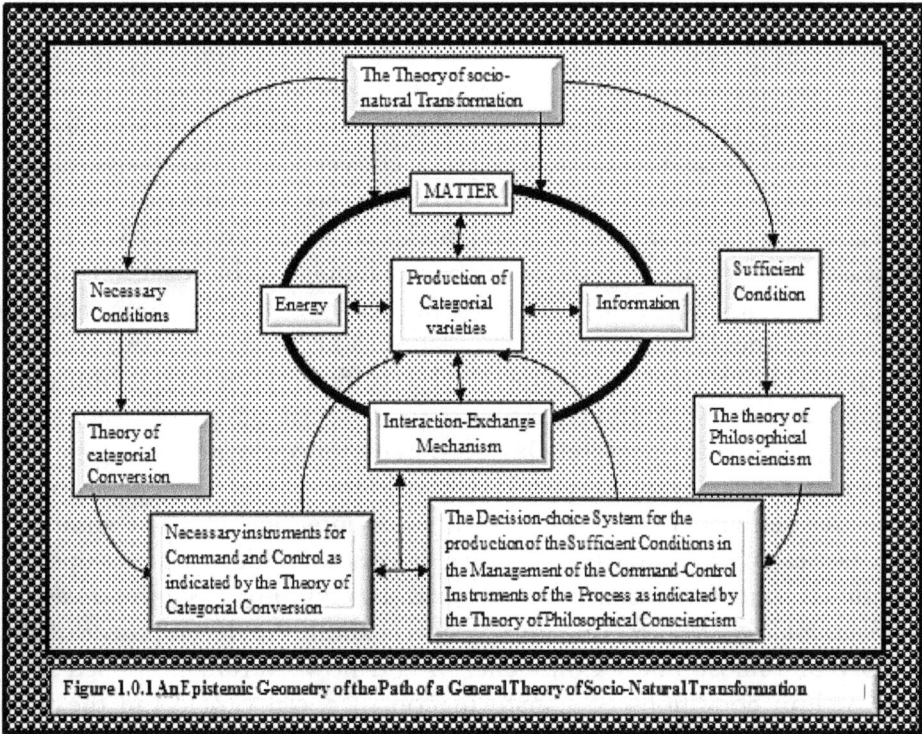

Figure 1.0.1 An Epistemic Geometry of the Path of a General Theory of Socio-Natural Transformation

The active and positive social transformations are seen in the Nkrumah's conceptual framework as composing of the development of intellectual guidance, and the development of a set of positive actions (social programs) consistent with the intellectual guidance for transforming categories of unwanted social conditions to categories of desired social conditions. These social programs may be seen as elements in the policy space that contains the *program set* $\mathbb{P}$. The implementation of the elements in the program set constitutes the *action set* $\mathbb{A}$. In this respect, the guidance must be practically related to the unity of thought and practice which may alternatively be expressed as the unity of theory and application where every application has an explicit or implicit theory behind it, and every theory points to a possible direction of application. Every practice has a theory behind it and every theory has an actual practice or potential practice. The thought, practice, direction of application and social action, individually finds expression in human intentionality and preferences that drive the social decision-choice processes, tactics and strategies of social action towards social goals and

73

objectives as indicated by the social intentionality. The intentionality and preferences provide information that acts as input into the decision-choice process to maintain or alter the category in which social elements belong. The general necessary conditions of *categorial convertibility* are derived under the theory of categorial conversion. The convertibility conditions are defined by a set of conditions of categorial moment, categorial transfer function and categorial transversality. These are necessary instruments that must be manipulated by decision-choice actions to bring about socio-natural transformations. The knowledge of these instruments for flexible manipulation is necessary but not sufficient to bring about transformations. The sufficient conditions must be created under the cultural confines of the social set-up by command-control managerial decision-choice system must design strategies and tactics that are relevant to specific national interest and the general social-objective set within the conditions of social intentionality. The strategies and tactics for fulfilling these sufficient conditions in social set-ups for continual transformation are indicated by Nkrumah as the *theory of philosophical Consciencism*. This monograph is about the development of the theory of Philosophical Consciencism and how Nkrumah applied it to the African conditions of transformation of social varieties in the time domain. Social transformation may not lead to the desired result due to competition of negative and positive forces in the social actual-potential polarity. In view of this, the overriding importance of the theory of Categorial Conversion is to isolate and make explicit the necessary command-control instruments that may be manipulated in the transformation process of varieties. These instruments are the exact controllers under ontological decision-choice actions. They are, however, the fuzzy controllers under epistemological decision-choice actions. The overriding importance of the theory of Philosophical Consciencism in support of the theory of Categorial Conversion is to create a wining organization of decision-choice systems for ontological and epistemological actions with winning sets of strategies and tactics to bring about the destruction of undesired actual variety and bring about the creation of the desired potential variety at each time point. As discussed, the complete theory of transformation is made up of the theory Categorial Conversion and the theory of Philosophical Consciencism. The two sub-theories may be viewed in terms of explanatory and prescriptive sciences.

1.1 Information Requirements, the Decision-Choice Process and the Theory of Philosophical Consciencism

The practice of Nkrumaism where thought and practice are unified in the decision-choice system may be seen in terms of Nkrumah's socio-political actions. Generally, each selected theory or thought for practice is expressed as prescriptive propositions that are aggregated into a social *prescriptive rationality* which may be related to either explanatory theory embedded in an explanatory science or related to prescriptive theory which is embedded in prescriptive science. The prescriptive rationality derived on the basis of explanatory science that contains explanatory theory and philosophy is called *explanatory-theory-based prescriptive rationality*. Similarly, prescriptive rationality derived on the basis of prescriptive science which contains prescriptive theory and philosophy is called *prescriptive-theory-based prescriptive rationality*. The explanatory science and the corresponding prescriptive rules are empirically derived on the average on the basis of explanatory theory. The prescriptive science and the corresponding prescriptive rules are axiomatically derived, on the average, on the basis of prescriptive theory [R3.7] [R15.14] [R3.10][R3.13][R13.7] [R13.13]. Both approaches have different areas of effective application within Nkrumaism and its applied framework of social transformation. The explanatory-theory-based prescriptive rationality is about the actual while the prescriptive-theory-based prescriptive rationality is about the actualization of potential.

The explanatory-theory based prescriptive rationality is used to destroy the conditions of oppression and then to transform societies which are either subjected from within such as domestic dictatorships or from without, such as imperialism, colonialism, neocolonialism and imperial occupation. The application of this type of prescriptive rationality is to bring about an effective change of the control of decision-choice sovereignty with which the lives of the people under subjugation have no possibility of improvement in accord with their collective preferences and will. The change in the control of the decision-choice sovereignty is about the acquisition of political freedom to acquire political power that helps to bring about economic and legal freedoms and power. It is on this basis that Nkrumah advices that: *political power is the inescapable prerequisite to economic power* [R1.207, p.78]. This advice is further strengthened by an additional statement:

Seek ye first the political kingdom became the principal slogan of the CPP, for without political independence none of our plans for social and economic development could be put into effect [R1.202, p.50].

Nkrumah's advice of *seek ye first the political kingdom* is more powerful in understanding sovereignty, freedom and justice than any political theory can analytically contain. This statement reflects the recognition of the fundamental institutional trinity of the political structure, economic structure and the legal structure and power trinity of political power, economic power and legal power for whose analytical potency has been discussed in [R3.12][R13.8][R13.9].

Nkrumah clearly came to understand the fundamental relationships among economics, politics and law, and how they affect the distribution of social power in relation to individuals and the collective, for the construction or dismantling of a unitary state. He came to understand that by the social construct, the ultimate social power is vested in the political structure that is consistent with *seek ye first the political kingdom*. He understood that when the political power is controlled, it can be used to enact laws, rules and regulations that will change the current legal structure which then alters the conditions in the economic structure on the basis of economic actions which are individually and collectively undertaken. The laws may also affect the individual and collective decisions in the legal and political structures in terms of social participation, freedom and justice within the individual-collective duality of the social set-up. The conditions of the social participation, freedom and justice within the individual-collective duality provide cost-benefit calculations in support of configurations of decision-choice actions by the individual and the collective. The organizational structure presents important interrelationships among society's freedom and justice on a pyramidal structure, and politics, law and economics constituting another pyramidal structure. The organizational structure is shown in Figure 1.1.1 with government and governance at the center. The three structures of politics, law and economics provide three sovereignties that constitute the social power system in which the political power constitutes the primary category that institutionally gives rise to the legal and economic powers as derivatives. The exercise of the social power system and the distribution of the power are controlled by on information-knowledge system which is generated by the content of the theory of Philosophical Consciencism. The general concepts of information, knowledge and information-knowledge system are discussed in [KKD, Gen Theory of info]. The relational structure of information and transformation dynamics of varieties is also discussed in [KKD, Info-dynamics].

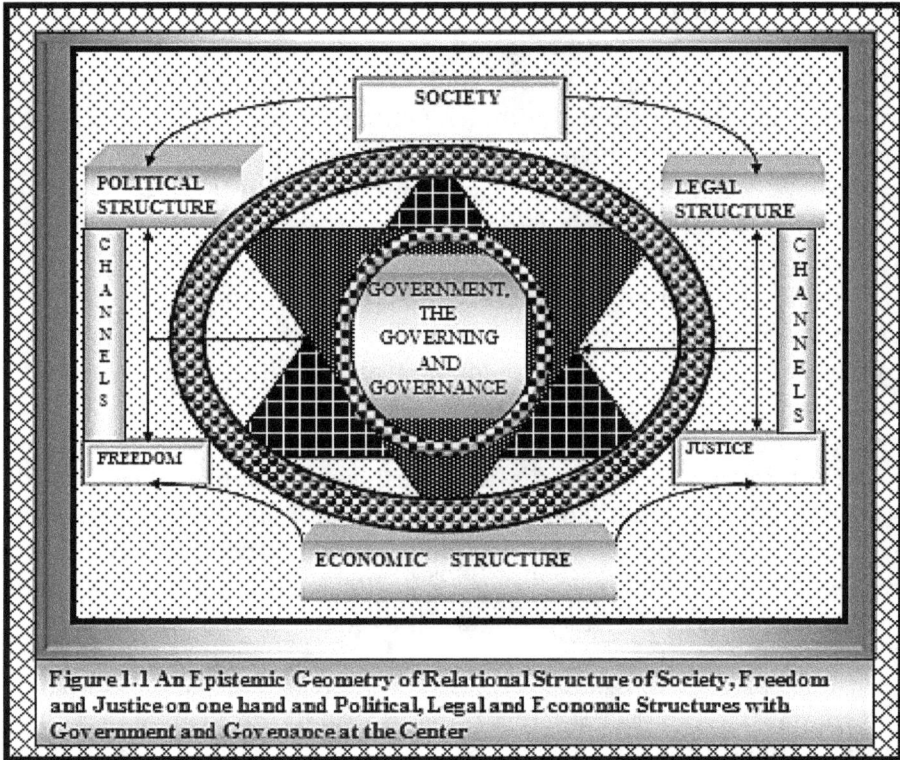

Figure 1.1 An Epistemic Geometry of Relational Structure of Society, Freedom and Justice on one hand and Political, Legal and Economic Structures with Government and Governance at the Center

The relational structure is complexly connected to freedom, justice and independence which must be related to the social states in the individual-community duality for social progress and stability. Here we must observe that in a social collectivity, freedom without law and rules may create anarchy, while unjust laws may take away freedom, institute injustices, and create oppression and possibilities of social instabilities that will foster social repression by force. The question then, boils down to how to create an optimal balance between law and freedom that will maintain justice within the individual-community duality given any social pole of the actual-potential polarity. Foreign control of the political structure through colonialism, or neo-colonialism either directly or indirectly controls the national sovereignty and political power. This

77

disenfranchises the citizens and reduces them to puppetry behavior in support of the foreign interest and imperial aspiration. Alternatively, a domestic control of the political structure leads to the control of national sovereignty and political power that can be used to control both the legal and economic powers for the interest of the domestic population. The external or internal control of the political structure can be used to consolidate the political power and shape the institutional political arrangements. It is within the understanding of this relational structure that Nkrumah's statement must be understood and related to the national decision-choice sovereignty and the control of the decision system that affects Africa's national interest, social vision and citizens' protection and social participation.

The application of Nkrumaism sees freedom and justice as seeds within independence that must geminate to free the African masses from imperialism, colonialism and neocolonialism. When independence is won, freedom and justice are mere seeds of potentials that must be watered to germination, matured to growth, guided for sustainability, and protected for the people. The analytical justification is found in the postulate of categorial convertibility in Nkrumaism as it has been explained in [R1.90b] where freedom and justice are Africentric traditions as they have been discussed in [R1.92, pp.82-138]. It is on the basis of this postulate that led Nkrumah to state:

> We have to be constantly on the alert, for we are steadfastly resolved that our freedom shall never be betrayed. And this freedom of ours to build our economics, stands open to danger just as long as a single country on the continent remains fettered by colonial rule and just as long as there exist on African soil puppet governments manipulated from afar. Our freedom stands open to danger just as long as the independent states of Africa remain apart [R1.202, p. xvii].

This statement by Nkrumah is a strong recognition of the properties of duality and polarity with continua under the conditions of categorial-conversion of reversibility. Categorial conversion works within duality and polarity under the principles of continuum and unity. The duality in all forms may be viewed as consisting of negative dual and positive dual that reside in continuum and unity. Similarly, polarity is viewed as composing of actual pole and potential pole existing in continuum and unity as have been discussed in [R1.90b] [R1.92]. The negative and positive duals and the actual and potential poles have different representations and interpretations in different applications using the intellectual framework of Nkrumaism. Nkrumah's application of the theory of categorial conversion is with social polarity and duality with special reference to the African conditions. The negative and the positive exist in a mutual continual negation which is extended to the actual and potential poles in continuum and unity over the path of categorial enveloping. Every successful categorial conversion over the conflict of the previous polarity leads to a new relational structure of polarity and duality with different conflicts that generate different forces of mutual categorial negation. The logical frame of the operations of polarity and duality within the framework of Nkrumaism is the recognition that the polarity exists as actual and potential poles and the duality exists as negative and positive duals which are linked together by a process to establish an infinite continuity of change in nature and society in relational continuum and unity. It is within this logical and analytical position that the statement that *the only thing permanent is change* acquires philosophical and mathematical credibility in nature and society. Similarly, it is within this logical and analytical claim that digital or discrete is a cognitive approximation to analog or continuum acquires logical and mathematical credibility.

Within Nkrumaism, every categorial conversion of a social set-up is a conversion of the actual (old social order) and is in favor of a potential (a new social order). When a potential is actualized, it sheds off the conditions of its potentiality and acquires the conditions of the actual to become a new actual to give rise to a new actual-potential polarity. The actualized potential becomes a target of categorial struggle for a categorial conversion which may be negative or positive conversion. The negative categorial conversion is consistent with the principle of categorial reversibility for backward qualitative motion while the positive categorial conversion is consistent with the principle of categorial non-reversibility for forward qualitative motion. Each case is defined by a set of conditions that places it in the category of its belonging as well as possible convertibility. Every actual contains remnants of the previous category as well as the seeds of potential categories to be derived since every duality contains negative and positive characteristic sets. There are conditions which allow the mutual existence of the socio-natural actual and the socio-natural potential. These conditions are different for social actual-potential polarity and natural actual-potential polarity. The general logical foundations for their categorial conversions are the same but the forces that create the conditions of convertibility are different. The conditions of the mutual existence of the actual and potential create temporary equilibrium states in the actual-potential balances for *categorial equilibrium* in the polarity depending on the relative set-relation of the duals in the polarity. There are three different sets under dynamic conflicts in continuum and unity of international political economy for any state and time in duality relative to the conditions of the poles in the relevant polarity.

1.2 Imperialism, Neocolonialism, African Emancipation, Categorial Conversion and System Dynamics

Let us relate the conditions of these three categories of imperialism, neocolonialism and African emancipation to the transformational dynamics of colonialism. National independence emerges from colonialism and imperialism which, by the principle of categorial reversibility may emerge from independence as neo-imperialism and neocolonialism with disguised characteristics to turn independence of the African states into a mirage in the desert. The negation of colonialism to independence, and the negation of the negation of independence to neocolonialism were clearly understood by Nkrumah under the application of the epistemic conditions of the theory of categorial conversion. In other words, just as independence emerges out of colonialism and imperial subjugation through a process of positive categorial conversion, neocolonialism and neo-imperialism emerge out of independence through a process of negative categorial conversion. The recognition of these new faces requires social and cognitive vigilance in terms of recognition of their characteristics. It is this logical presence of the double negation as generally seen but with special reference to the African politico-economic conditions that motivated Nkrumah to tackle the morphology of neocolonialism to expose its inner mechanism [R1.204] and to offer an internally lasting and sustainable solution through his work on African Unity [R1.2O2]. The logical structure of this work on the basis of the theory of categorial conversion has also been extended in [R1.191]. Similarly, by the principle of non-reversibility in the theory of categorial conversion, complete emancipation composed of economic freedom and nation-building may emerge from national independence through the process offered by the theory of categorial conversion as is fully developed in [R1.190b] by building on the

Africentric conceptual system with the embedment Nkrumah's basic conceptual system. Neocolonialism and complete emancipation are important seeds in the colonialism-independence polarity waiting to geminate within national independence when it is attained. Which one of the seeds will geminate depends on the socio-economic environment, the peoples' ideological awareness and the counter insurgency of imperialism, its *deceptive information structure* and the peoples' acceptance. After decolonization the colonialism-independence polarity is transformed into two possible new actual-potential polarities of independence-emancipation polarity or independence-neocolonialism polarity. Here, independence is the actual and emancipation and neocolonialism are the potential. Whether national political independence will be transformed to complete emancipation or neocolonialism will depend on conditions of how positive and negative actions are generated from within the society. It is on these conditions of double negation in categorial conversion that motivated Nkrumah to state that:

> There is no force, however formidable, that a united people cannot overcome [R1.207, p.107]. It is by the people's effort that colonialism is routed, it is by the sweat of the people's brow that nations are built. The people are the reality of national greatness. It is the people who suffer the depredations and indignities of colonialism, and the people must not be insulted by dangerous flirtations with neo-colonialism [R1.203, p103].

It may also be noted that colonialism of all forms emerges from national independence just as national independence emerges from colonialism of one form or the other. The situation may be seen in the conditions of freedom-oppression duality of justice–injustice duality with continuum and relational unity. When national independence is attained by decolonization, it ascends to the ladder of actuality. All other elements stand as potential relative to the national independence as the actual. Given the set of conditions of the actual, there are three different sets of conditions under conflict around the present actual which may be examined in unity for any given state and time for the understanding of transformational dynamics of the social set-up. These conditions create temporary equilibrium states in the polarity depending on the relative relational structure of the dual of the dualities in each pole of the actual-potential polarity. This point of activity in the actual-potential polarity that exists in transitional equilibrium-disequilibrium duality has been theoretically illustrated in the monograph [R1.90b]. It is analytically and practically useful to illustrate this point with a politico-economic institutional arrangement in the social set-up.

For example, national independence emerges from colonialism and imperialism which by the principles of *categorial reversibility* may emerge from independence as imperialism and neocolonialism with a designated characteristic set to turn national independence into a mirage. The imperialism and neocolonialism that emerge from the decolonized process of independence present new faces and new characteristics in appearance but with acute exploitive behavior. The recognition of these new faces and the aggressive underlying behavior by the use of soft power require a social and cognitive vigilance in terms of their characteristics. Similarly, by the principle of non-reversibility, the complete emancipation composed of economic freedom and nation-building may emerge from the national independence on the basis of the peoples' organic action within the negative-positive duality. When national independence is attained by whatever means, complete independence and neocolonialism are important seeds in the conditions of national independence. Let us keep in mind the logic of continual change where there is only dominance of either a negative or positive characteristic set but not complete elimination. There is always a residual of either the set of the negative characteristics or the set of the positive characteristics. The dominance of a positive characteristic set implies the existence of reminisces of the negative characteristics set. Similarly, the dominance of a negative characteristic set implies the existence of reminisces of the positive characteristics set. They are waiting to geminate under appropriate socio-political conditions while national independence sits on temporary and unstable equilibrium. The conditions of temporary equilibrium must be maintained from within the social set-up. The conditions for change must be created from within to usher in an appropriate social environment that will be conducive in nurturing the appropriate categorial moment for appropriate categorial conversion.

The conditions of temporary equilibrium are maintained by a set of neutral actions. The conditions of neocolonialism are created by a set of negative actions while the conditions for complete emancipation are created by a set of positive actions from within the social set-up. The imperial external forces seek to create negative conditions from within and from without to generate an appropriate categorial moment to convert the national independence to a neocolonial state that denies life, liberty, justice and happiness under the framework of the will of the people in the neocolonial state. This is the negative categorial moment for reversibility of independence into a colony with a new face, new methods of exploitation and dehumanization. The success in creating a neocolonial state is attained when the process leads to the creation of the negative *Nkrumah delta*. This is the postulate of reversibility that is explained in [R1.90b] [R3.13]. Let us note that the postulate of categorial reversibility does not contradict Nkrumah's statement: *Once a colony has become nominally independent it is no longer possible as it was in the last century, to reverse the process. Existing colonies may linger on, but no new colonies will be created* [R1.204, p. ix]. He clarifies the reversibility situation by saying that *colonialism has achieved a new guise. It has become neo-colonialism, the last stage of imperialism* [R1.204, p.31]. The explanation lies in the behavior of the actual-potential polarity and the respective dualities in the poles. When political colonialism is destroyed, it becomes a potential with a seed in the independence. The independence becomes an actual with a seed in the potential bringing into focus a new contradiction, tension, struggle and politico-economic dynamics. The interactive processes between independence and colonialism have a central managing-unit power.

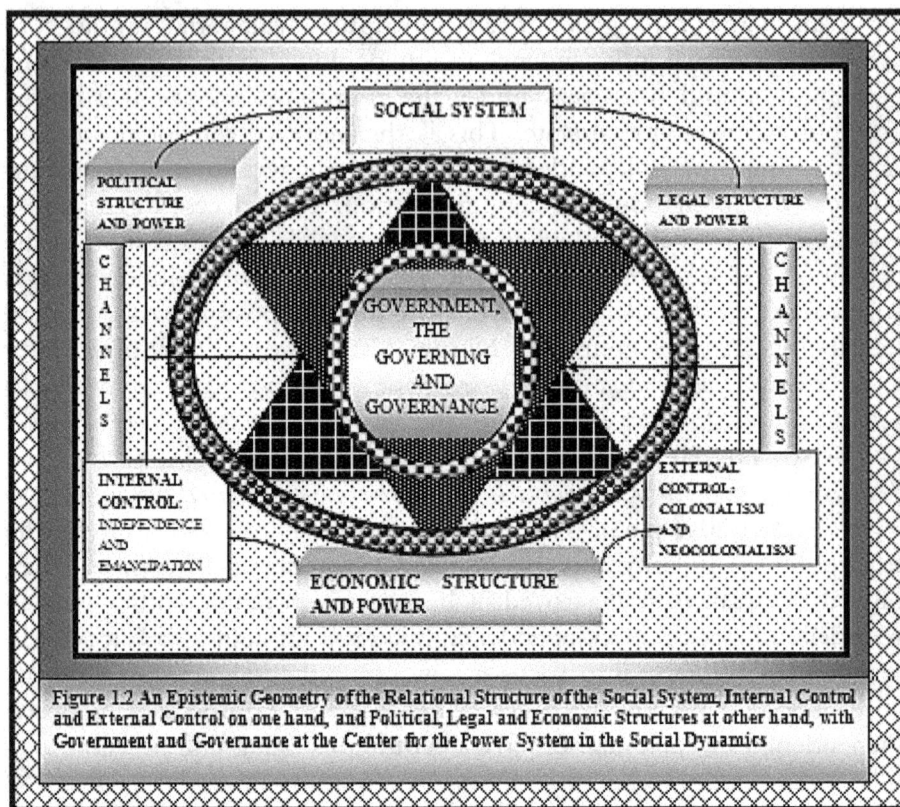

Figure 1.2 An Epistemic Geometry of the Relational Structure of the Social System, Internal Control and External Control on one hand, and Political, Legal and Economic Structures at other hand, with Government and Governance at the Center for the Power System in the Social Dynamics

The central managing power may be internal or external which constitutes controls over the social decision-choice system, with the production of cost-benefit configuration and cost-benefit distribution. External control distributes costs to the colonized and benefits to the colonizer. Internal control distributes both costs and benefits to the citizens. It is this asymmetry of cost-benefit creation and distribution that fuels the fire of conflict and motivates national liberation forces.

The national liberating forces, on the other hand, seek to create positive conditions from within to generate appropriate categorial moment to convert the state of national independence to complete emancipation that will ensure life, justice and liberty of the people of independent states. All the external forces supporting these internal forces under the principle of non-compromise are welcomed with suspicion. The imperial external forces and national liberation forces are opposites in duality with the polarity, where complete emancipation constitutes a pole and neocolonialism constitutes the other pole forming

the *emancipation-neocolonialism polarity.* These opposites exist under
continual conflict and in relational continuum and unity. Inside each pole
of emancipation-neocolonialism polarity, there is a duality composed of a
set of emancipation characteristics and neocolonial characteristics which
are inter-supportive of their existence with mutual negation through the
development of a categorial moment to create a transfer function to set
the potential against the actual. The matter containing the forces of
change is the people as the vanguard of continual change. The conflicts
are generated by opposing preferences and ideology of the people in the
opposing duals in relation to freedom and justice under the multiplicity
of cultural confines. Socio-political development implies a destruction-
construction process, where an existing social variety is destroyed and
moved to the potential space, and a potential social variety is actualized
with a qualitative movement from the potential space to the actual space
through the transformation-substitution dynamics.

Let us keep in mind the full epistemic view of the principle of
opposites as it applies to social polarities in relation to the actual-
potential polarity. Actual emerges from the potential and potential
emerges from the actual as one follows the evolution of kind.
Colonialism emerges from independence. There are a set of conditions
that makes it possible. First, the conditions of independence have to be
destroyed by reducing its characteristics to an ineffective minimum by
introducing into the African social organs and reality a categorial
moment which allowed the categorial conversion from independence to
colonialism by altering the relative balance between the forces of external
domination and internal stability, between the forces of freedom and the
forces of oppression, and between the forces of justice and the forces of
injustice. The destruction of African territorial and socio-political
independence and the emergence of colonialism did not just happen by
accident which is discounted in the Africentric conceptual system
[R1.92]. A categorial moment, composed of energy, power and force was
created by competing imperial European countries with resource-seeking
aspirations through arms robbery. The creation of the social energy,
power and force was done in a number of ways from without and from
within. The external process was composed of killing, terror and
disrespect for humanity coupled with mass execution and many other
instruments of coercion, that is, instruments of violence and fear. The
internal process involved deception, propaganda, divide-and-rule, and
internal recruitment of stooges through dubious biblical claims and

empty promises that reside in some place called Heaven. The external and internal processes by the members of the imperial order for creating the categorial moment were amplified by the creation of various types of institutions of oppression through which policies were created to advance conversion from original African independence to colonialism with imperial occupation and control of the people, and the capture of national sovereignties supported by institutional configurations through which the imperialist ideology and policies are transmitted. Under the principle of opposites, one is dealing with the conflict between domestic and foreign control of domestic national sovereignty where such conflict resides in justice-injustice duality. The imperialist institutions were appropriate and optimally created to bring about the needed categorial moment for the categorial conversions from an independence pole to a colonialism pole through the activities of the imperialists. In other words, by creating an appropriate decision-choice system for managing the set of fuzzy social controllers the imperialists were able to manufacture a categorial moment to destroy the existing variety of the African social order and created a new social variety of the African social order called colonialism from the potential space. The foundations of role of information in the interplay of decision-choice actions and the management of the fuzzy social controllers in destroying and creation of varieties are discussed in the monograph on general theory of information [R9.12]. The general important point to keep in mind is that the information about the past and present are in stock-flow disequilibrium.

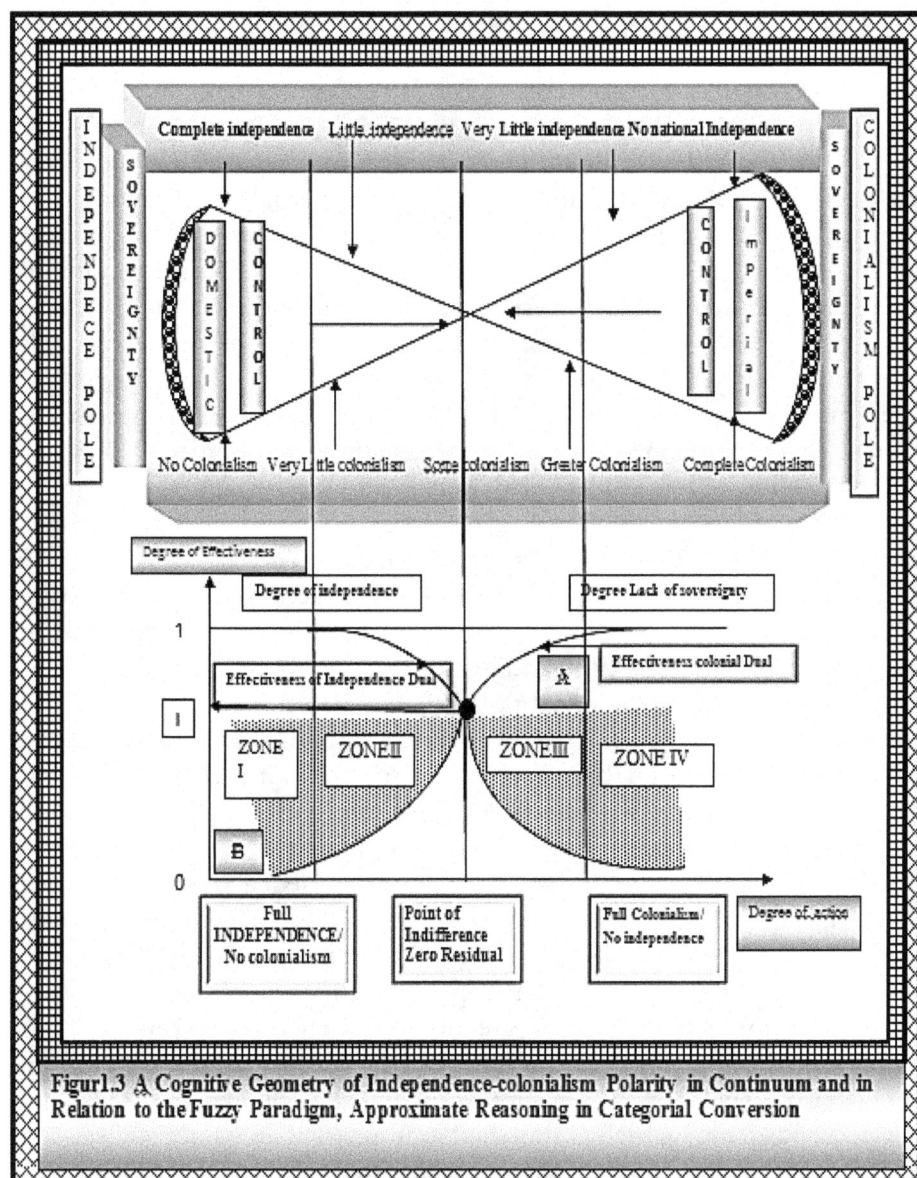

Figur1.3 A Cognitive Geometry of Independence-colonialism Polarity in Continuum and in Relation to the Fuzzy Paradigm, Approximate Reasoning in Categorial Conversion

The European imperialists transformed Africa's sovereignties and traditional independence into European colonialism and loss of traditional sovereignties, where national decisions were imposed on the Africans from outside through the seizure of the political structure and the changes in the legal structure and order to promote the imperialists' interests, where the Africans' powers to decide for themselves were replaced by the imperialists' powers to decide. The structure of the relational continuum and unity of independence-colonialism polarity, zones of control and sovereignty is presented as cognitive geometry in Figure 1.3. By understanding the mechanism of categorial conversion on the part of the people of imperial states for negation of independence, a counter mechanism may be undertaken for the negation of colonialism to a new independence different from that of African's traditional independence. It is on this basis that Nkrumah spent a lot of time to study and to write about colonial freedom and justice as delivered by the imperial system [R1.197][R1.200][R1.204][R1.206]. The same conditions in the understanding of colonialism and imperialism by Nkrumah also motivated other members of anti-imperialist minds [R1.9][R1.9][R1.63] [R1.102]. Here one is dealing with social actual-potential polarity where the traditional independence is the actual pole and also the primary category. In all socio-economic systems, care must always be exercised in defining the opposites and the corresponding actions. These opposites must be related to decision-choice actions towards redemption or oppression, towards freedom and justice or towards law and order.

In this framework of social polarity, colonialism is the potential pole and a derivative from the actual. The initial traditional independence pole is defined by a duality which is specified in terms of a characteristic set. The characteristic set is divided into positive and negative sub-sets in conflict with each other, as well as give identity to the duality as positive duality that defines the identity of the independence pole in the sense that the positive characteristic sub-set dominates the negative characteristic set. The negative characteristic subset produces negative action on the aggregate, and the positive characteristic set produces positive action on the aggregate. Independence is implied by the condition that the positive characteristic subset dominates the negative characteristic sub-set in the positive duality. The colonialism pole is also defined by a duality which is specified in terms of a characteristic set. The characteristic set is also divided into positive and negative sub-sets in conflict with each other, as well as give identity to the duality as negative duality that defines the identity of the colonialism pole, in the sense that the negative characteristic sub-set dominates the positive characteristic sub-set. The negative characteristic sub-set produces negative action on the aggregate and the positive characteristic set produces positive action on the aggregate. Colonialism is implied by the condition that the negative characteristic sub-set and negative action dominate the positive characteristic sub-set and positive action in the negative duality.

Colonialism is actualized if the negative forces and the negative action of the colonial pole, on the net, outweigh the net positive forces and positive action of the independence pole of the independence-colonialism polarity. The actualization of colonialism reverses the conditions of dominance in the polarity. Independence is negated and pushed to be a potential while colonialism becomes the actual defining new conditions of struggle by role reversal under the principle of opposites [R1.90b] [R1.92]. When independence is negated the existing social variety is destroyed, and in place is a creation of a new social variety called colonialism from the potential space. To claim national independence requires the act of decolonization to assert control over national sovereignty and over the national decision-choice system. This implies taking control of the national political structure that holds the decision-making power which can be made to control the legal structure and the economic structure by defining the national interest, social vision and the goal-objective formation with a winning *action set* that will support the national aspirations defined within the social vision. The process works in categories and transformation of categories, where colonialism is seen as a category and so also are independence, neocolonialism and emancipation. One may view independence and emancipation as opposites where independence and colonialism constitute a social polarity, and neocolonialism and emancipation constitute another social polarity These polarities have different characteristics but are similar to slave-master polarity, in concept where the sovereignty of the slave in decision making is controlled by the power of the master with violence or non-violence. A slave is a slave just as a country is a colony because he or she has lost his or her sovereignty which implies a loss of determination of one's destiny.

The underlying theory and the logic of categorial conversion as has been presented in [R1.90b][R15.15] work with the notion of actual-potential polarity where each pole has a residing duality that is made up of negative and positive duals. Every social category exists as either social actual or social potential to constitute a social polarity in relational continuum and unity. Both the actual and the potential are under the internal behavioral activities of the positive and negative characteristic sub-sets seeking to negate each other. The actual pole through its duality seeks to strengthen and maintain itself as actual by negating some of the characteristics of the potential contained in it. The potential pole through its residing duality seeks to strengthen and to unseat the actual by negating some of the characteristics of the actual contained in it. In this logical frame every actual is past-present connected where the present always resides in the past. Similarly, the potential is present-future connected where the future always resides in the present, and hence by logical extension, the future resides in the past as projected by the *Sankofa-Anoma* philosophical thought of the Akan people of Ghana [R15.13]. The present reality is a derivative from the past reality which is also its primary category. The present reality is a primary category for the future reality which is currently potential residing in the present. The *sankofa-anoma* philosophical thought is always related to the *asantrofi-anoma rationality which simply* says that every decision-choice action is cost-benefit defined and that one cannot choose the benefits without the costs. In other words, the conditions of the past are cost-benefit defined, the conditions of the present are cost-benefit defined and similarly the conditions of the future are cost-benefit defined. These conditions exist in relational continuum and unity in the primary category of social essence and generate forces of transformation of social polarities in derived categories. In any social set-up, the *sankofa-anoma* philosophical thought, and the *asantrofi-anoma rationality* find logical expressions in *anoma-kokone-kone problem* and *funtum-mereku-denkyem-mereku problem*. The *anoma-kokone-kone problem* defines the existence of individual-community duality with conflict in relational continuum, where the individual has no identity without the community, and the community is not defined without the individuals. The *funtum-mereku-denkyem-mereku problem* defines the existence of social dualities with a family of conflicts in a relationally inseparable unity. The above relational structure is required in explaining the dynamics of actual-potential polarity in destruction and creation of social varieties.

In a colonized territory in Africa, colonialism is the present reality while decolonization is a future reality. When decolonization is achieved and completed, it takes the position of present reality while colonialism is kicked into the potential and becomes the past reality as well as it becomes a contest for the future reality in a different form that can be actualized if the appropriate conditions are created with the colonized-colonizer polarity under the historic process as seen by Nkrumah. The actualization of decolonization leads to two possible socio-political polarities. The two polarities emerge where after decolonization, national independence with sovereignty is claimed. In the claimed independence are the seeds of emancipation and neocolonialism. One of them is *independence-emancipation polarity*, while the other is *independence-neocolonialism polarity* with corresponding systems of dualities defining the conflicts on the basis of the historic actors. The choice of one of these social polarities will shape the path of African socio-political history in terms of polar conflict resolution as seen in Nkrumah's conceptual system. Nkrumah's choice is the *independence-emancipation polarity* with strong aversion to the *independence-neocolonialism polarity* at all cost. The choice that the African countries have made individually and collectively, directly or indirectly has been the independence-neocolonialism polarity which has redirected the history of Africa into another domain of servitude [R1.91] [R1.204] [R1.200]. The result is that, currently, the neocolonialism pole is the reality and existing socio-political variety while true independence and emancipation are epistemic elements of socio-political variety in the potential space with independence residing in a desert mirage. The two analytical paths may be illustrated with epistemic geometries. The first involves independence-neocolonialism polarity while the second involves independence-emancipation polarity. The path of categorial conversion of political polarity is shown as an epistemic geometry in Figure 1.3 where relational continuum and unity of positive and negative forces are connected to generate social action.

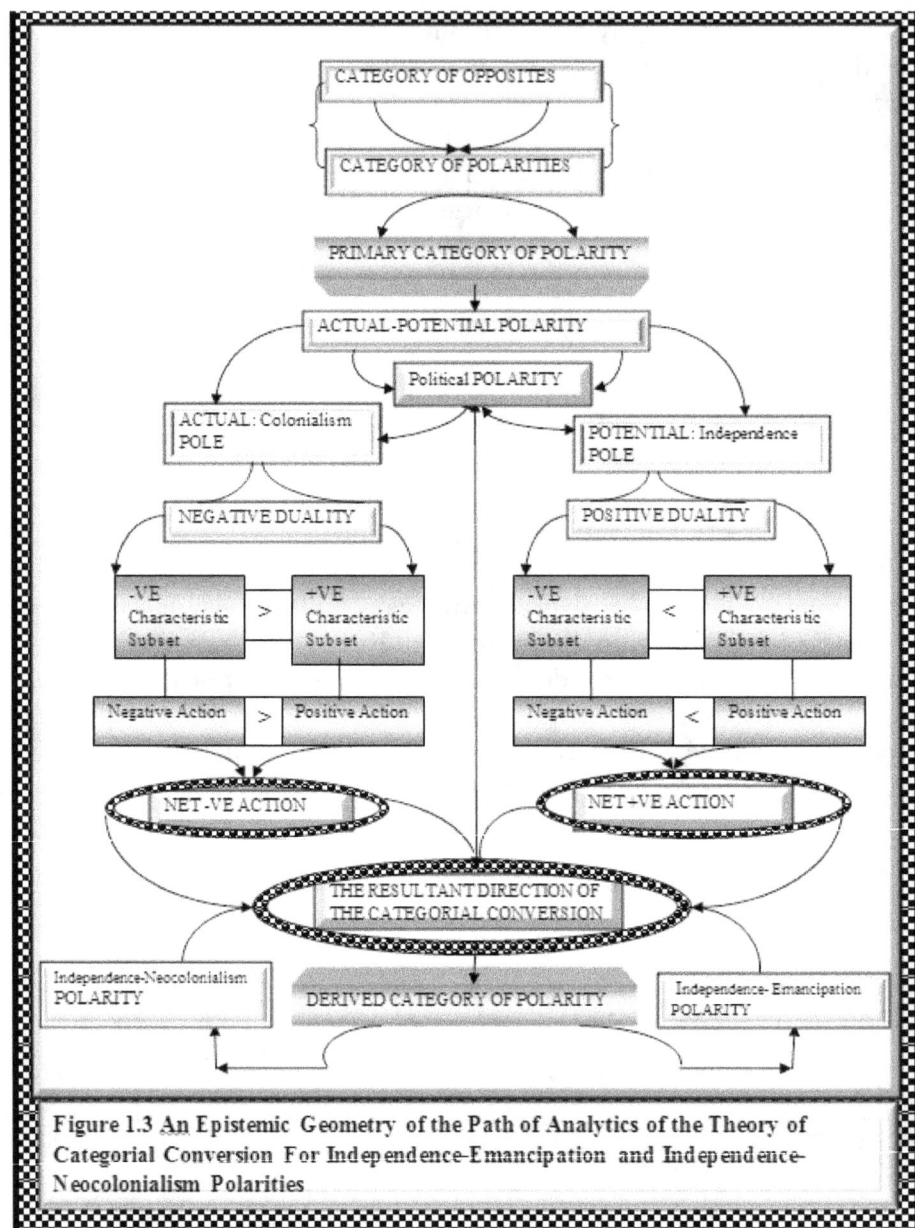

Figure 1.3 An Epistemic Geometry of the Path of Analytics of the Theory of Categorial Conversion For Independence-Emancipation and Independence-Neocolonialism Polarities

The conceptual system of Nkrumaism is to fashion efficient tools to be used to empower the characteristics of emancipation to overwhelm the characteristics of neocolonialism. This must be done through the conditions of categorial conversion by creating favorable conditions of

92

categorial moment to generate a qualitative motion for a *categorial transfer function* by changing the balance between neocolonialism and emancipation in favor of the national emancipation composed of socio-economic development, social-vision setting, sovereignty defense and national-interest formation. It is by the efforts of the people that colonialism, injustice and oppression are rooted in the society and the people's national sovereignty is taken away. The same people hold the power to reverse the process because they constitute the backbone of positive action from within to change their socio-political conditions. They are also the backbone of negative action from within to allow the existence of their suffering. The people are, therefore, the *matter* whose identity is revealed by the set of characteristics with a general production of *energy*. The set of characteristics are divided into positive and negative sub-sets of characteristics which produce positive and negative energies that are translated into positive and negative actions for categorial conversion of social polarity under an appropriate *information structure*. The people generate the required energy for creating the conditions for conversion of social polarities through their formation of negative and positive characteristic set and the corresponding negative and positive actions under the principle of cost-benefit rationality. Here, the costs and benefits are defined in terms of real characteristics. The production of social action must be organized to create a winning force that will set the willed potential against an unwanted actual if genuine independence and emancipation are to be claimed by the Africans against independence and neocolonialism in a fantasy of a mirage . In the sequential analysis, attention must be paid to the relational structure of matter, energy and information. The categorial conversion is such that the people constitute the natural matter, they produce information and energy to bring about qualitative motion. It is on the basis of this conceptual structure that categorial conversion is linked to Philosophical Consciencism as its epistemic vanguard in the dynamics of social actual-potential polarities for destroying and creating socio-political varieties. In fact it is through this logical structure that one can explain qualitative motion. It is also through this logical structure that one can construct theories of economic development, political development and legal development where every stage present time point of actual variety which is ready for destruction and replacement of a potential variety.

In the dynamic action processes of relational behavior of social polarities and the supporting system of dualities, Nkrumah was confronted with two formidable problems of African society under the scientific application of the theory of categorial conversion. The first problem was the categorial conversion of *political polarity* which was *colonialism-independence polarity*. Here, colonialism was the actual and independence was the potential that needed to be actualized by selecting appropriate social controllers through the people's organized positive characteristics to empower the social positive action from within. The second problem was and still is the categorial conversion of economic polarity given the legal process under national independence and decision-choice sovereignty. In the second problem, socioeconomic backwardness is the actual and the socioeconomic development is the potential which must be actualized by the people's organized positive characteristics to empower the social positive action from within to move the social setup through stages of socioeconomic development. In other words, after decolonization the African social system must solve the problem of *economic polarity* which is *underdevelopment-development polarity*. The sequence of the problems faced by Africa as seen in its organic form under the application of the theory of categorial conversion is *the categorial-conversion problem of political polarity* and the *categorial-conversion problem of economic polarity* which are relationally linked by the *categorial-conversion problem of legal polarity*. The twin problems that have been pointed out is captured by an Nkrumah's statement: *Every movement for independence in a colonial situation contains two elements; the demand for political freedom and the revolt against poverty and exploitation* [R1.202, p. 51]. We have discussed the first problem of categorial conversion of political polarity of internal-external control. The solution to the problem of the political polarity offers a direct framework to create the convertibility conditions to the solution of the problem of legal polarity and the indirect framework to create the convertibility conditions to the solution of the problem of economic polarity. The solution to the second problem of categorial conversion of the economic polarity is the problem of nation building to which a reflective attention is now turned. It is useful to keep in mind that social variety at any time point is a composite aggregate of political, economic and legal varieties which are under cultural and institutional power tension from within. It is this power tension that provide energy for qualitative self-motion of varieties.

1.3 Nation-Building, Socio-Economic Polarity, Categorial Conversion and Command-Control Dynamics

Decolonization bestows on the previous colonized people the social decision-choice sovereignty to manage the affairs of the country and its people. It bestows on the people the natural right to participate in the social affairs of the nation and the choice to select or not to select the members of the decision-making core to follow any system they choose to construct their society and manage the nature of its formation in accordance with their preferences and will. It also bestows on them the right to participate which was denied to them under colonialism, occupation and imperial order. Decolonization, leading to national independence, bestows freedom and responsibility on the people when they control the political structure which implies the control of the national decision-making power over political, legal and economic structures. It is through this relational understanding that led Nkrumah to explicitly state: *Seek ye first the political kingdom* [R1.202, p50]. This statement is amplified further by the following statements: *Political power is the inescapable prerequisite to economic and social power* [R1.207 p.78]. *Political independence is only a means to an end. Its value lies in its being used to create new economic, social and cultural conditions which colonialism and imperialism have denied us for so long* [R1.207, p.80]. *Independence must never be considered as an end in itself but as a stage, the very first stage of the people's revolutionary struggle* [Axioms, p83] *When independence has been gained, Positive Action requires a new orientation away from the sheer destruction of colonialism and toward national reconstruction* [R1.203, p.105]. To speak of nation-building, we must understand the concept and content of a nation. The concept and content of the nation must be related to stages of energy, motion and intra-inter stage transfers.

Definition 1.3.1: Nation

A nation is conceived as a union of qualitative and quantitative characteristic sets of political, legal and economic structures, where the qualitative and quantitative characteristic sets collectively define the identity and character of the nation. The qualitative-quantitative characteristic sets are also a union of negative and positive characteristic sub-sets with structured relative qualitative-quantitative dispositions

95

which then define the human disposition given its population, climatological and geomorphological conditions.

Definition I.3.2: Nation-building

Nation building is a transformation of a nation's political, legal and economic structures over various stages of belonging. Each stage is identified by an organic set of qualitative-quantitative characteristics that place it in a category of dynamic enveloping of the actual-potential polarity where the nations time-point identity is a variety waiting to be transformed.

Definition 1.3.3: Nation-building Stages

The identity of a stage of a nation building is fixed by a set of qualitative and quantitative characteristics where every stage belongs to a category and where each stage is distinguished from other stages by the relative negative-positive characteristic set defined in terms of socio-cultural institutional configuration that encompasses political, legal and economic allocation, production and distribution.

Proposition I.3.1

Given the quantitative disposition, every stage category is fixed by its qualitative disposition in the nation-building process and every nation-building stage is a derived category by categorial conversion from the previous stage which serves as its primary category.

Definition 1.3.4: National Historical Process

The collection of all the stage categories from the past through the present to the future constitutes the enveloping of the categorial conversion of the national historical process of the actual-potential polarities with a relevant supporting set of dualities of the nation-building process. It is the *sankofa-anoma enveloping* of categorial conversions under an *asantrofi-anoma rationality* in the decision-choice space.

Note 1.3.1

The nation-building process is basically a continual transformation of characteristics and relative negative-positive characteristic set. Nation building is a continual transformation of the social system composed of its qualitative and quantitative dispositions and the institutional configuration that holds them through stages. Every socio-economic stage, just like every socio-political stage, has an identity defined by a set of characteristics that places it in a category. Every stage is a derived category that also serves as a primary category for the subsequent categories. A movement from one stage to the other is a categorial conversion induced by the social action of the people. Such social action, at the aggregate level may be intentional or unintentional. It may also be internally or externally induced. When it is internally induced with intentionality, the categorial conversion follows the will of the people with sustainability into the future except violent interruption. Every sustainable categorial conversion is an internally induced transformation of the characteristics of the relative combinations of political, legal and economic structures [R15.15] [R13.8] [R13.9]. When it is internally induced without intentionality, the transformation follows an unwilled path of categorial conversion. The will of the people is violated when the transformation is externally induced with external intentionality or un-intentionality. Here, the development is not of the people, by the people and for the people because the people have lost their sovereignty and the will to their destiny. The will of the people to nation building to define their evolving history and destiny is encapsulated in their culture and the ideological conditions of their sovereign will. The same will of the people to be independent from foreign control and claim their sovereignty in the decision-choice process rests also on culture and ideology [R3.7]. It is this sovereign will that creates the internal forces of social transformations of one social variety to another social variety without external impression. It is this sovereign that must be manipulated by the management of command-control instruments.

The stages and the path of categorial conversion are illustrated by epistemic geometries in Figure 1.4 and Figure 1.5 where the systems are viewed as internally closed for internal decision-choice actions. Theoretically, each stage is either an actual or potential where every actual stage has a corresponding potential that constitutes actual-potential polarity. Here, the actual-potential polarities are defined by the characteristics of degrees of development and underdevelopment. The degree of development of the actual constitutes the actual pole while the degree of development of the potential constitutes the potential pole. The degree of development of the actual is assumed to be less than the degree of development of the potential when a transformation is sought. In other words, the actual variety is less socially preferred in comparison with a set of potential varieties from which a possible constitute may be created. The creation of the categorial convertibility conditions and the effectiveness of the categorial conversion in the solution to both the problems of political and economic polarities are substantially induced by culture and ideology which also influence culture and social actions in the political and economic structures in terms of the management of the necessary social instruments for the command and control process. Let us now turn our attention to the role that culture and ideology play in categorial conversion.

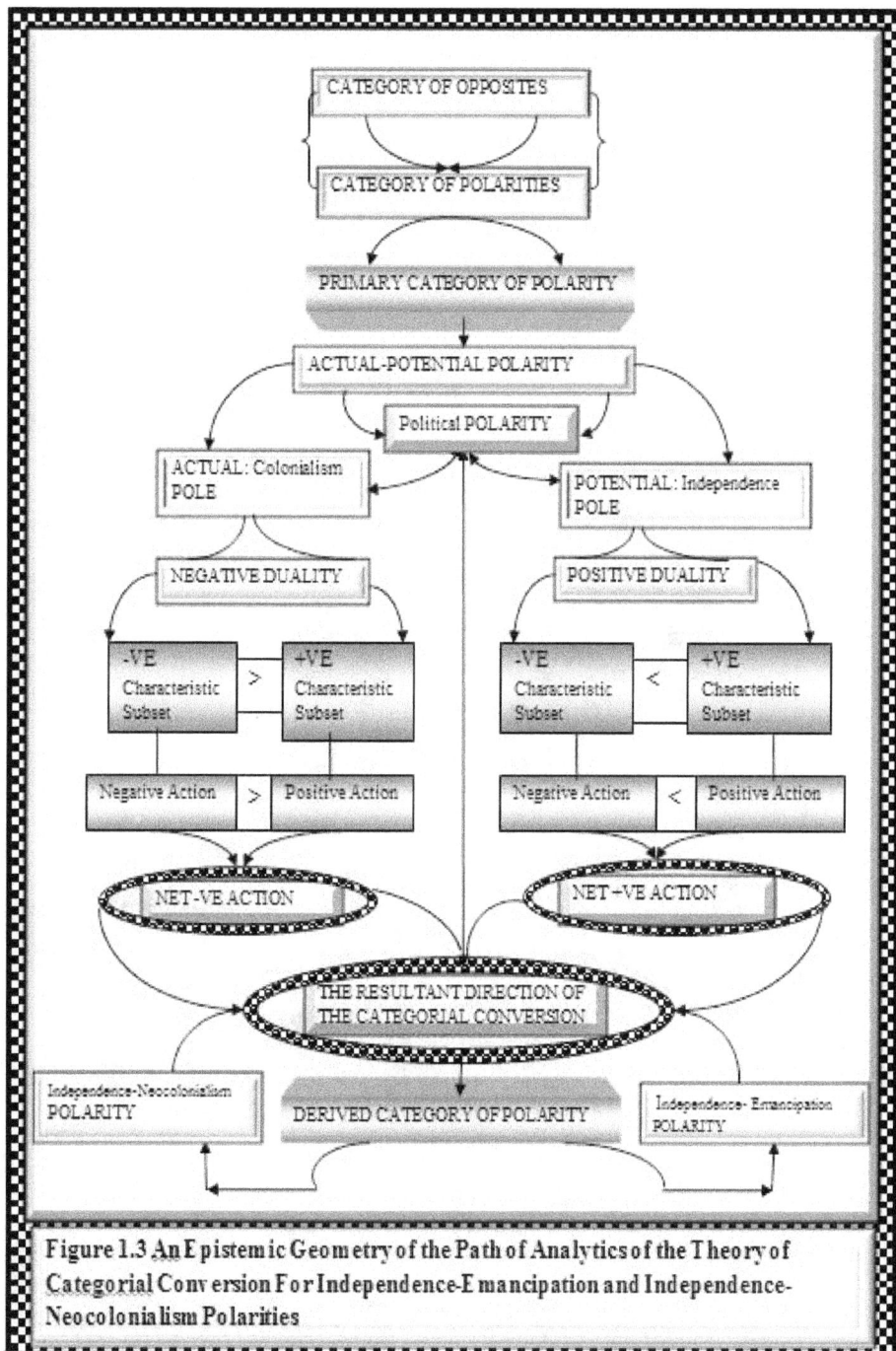

Figure 1.3 An Epistemic Geometry of the Path of Analytics of the Theory of Categorial Conversion For Independence-Emancipation and Independence-Neocolonialism Polarities

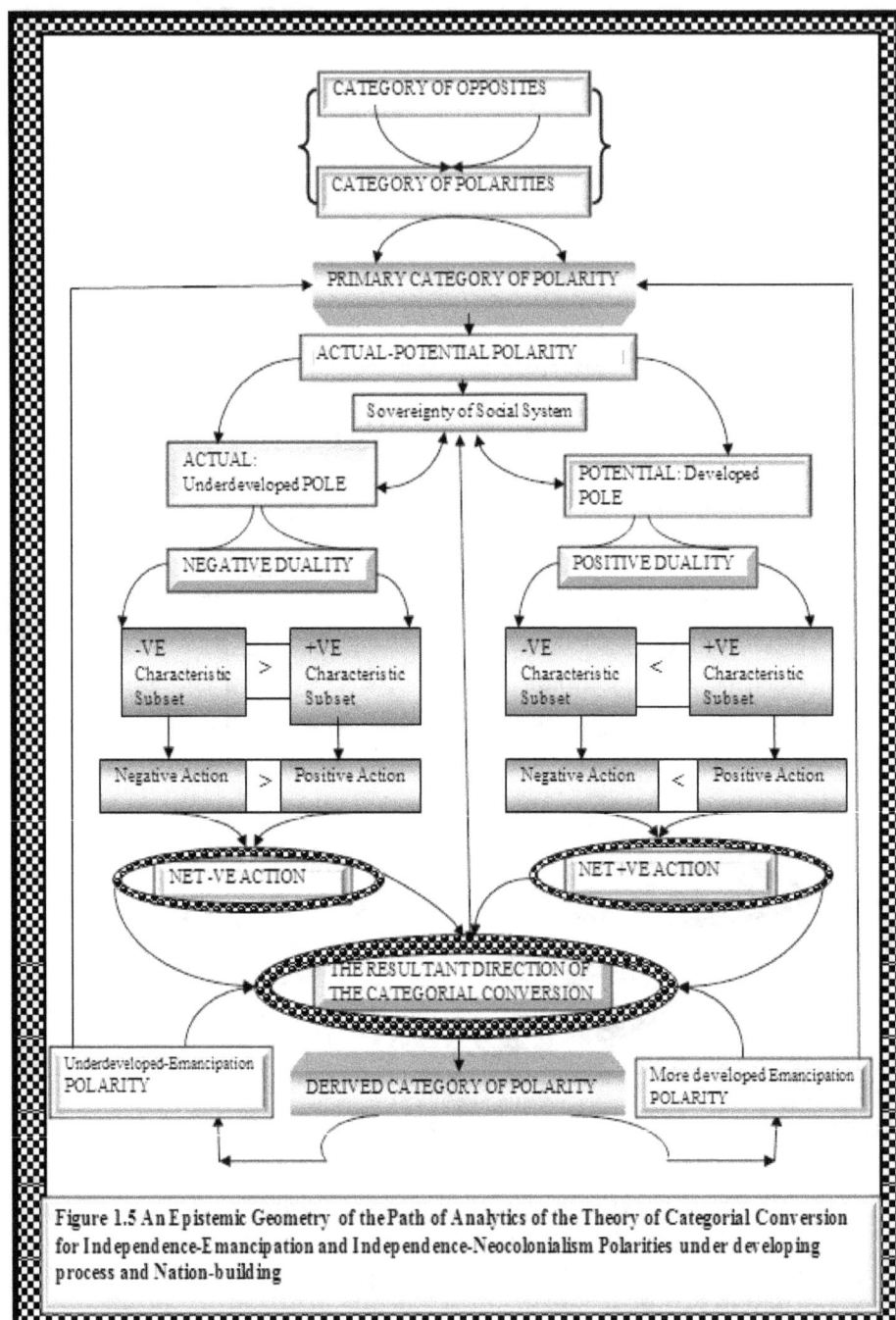

Figure 1.5 An Epistemic Geometry of the Path of Analytics of the Theory of Categorial Conversion for Independence-Emancipation and Independence-Neocolonialism Polarities under developing process and Nation-building

1.4 The General Concept of Philosophical Consciencism in Managements of Command-Control of Socio-Natural System Dynamics

The theory of categorial conversion provides us with a general dynamical mechanism of qualitative and quantitative dispositions of socio-natural elements defined in terms of categories of being indicating the necessary conditions of convertibility. The central core of the mechanism is composed of *categorial moment, categorial transfer function* and *categorial transversality conditions* in system dynamics. The categorial moment relates to intra-categorial conversion of internal qualitative disposition of categorial elements defined in terms of varieties at each time point. The categorial transfer function relates to intra-categorial conversion of qualitative-quantitative dispositions of categorial elements also defined in terms of inter-temporal varieties. The categorial transversality conditions relate to the completion of intra and inter categorial conversion of categorial elements of varieties. These three are in relational continuum and unity. They constitute the necessary organic convertibility conditions for categorial conversion of categorial elemental varieties of the universal system. The set of convertibility conditions must be created by socio-natural processes that take place within socio-natural polarities through the behavior of the corresponding dualities whose identities are defined by relative positive-negative characteristic sets.

The socio-natural processes in creating the convertibility conditions are the works of information-decision-choice interactive systems guided by socio-natural principles in matter-energy space. There are two important sets of polarities under categorial conversion that are available for epistemic actions in the epistemological space. They are the sets of natural actual-potential polarities and social actual-potential polarities, and corresponding to them we have natural categorial conversions and social categorial conversions. In natural actual-potential polarities, the categorial convertibility principles are embedded in the natural ontological design order of natural matter-energy relational structures. The epistemic discovery of matter-energy relational structures and the implied principles for categorial conversions provide the justification of searches for scientific knowledge as well as constitute the foundations of the development of different knowledge sectors and all actual and potential engineering sciences. The necessary conditions from the theory of categorial conversion indicate what command-control instruments must be manipulated to create sufficient condition for convertibility of varieties. The manipulations are done by *decision-choice systems* with socio-natural intentionalities that are consistent with the internal structures of the socio-natural varieties. The understanding of the general process is that the socio-natural varieties may be cognitively partitioned into a set of natural varieties and a set of social varieties. The sufficient conditions for transforming the natural varieties are manufactured by natural decision-choice system under appropriate conditions that constitute general principles of natural actions. The sufficient conditions for transforming the social varieties are manufactured by social decision-choice system under appropriate internal conditions that also constitute general principles of social actions .The set of the general principles which guides natural categorial conversions shall be referred to as *natural Philosophical Consciencism*. This natural Philosophical Consciencism is born out of ontological crises in the nature of categorial existence within elemental dualities by conflicts generated from internal forces of negation by competing negative and positive characteristic sets, to induce the direction of the resultant force for categorial conversion of natural actual-potential polarity. There are as many types of natural Philosophical Consciencism as there are elemental categories of varieties. There is an important thing that must be kept in mind, and that is for any phenomenon and object, there is always one categorial reality and infinite number of categorial potential varieties that will serve as a replacement under the transformation-substitution principles. The nature of the actualized potential variety in place of the existing actual variety will depend on how the existing actual is destroyed. It is useful to keep in mind that transformation of varieties is problem-solution duality, where each solution creates a new problem. The nature and complexity of a new problem emerging out of a solution will always depend on the method and manner for solving the problem in all natural problem-solution processes. This is also the case in the social space regarding the dynamics of the problem-solution process under the transformation-substitution principles within the behavior of the actual-potential polarity. In general, a problem is an actual, and when it is solved by a transformation, a vacuum is created in the space of the actual. The type of the solution brought by the transformation is among many elements in the potential space that is actualized to fill the vacuum as a substitute. In this analytical framework, a problem must be seen as an actual variety while a solution comes from many potential varieties, each one of them brings its own problem under the principles of duality, where each problem has a corresponding solution, and each solution has a corresponding problem in a relational continuum and unity.

In social actual-potential polarities, the set of convertibility principles is embedded in the epistemic decision-choice order of human matter-energy relational structures. The creation of the forces from matter-energy relational structures on the basis of social decision-choice activities is the justification of a search for knowledge and the discovery of decision-choice rationality of human action. The set of the principles which guides the human decision-choice actions to bring about social categorial conversions relevant to a particular social society shall in general be referred to as *social Philosophical Consciencism.* Social Philosophical Consciencism is born out of epistemic crisis in the social categorial existence (for example, colonialism-independence polarity or slavery-freedom polarity) of social elemental duality; that is, the crises of the conscience of the people relative to their collective and individual personalities as valued through perceptions of freedom and justice which may have defined net real values in the general cost-benefit space. There are as many types of social Philosophical Consciencism as there are societies.

The categorial convertibility conditions and the processes in creating them may be set up in set-theoretic terms. Let $\mathbb{A}$ be a set of natural categorial conversions with a generic element $\alpha \in \mathbb{A}$ and with an infinite index set $\mathbb{I}_{\mathbb{A}}^{\infty}$. Let the corresponding set of natural Philosophical Consciencism be $\mathbb{N}$ with a generic element $\nu \in \mathbb{N}$ and with an infinite index set $\mathbb{I}_{\mathbb{N}}^{\infty}$. Similarly, let $\mathbb{B}$ be a set social categorial conversions with a generic element $\beta \in \mathbb{B}$ and with an infinite index set $\mathbb{I}_{\mathbb{B}}^{\infty}$. Let the corresponding set of social Philosophical Consciencism be $\mathbb{S}$ with a generic element $\sigma \in \mathbb{S}$ and with infinite index set $\mathbb{I}_{\mathbb{S}}^{\infty}$. The general convertibility conditions as abstracted from the general theory of categorial conversion may be specified in terms of set-theoretic representation. Let Γ be a set of measures of categorial moments with a generic element $\gamma \in \Gamma$ and with an infinite index set $\mathbb{I}_{\Gamma}^{\infty}$. Similarly, let $\mathbb{T}$ be a set of transfer functions with a generic element $\tau \in \mathbb{T}$ and with an infinite index set $\mathbb{I}_{\mathbb{T}}^{\infty}$. Finally, let $\mathbb{G}$ be a set of sets of categorial transversality conditions with a generic element $\varphi \in \mathbb{G}$ and with infinite an index set $\mathbb{I}_{\mathbb{G}}^{\infty}$. These are the set-theoretic structure for the construction of the theories of the categorial-conversion and Philosophical Consciencism.

The infiniteness of the sets has an important meaning regarding ontological and epistemological existence and processes which are continually ongoing in relational continua and unity. The infiniteness of the ontological space and categorial conversion processes from within nature ensure the permanency of the universe with relational continuum and unity. The categorial conversion processes within the ontological space produce never-ending arrangements and rearrangements of natural varieties and categorial varieties with continual elemental migrations among categories, where the set of processes of the emergence of the new and the disappearance of the old is a general law of nature under a stability of multiplicity of rhythmic forces. It is through this understanding that provides the support of the Africentric assertion that the universal system has no beginning and no end represented by the all *Seeing Eye* (*Abode Sante*) as an *Adinkra* symbolic representation of the idea that nobody of human existence saw the Beginning of creation and nobody will ever see its End, for the End resides in the Beginning which also resides in the End [R1.124], [R1.219][R1.224][R1.243]. The End and the Beginning, therefore, mutually create themselves as well as mutually negate themselves in ontological actual-potential polarity. The universe is a beginning-end duality within the nothingness-somethingness polarity where darkness-light duality resides in each pole of the system of actual-potential transformation processes. The infiniteness generates the permanency of the universe and the permanency of continual categorial conversions where some categories are destroyed by modification and transformation, and new ones are created within the universally unified structure under substitution-transformation duality with internal processes. The system, as set up with categories and categorial conversions, does not support any conceptual idea of an expansion of the universe. It, of course supports expansion of the infinite set of old and new varieties in the actual potential spaces where there are destructions of actual varieties and creation of new actual varieties and categorial varieties from the potential space while the universal continuum and unity are maintained within the infinitely closed universe. In this framework, it is our knowledge set that is continually expanding in relation to the expanding set of varieties and categorial varieties that constitute the continual expansion of the ontological information set.

The infiniteness, seen at the level of epistemology presents limitativeness and limitationality of cognitive agents in terms of qualitative and quantitative dispositions iof information, information-processing capacity and knowing, where epistemological categories are temporarily but not permanently fixed in the logical process within the construction-reduction duality for a self-exiting and self-correction knowledge system [R3.10][R3.13]. In this discussion, the set-representation of the relational structure among categorial conversion, categorial convertibility conditions and Philosophical Consciencism is summarized in Table 1.

TABLE 1: SET-THEORETIC SYMBOLS AND CORRESPONDING CONCEPTS

CONCEPT	SET REPRESENTATION NATURAL POLARITIES	SET REPRESENTATION SOCIAL POLARITIES	CONVERTIBILITY CONDITIONS
Measures of Categorial Moments	$\Gamma_N, \; \gamma \in \Gamma_N,$	$\Gamma_S, \; \gamma \in \Gamma_S,$	
Index Sets	$I_N^x, \; I_{\Gamma_N}^x$	$I_S^x, \; I_{\Gamma_S}^x$	
Categorial Transfer Functions	$\mathbb{T}_N, \; \tau \in \mathbb{T}_N,$	$\mathbb{T}_S, \; \tau \in \mathbb{T}_S,$	
Index Sets	$I_N^x, \; I_{\tau_N}^x$	$I_S^x, \; I_{\tau_S}^x$	
Categorial Transversality Conditions	$G_N, \; \varphi \in G_N,$	$G_S, \; \varphi \in G_S,$	
Index Sets	$I_N^x, \; I_{G_N}^x$	$I_S^x, \; I_{G_S}^x$	
Categorial Conversion	$A_N, \; \alpha \in A_N,$	$B_N, \; \beta \in B_N,$	
Index Sets	$I_N^x, \; I_{A_N}^x$	$I_N^x, \; I_{B_N}^x$	Necessary Conditions
Philosophical Consciencism	$N, \; \upsilon \in N,$	$S, \; \sigma \in S,$	Sufficient Conditions
Index Set	I_N^x	I_S^x	

The system that is portrayed by the table of set-theoretic representation has two parts. One part concerns categorial dynamics of natural and social categories and conditions of convertibility. This is the main occupation of the theory of categorial conversion. It is about the study of the dynamical continuum and unity of the qualitative and quantitative motions of natural actual-potential polarities of $\mathbb{A}_N$ and $\mathbb{N}$ and its logical extension to $\mathbb{B}_S$ and $\mathbb{S}$ The other part concerns the understanding of the manufacturing process of forces that bring about the convertibility conditions of natural categories, and how such understanding may help in the understanding of socioeconomic transformation as well as the use of such understanding to bring about social categorial conversion. This is the main concern of the theory of Philosophical Consciencism as the foundation of the practice of Nkrumaism. The theory is to help us to create a framework for action under conditions of social resistance and change within social continuum and unity of the qualitative and quantitative dispositions of social categories defined where the identity of social categories like the identity of natural categories are seen in terms of actual-potential polarities, and where the identity of each pole is defined by the residing negative or positive duality whose existence is made possible by negative and positive characteristic subsets in mutual negation by producing negative-positive forces of conflict in tactical and strategic actions in the game space of mutual negations and categorial conversions. For consistency of logical development and understanding, two types of Philosophical Consciencism of *natural Philosophical Consciencism and Social Philosophical Consciencism* have been introduced. The theoretical development in this monograph will be centered on *social Philosophical Consciencism*. The development of the complete framework of natural Philosophical Consciencism is left to scientific and technological theories to provide.

1.5 The Basic Epistemic Foundation of the Theory of Philosophical Consciencism in System Dynamics

Philosophical Consciencism projects an integrated system of social philosophy and ideology on the basis of which social convertibility conditions are created by mimicking the convertibility conditions of forces at work in natural categorial conversion of natural actual-potential polarities for categorial conversions of social actual-potential polarities. The effective strength of Philosophical Consciencism is seen in its guidance to craft a system of ideology, principles, social vision, national interest, social goal-objective set, programs and social actions that are appropriate to a particular social formation. It is a guide to create a relevant social decision-choice system that is fully compatible with the socio-cultural confines of a particular society to bring about the sufficient conditions needed for the transformations of social varieties and categorial varieties.

1.5.1 Ideology, Principles and Programs in Social Systemicity

From the viewpoint of Philosophical Consciencism, there are ideologies, principles, social vision, national interests, social-goal-objective formation, social programs and actions existing in relational continuum and unity as seen under the principles of opposites. In this relationality of thought, ideology may be viewed as an intellectually derived category from a primary category of philosophy. *Ideology* is a belief system derived from a particular *philosophy* which serves as its logically justification support. The philosophy with the implied ideology in the social decision-choice system for social transformations is Philosophical Consciencism which when derived from the African experiential information structure acquires a content of Nkrumah's conceptual system for Africa's emancipation. On the basis of an ideology and the philosophical support, social *vision* and *national interests* are established to define the path of categorial conversions implied in categorial dynamics of social *actual-potential polarities* for the disappearances of actual varieties and the emergence of potential varieties. From the ideological system, a *set of principles* is structured to define the essences of the ideology and the conceptual core of its philosophical support. Such principles are intended to simplify the individual and collective decision-choice actions in the tactical and strategic game space for establishing *social programs* for *social actions* to accomplish the social vision and national interest with the guidance of the set of principles.

In this respect, the social programs are subject to continual evaluation and revision to assess the disparity in the actual-potential polarity. The revision is motivated by the assessed distance between the social actual and preferred social potential in the game space of social vision and national interest that are derived from the philosophy and ideology. In this conceptual framework, a change in the principle means a change in the ideology which also means a change in the supporting philosophy and hence changes in the social vision and national interest that are derived from the philosophy and the implied ideology. It is on this relational framework of concept revolution and transformation on the basis of social will and principles that Nkrumah preset the quote of Mazzini:

> Every true revolution is a programme; and derived from a new, general, positive and organic principle. The first thing necessary is to accept that principle. Its development must then be confined to men who are believers in it, and emancipated from every tie or connection with any principle of opposite nature.

It is clearly stated that the programs are derivates of the principles that presents the constancy and stability of the ideology and its supporting philosophy. Given the principles, the social programs and actions may change since they are instruments to accomplish social vision and national interest implied by the ideology and its philosophical support. A change in the social programs merely indicates the existence of an unwanted distance between the actual and potential poles of the actual-potential polarity in reflection of the activities of the negative-positive forces in the social dualities. A change in social programs points to the assessed relative effects of the conflicts in the negative and positive forces in the dynamic game space of the actual-potential polarity in accord with the ideology and supporting philosophy. In this discussion, I have argued and will continue to argue that Philosophical Consciencism constructed from the African experiential information structure with tools from the African Conceptual system is pro-African and is completely opposed to philosophical Consciencism constructed from any experiential information structure of a non-African experiential information structure with logical tools derived from colonialism, neocolonialism and imperialism constructed to project anti-African philosophy. This argument will be further enhanced in the chapters to come.

1.5.2 The Basic Foundation of Philosophical Consciencism in the Dynamics of Social Systemicity

The following basic analytical sequence constitutes the epistemic foundation of the development of the theory of Philosophical Consciencism with African content:

1. Every ontological element belongs to a category where each category is defined by a characteristic set in negative and positive combination to define its qualitative and quantitative dispositions as a duality under relational continuum and unity. The ontological elements are the ontological varieties and the category in which each variety belongs is the categorial variety

2. Sustainable change is from within the elemental categories involving the dynamics of the qualitative and quantitative disposition of the elements (varieties) and is induced by forces from within.

3. The forces of change are internally manufactured by the struggles and contradictions of the negative and positive characteristic sub-sets that define the duality where activities of these conflicting forces produce convertibility conditions of categorial moments, categorial transfer functions and convertibility conditions in each actual-potential polarity. The conditions of categorial moments, categorial transfer functions and convertibility conditions in each actual-potential polarity constitute the necessary conditions for transformation.

4. Every point of change is described by an actual-potential polarity where each pole has a residing duality that is either negative or positive which is viewed as general notions where the negative or positive duality acquires its identity through the nature of the relative negative-positive characteristic set.

5. The creation of convertibility conditions is through the Philosophical Consciencism developed to affect the behavior in the decision-choice space to alter the relational terms of the poles in the actual-potential polarity. This is the manufacturing of the sufficient conditions of the transformation of varieties and categorial varieties.

109

CHAPTER TWO

CATEGORIAL CONVERSION, CULTURE, PHILOSOPHICAL CONSCIENCISM AND NKRUMAIST SOCIAL DYNAMICS

Social categorial conversions are transformations of social polarity from within the social setup. The matter and energy required for these transformations are people and their social actions in the national decision-choice spaces with specific experiential information that define the distribution of national identities. These social actions are the products of individuals in aggregative form guided by some form of individual and collective rationalities. The individual rationalities may substantially conflict with one another to generate and produce an unwanted direction of categorial conversion of the social set-up. It is here that *institutional organization* of individual rationalities into a collective rationality, *unification* of individual preferences into a collective preference, *mobilization* of individual forces into a collective force, and an *integration* of individual actions into an effective collective action become essential strategies in working pathways in the application of the theory of categorial conversion to social polarities. Let us keep in mind that an individual exists as a duality with relative combinations of positive and negative characteristic sub-sets that define the individual identities as well as generates positive and negative actions respectively. The qualitative dynamics of the relative negative-positive characteristic sets and the corresponding relative negative-positive action present the identities of the individuals in the collective and in relation to the direction of the categorial conversion. The qualitative dynamics of the individual relative negative-positive action are under the dynamics of the cultural and ideological contents of the social set-up as well as the changing nature of social information. In the society, there is the cultural content and its growth that provide the accumulated *cultural capital* and its rate of accumulation as *cultural investment* in the society. There is also ideological content and its growth that also provide the *ideological capital* and the rate of *ideological investment*. Together, the social set-up is identified by the characteristics of the cultural and ideological configurations that exert preponderating effects on the structural dynamics of the individual and collective qualitative and quantitative dispositions to bring about social categorial conversions. Let us deal with the individual roles of culture and ideology and then their interactive effects in the theory of categorial conversion.

2.1 Culture, Categorial Conversion and Philosophical Consciencism

The role of culture in defining boundaries of information structures, its needs and uses in socioeconomic decision-choice processes have been pointed out in various discussions [R3.7] [R3.10]. The same culture influences the development of social thought and methods of obtaining it. Social and individual preferences are established by a defined social environment within any social formation that draws parametric boundaries of acceptability of individual and social actions given human nature. The culture presents a social environment that establishes the social and individual preferences to define values of socio-natural elements in the social formation given human nature whatever it may be. Every social formation is composed of three structures of economics, politics and law which project various decision-choice problems and challenges in their respective environments. The decision-choice problems considered either individually or collectively must be dealt with within the existing culture at any point in time and over the path of collective and individual social existence. It is the sequence of solutions to the sequence of problems that shapes the path of national history as well as changes in the path of the national history. Generally, it is the sequence of solutions to social problems of the *past*, which generates a sequence of social problems *present*, which then requires the development of new sequence of solutions to create a sequence of social problems of the *future* in the history of human time trinity of the past-present-future configuration of *sankofa-anoma* tradition. This past-present-future dynamics of the social problem-solution processes is decision-choice dependent relative to the general dynamics of actual-potential polarity, where each social problem constitutes an actual pole and the corresponding social solution constitutes the potential pole in social transformation composed of categorial conversion and social philosophical Consciencism.

Culture, therefore, helps to define the decision-choice transformative toolboxes of society at both levels of the qualitative dispositions and quantitative dispositions of individuals and the collective in the decision-choice space. Culture also performs the two roles of resistance to foreign oppression as well as galvanizer for the transformation of political polarities. At the level of transformation, it is this powerful role of culture in the collective national life of a people over time that led Abraham to state:

> All EVENTS OF LARGE significance take place within the setting of some culture, and indeed derive their significance from the culture in which they find themselves. It could therefore happen, and does indeed happen, that the same event, occurring as it were between the frontiers of two different cultures, should be invested with differing significance, with different capacities for determining the direction of policies arising therefrom. This immediately raises problems for a number of disciplines including, above all, history and social anthropology. The writing of the history of one culture from the milieu of another culture, which is not – relevant to the events and situations concerned – isomorphic, raises serious questions of cultural bias and distortion. It does not necessary offer objectivity, where it touches evaluation of facts and events, a cultural alien can only offer an alternative set of prejudices [R1.1, p. 11].

At the level of resistance against foreign domination, the recognition of the powerful role of culture as a defense of people's existence and national life led Cabral to state:

> Culture is simultaneously the fruit of a people's history and a determinant of history, by the positive and negative influence which it exerts on the evolution of relationships between man and his environment, among men or groups of men within a society, as well as among different societies [R1.60a, p.41].

In this respect, a question may be asked: What is culture and in what context is it being used? This question will be answered under the theory of Philosophical Consciencism and then related to the creation of categorial convertibility conditions composed of categorial moment, categorial transfer functions and categorial transversality conditions for the needed categorial conversion of social polarities. The categorial convertibility conditions are the sufficient conditions of transformation, and the creation of the set of sufficiency conditions is the result of the culture that gives rise to efficient development of Philosophical Consciencism under a particular experiential information structure Let us keep in mind that the theory of categorial conversion is to derive necessary convertibility conditions of actual-potential polarities. The theory of Philosophical Consciencism is to derive tactical and strategic actions through the management of the necessary command-control social instruments to create the set of sufficient convertibility conditions to bring about categorial conversions. In this way, the theories of categorial conversion and Philosophical Consciencism reside relationally in logical continuum and unity in understanding the general mechanism of socio-natural transformations.

The theory of Categorial Conversion is about the understanding of the dynamic nature of existence and identities of the needed command-control instruments as the *necessary conditions* for transformation, while the theory of Philosophical Consciencism is about the decision-choice understanding of the mode of change in complexity, and how to use this understanding to design relevant strategies and tactics as *sufficient conditions* to resist or bring about polar transformations. It is here that Abraham's quotation above *that the same event, occurring as it were between the frontiers of two different cultures, should be invested with differing significance, with different capacities for determining the direction of policies arising therefrom* [R1.1, p.11]. It is also here that the development of the theory of Philosophical Consciencism acquires its ultimate relevance and utility for understanding and engineering social transformations. The theory of Categorial Conversion is an *explanatory theory* about the necessary conditions in the theory of socio-natural transformation. The theory of natural Philosophical Consciencism at the level of nature may be viewed as an explanatory theory about the sufficient conditions in the theory of natural transformation. The theory of social Philosophical Consciencism is basically a *prescriptive theory* about the creation of sufficient conditions in the theory of social transformation. The conditions under which a prescriptive theory and science are viewed in terms of explanatory theory and science, and the explanatory theory and science are viewed in terms of prescriptive theory and science are discussed in [R3.7] [R3.13]

Definition 2.1: Culture Verbal

Culture may be seen as a totality of human past, present and future experiences within a social environment of its resistance where culture and social environment are mutually interdependent as well as mutually negating given the natural environment, where the totality of human

114

past, present and future experiences constitutes the experiential information set which is continually expanding as some old experiential varieties give way and new experiential varieties are formed by acquaintances.

Note 2.1

There are two sides of culture. There is the set of qualitative characteristics given its quantitative disposition. There is also a set of quantitative characteristics given its qualitative disposition. The characteristics are partitioned into sub-sets of negative and positive characteristic sub-sets in a relational continuum and unity in such a way that every negative characteristic sub-set has a supporting positive characteristic sub-set and vice versa. The union of the two presents not only the unity of the culture but also the identity of the culture and the character of the nation. These negative and positive characteristic sub-sets are in relational tension and in the process of continual mutual negation through their influence on methods of cognition and human thought in relation to conversions of social polarities. By its influence on thought and social practice, culture can emancipate individual and collective minds of a people to critically see the relational structure of social actual-potential polarities and the relative cost-benefit configurations of the corresponding actual and potential poles with supporting systems of dualities. Every social formation has an evolving cultural configuration under continual tension that produces forces of it evolution or revolution through its effects on cognition, thinking, information processing capacity and the individual and collective decision-choice rationality. Culture is collectively democratic as well as dictatorial on its restraining ability on the individual behavioral rationality under rules of commons (social goods). Changes in the qualitative-quantitative characteristics lead to changes in the cultural configuration and hence its preponderating effect on human action. Changes in qualitative and quantitative expositions are revealed by the expressions of motion in terms of directional movement. Analytically, there are a lot of discussions on quantitative motion as seen in space-time relation. However, there are few, if any, discussions on *qualitative motion* in the knowledge-production system. It must, therefore, be made clear what is meant by *qualitative movement* and the supporting *qualitative equation of motion*. Qualitative motion and its analytical dynamics require the development of qualitative mathematics that go beyond the simple structure of negative, neutral and positive [R9.11][R9.12][12.6][R12.69][R 12.13][R12.14].

Definition 2.2: Qualitative motion

A qualitative movement of an element in a category to another is inter-categorial transformation that is called a *categorial conversion* that is induced by a qualitative motion by intra-categorial changes of the relational structure of the negative and positive characteristic sub-sets in elemental categories to produce an inter-categorial transfer function.

Note: 2.2

Every culture is identified by its qualitative disposition which reflects the internal characteristics and the relative composition of its negative-positive characteristic structure. It must also be kept in mind that every qualitative disposition of any element has a supporting quantitative disposition in a dualistic setting. Much of this discussion is offered in [R1.91][R1.92][R3.7]. Categorial conversion appears in two ways as inter-categorial conversion and intra-categorial conversion. The inter-categorial conversion is concerned with transformation in the qualitative disposition. The intra-categorial conversion is concerned with changes in quantitative disposition.

Proposition 2.1

Every qualitative motion in a social set-up is propelled by a *social categorial moment* that obtains its energy from the conflicts and contradictions in the individual preferences in the collective decision-choice space for any given institutional configuration and social information structure.

The preferences are the reflection of the material existence of the people in terms of matter. The institutional configuration provides us with the internal organizational arrangements of the members that allow them to constitute a social unity whose stability is produced by a cultural unity [R1.84]. The information structure is the complex work of communications that induce the direction of individual decision-choice action and the resultant direction of the collective decision-choice action under cultural rationality. Social and individual preferences are established by a defined social environment within any social formation given human nature. This social formation is composed of three structures of economics, politics and law that project social decision-choice problems of integrated environments that must be dealt with within the existing culture at any point of time and over the path of social existence.

As the social preferences that capture the individual preferences change, new conflicts arise that require restructuring of the existing institutional configuration and the collective value system. These conflicts are cultural which express themselves through the institutional dynamics that install categorial quality at each state as a temporary equilibrium. Each equilibrium state over the categorial enveloping of the nation building establishes an interaction between alteration of social circumstances and the contents of individual and collective consciousness. The circumstances are not permanent at each temporary state. They can be altered by categorial conversion that is induced by the people who are guided by thought and motivated by will to act to alter their social circumstances. The will to act to change social circumstances is propelled by the changes in the content of collective and individual consciousness which resides in the people's culture. The culture is the creation of the people by the people and for the people which directs them to accept or reject their circumstances. It is the acceptance-rejection decision-choice actions of the people's circumstances that define the path of the people's history in terms of decision-choice outcomes. Here, culture defines the fruits of the people's historic struggle against nature and in relation to collective existence and the stability that may be required to increase its strength and reduce its weakness. It is also a determinant of such historical struggle.

The people have no claim to their culture until they have mastered its content and transformative force which it exerts on their dignity, sovereignty, organization and welfare. Culture is dualistically defined by the constituent positive and negative characteristics which claim its identity. Culture, therefore, has simultaneously positive and negative influences on the categorial-conversion process by either affirming the people's social circumstances or by rejecting them through the understanding of their cost-benefit implications. Every circumstance is an actual that is defined by ontological characteristics and belongs to the ontological space. The understanding of the circumstance is defined by an epistemological characteristic set and belongs to the epistemological space through the creation of thought for its explanation given the information structure. The thought belongs to the category of explanatory theory and explanatory science. The changing of the circumstance is a potential that resides in the epistemological space given human thought and social information structure [R9.11]. The actualization of this potential is a decision-choice action that must be guided by thought in the cultural environment with judicious applications in the selection of a cost-benefit combination in the cost-benefit duality, which must be related to the conversion of the actual-potential polarity. The conscious decision-choice activities to actualize the potential belong to a thought process which in turn belongs to the category of prescriptive theory and science. The placing of these conceptual elements in either the ontological space or epistemological space is important to the understanding of the constructs of explanatory and prescriptive theories within construction-reduction duality in the process of categorial conversions of social polarities. Such placements force one into judicious application of the methods of characteristics-based analytics in the information space to identify relevant social categories. In this respect, we need a clear conceptual understanding among the ontological space, epistemological space, characteristic sets and socio-natural categories.

Proposition 2.2

The general ontological space $\mathbb{U}$ is defined by a set of ontological elements, Ω with a generic element $\omega \in \Omega$, with an infinite index set $\mathbb{L}^{\infty}$, and a corresponding set of ontological characteristics $\mathbb{X}$, with a generic element, $x \in \mathbb{X}$ which is equipped with an infinite index set $\mathbb{I}^{\infty}$. The ontological characteristic set $\mathbb{X}$ is partitioned into groups, $\mathbb{X}_i$, $i \in \mathbb{I}^{\infty}$ in order to establish similarities and differences among ontological elements of $x \in \Omega$. By characteristics-based analytics, there is a set of ontological categories $\mathbb{C}$ with generic elements, $\mathbb{C}_{ij} \in \mathbb{C}$ with an infinite index set $\mathbb{J}^{\infty}$ to which each element $\left(\omega_i, i \in \mathbb{I}^{\infty} \right)$ belongs.

It may be noted that there are many concepts and ideas in the proposition 2.2 that need some reflections and explanations. These concepts and ideas will not be discussed here, however, the interested reader is referred to see [R3.7][R3.10][R3.13].

2.2 DNA Reflections on Culture and Categorial Conversions in Social Systemicity

The discussion here will be limited to concepts and ideas that are relevant to the understanding of culture and Philosophical Consciencism and their effects on categorial conversion and within the practice of Nkrumaism, with special reference to complete African emancipation. A human being is a complete organism as well as a complete functioning dynamic system with various integrated parts. A human being has within its complete organism a complete division of labor in executing tasks in the production-consumption duality in relational continuum and unity. The identity of every human being is completely defined by a complete set of characteristics which presents itself as a union of negative and positive characteristic sub-sets within the social set-up. The unity of the negative and positive characteristic sub-sets of the human identity defines human existence in life as well as negates its existence life. Each of these characteristics may be viewed as deoxyribonucleic acid commonly referred to as the DNA which carries the relevant information about the organism on the basis of which decision-choice actions are undertaken. The introduction of the concept of DNA rests on the idea that culture is a living organism with matter, energy and information. From proposition $2.2, \mathbb{X} = \bigcup_{i \in \mathbb{I}^{\infty}} \mathbb{X}_i \qquad \mathbb{C} = \bigcup_{j \in \mathbb{J}^{\infty}} \mathbb{C}_{ij}$

where every $\mathbb{X}_i$ fixes and defines the DNA in $\mathbb{C}_{ij}, j \in \mathbb{J}^{\infty}$. For the nature of human culture, it is useful to consider the category $\mathbb{C}_{hj}, h \in \mathbb{I}^{\infty}$ and $j \in \mathbb{J}^{\infty}$. The term $\mathbb{C}_{hj}$ is the category of humans with a corresponding characteristic set $\mathbb{X}_h$ which defines the conditions of the human DNA. Every characteristic $x \in \mathbb{X}_h$ is a DNA where $\mathbb{X}_h$ is the genome. The complete collection of all DNA's with differential organizational structure constitutes the *human genome* that defines the identity of an ontological element that belongs to the category of humans. Let us keep in mind that the use of DNA and the genome is in relation to an ontological classificatory scheme, and hence it is in reference to a differential qualitative disposition in the ontological space whose information is mapped onto the epistemic space for analytical understanding and knowing. For the purpose of ontological identity, the

119

genome will be assumed to be the same for all humans under the principle of human ontological equality. This process presents a category that is linguistically defined as humans. Such an ontological equality is mapped onto the epistemological space under an epistemic action.

Culture comes with a people with a corresponding natural environment. It is, therefore, analytically useful to specify the corresponding natural environment $\mathbb{E}$ of human habitation which is described by a set of characteristics $\mathbf{X}^{\mathbb{E}}$. Let natural environmental characteristics $\mathbf{X}^{\mathbb{E}}$ be partitioned into sub-sets of characteristics that describe different natural environments in terms of climatological and geomorphological conditions. Let the index set of such partition be $\mathbb{Q}$ with generic element $q \in \mathbb{Q}$ such that $\mathbf{X}_q^{\mathbb{E}} \subset \mathbf{X}^{\mathbb{E}}$ defines the qth natural environment that is consistent with a social formation. Similarly, let the category of humans be $\mathbb{C}_h$, with $j \in \mathbb{J}^\infty$ suppressed, be partitioned by the natural environment such that for any group of the people with a social formation, there is a corresponding geographical environment that contains climatological C and geomorphological G characteristics of the form $\mathbf{X}_q^{\mathbb{E}} = \mathbf{X}_q^G \bigcup \mathbf{X}_q^C$ that defines the environment of geomorphology and climatology with particular demographic conditions. These concepts and symbols will be used to define culture.

Definition 2.3 Culture Set-theoretic

Culture $\mathbb{C}$ is a set of characteristics, which is collectively produced by the members in a social formation to reconcile their conflicts in the individual-community duality for individual-collective support, collective stability, security, communication and understanding given human nature in particular geographical locations under social consumption-production duality. Thus,

$$\mathbb{C} = \left\{ \mathbb{C}_\lambda = \varphi(h,q) \mid \lambda \in \Lambda,\ h \in \mathbb{H},\ g \in \mathbb{G} \right\} = \left(\mathbb{H} \otimes \mathbb{G} \mid \Lambda \right)$$

where Λ, an index , is a set of different cultures, $\mathbb{H}$ is a set of human characteristics and $\mathbb{G}$ is a set of geographical characteristics which is made up of geomorphological and climatological characteristics and φ is a transformative operator.

As defined, the characteristic set $\mathbb{G}$ specifies conditions of the natural environment and the characteristic set $\mathbb{H}$ specifies the conditions of the demographical environment. The product of the two characteristic sets $\mathbb{C} = \mathbb{G} \otimes \mathbb{H}$ defines the social environment for each category of the political economy of social formation and each category of the developmental stages within any category of any given political economy of a social system and its formation with a corresponding cultural characteristic set given an index set Λ of different societies. In this respect, accumulated cultural characteristics which include modal forms of communication are the national *cultural capital* of diverse forms which are the fruits of the social production through history. The increases in the national cultural capital constitutes the foundation of additions to the cultural characteristic set called the *cultural investment,* which helps to define the direction of the national history by influencing new individual and collective decision-choice activities. The teaching and learning of the culture and its progress is called the cultural education to maintain social stability of cultural unity. This is an important reason why Diop discussed the foundations of African Cultural unity [R1.84], why Nkrumah discussed the unity of African conscience [R1.203] and Dompere discussed a unified foundation of African intellectual unity required to understand the African cultural unity and unity of African conscience [R1.92]. The importance of the cultural capital, investment and education cannot be overemphasized, for they constitute the core factors to bring about national progress though their effects on the managerial decision-choice actions of command-control of the dynamics of social actual-potential polarities. Culture and the unity that it brings to the members of the social set-up form the basis of effective resistance against foreign aggression and domination by unifying the members into a collective defense. They also form the basis of freedom, justice and national progress through qualitative and quantitative transformations by the manipulations of the negative and positive characteristic sub-sets to produce the appropriate negative and positive actions that culture generates. By affecting the content of philosophy culture defines and establishes the boundaries of thought and acceptable areas of application. For this manipulation to work to produce the desired results, the people must accept and be united by the essential characteristics of the culture. The understanding of these elements of culture is important in understanding Nkrumah's concerns with the three sub-sets of the cultural characteristics of traditional African, Islamic African and Euro-Christian African characteristics that exist in relational disunity, where such characteristics must be reworked to produce a relational cultural unity with a relational continuum of the past-present-future historical dynamics of African society from the antiquity[R1.203].

Within the national cultural characteristics, there is a partition that satisfies the national population into groups and classes of different income levels with different ambitions that relate to their personal and national interests, personal vision and national vision. From the viewpoint of national transformations that relate to freedom, justice and complete emancipation, some of these groups may hold antagonistic cultural characteristics that merely relate to their personal interest and against the important elements of the collective social interest. Some of these groups may have common personal interests with the oppressive forces that are external and internal subjugation of the nation and her people. Just as a united culture may serve as a powerful instrument of national resistance; a disunited culture may serve as a powerful instrument to encourage external subjugation and oppression.

Culture, like any phenomenon, is encapsulated by a positive-negative duality as defined by its characteristics under the principle of relational continuum. In thinking about disunity-unity transformation, the phenomenon of interest is cultural polarity corresponding to actual-potential polarity where each pole has a residing duality defined in terms of characteristic sets. The phenomenon of duality is characterized by a positive-negative characteristic set and corresponding positive-negative action under conflict in relational continuum and unity. The phenomenon of polarity is characterized by positive duality that identifies the positive pole and negative duality which identifies the negative pole under conflict in relational continuum and unity. There is a category of social polarities and the category of natural polarities consistent with the African conceptual system. Here, the primary category of polarity is the actual-potential polarity where all other categories of polarities are derivatives. This conceptual system which is referred to as *polyrhythmicity* with *polyrhythmics* [R1.92] as its laws of thought, allows us to work on the *theory of characteristics-based analytics* in dealing with the production of the theories of category formation and categorial conversion through the methodological duality of construction-reduction processes. Culture, as a phenomenon, cannot be excused from the conceptual system and the interplays of elements defining the negative-positive duality.

The characteristic set of culture $\mathbb{C}$ is thus a partition into positive characteristic subset $\mathbb{C}^P$ and negative characteristic subset $\mathbb{C}^N$ such that $\mathbb{C} = \left(\mathbb{C}^P \bigcup \mathbb{C}^N\right) = \left(\mathbb{H} \otimes \mathbb{G}\right)$ for any given $\lambda \in \Lambda$. The culture is said to be progress-retarding if $\left(\#\mathbb{C}^N\right) > \left(\#\mathbb{C}^P\right)$ in which case the negative elements in the culture outweigh the positive elements in the cultural characteristic set. It is said to be progress-enhancing if $\left(\#\mathbb{C}^N\right) < \left(\#\mathbb{C}^P\right)$ in which case the positive elements in the culture outweigh the negative elements in the cultural characteristic set. These relationships hold for politico-economic transformations and stage-to-stage transformations through time. In an active search for a transformation of the actual-potential polarity, there arises a need to consciously introduce an action into the duality for the principle of negative reversal and maintain the dominance of the positive characteristic set for forward qualitative motion. The principle of positive reversal works through the creation of a social policy set, the implementation of which requires the casting of an appropriate institutional configuration through which the policies can be transmitted to bring about the required categorial moment to induce a transfer function between categories. This monograph, even though it is not directly about the theory of economic development, however, contains the analytical tools that are useful in the study and construct of such a theory [R15.14]. Economic development is about changes in the qualitative-quantitative dispositions regarding socio-economic varieties in the dynamics of actual-potential socioeconomic polarities, and this is also the case of all transformations of social actual-potential polarities. In other words, the theory of categorial conversion that provides the understanding of the necessary conditions, and the theory of Philosophical Consciencism that provides the understanding of sufficient conditions for transformation are general and that they unite to form the general theory of transformation in the quantity-quality space with neutrality of time.

2.3 Culture, Ideology and Categorial Conversion of Social Polarities in System Dynamics

The positive and negative characteristics of culture are revealed through tribal, ethnic, class, religious and other possible divisions that may be antagonistic to one another. The possible antagonistic relations among different divisions are generated by differential thoughts and beliefs that are held by each other to impose disunity in national interest and social vision. These differential thoughts are more or less differential sub-ideologies within the organic ideology. They generate temporary irreconcilable differences that affect the national interest formation, social-vision definition and the national character to create schizophrenia in the national rationality and decision-choice behavior toward nation building. They, further, create a system of conflicts that draws national and individual resources to pursue unproductive activities and social waste as well as tear asunder the social cohesion required for progress. For the categorial conversion process, the negative and positive characteristics of these sub-divisions must be identified, analyzed and unified for national progress. It is the recognition of these antagonistic relational structures of sub-categories of thought and their impact on national resources and social efforts that led Nkrumah to introduce the three segments of the African social experiential information structure and unstable culture which act as an important binding constraint on African social decision-choice actions after decolonization. The segments are identified as traditional Africa, Islamic Africa and Euro-Christian Africa. These three segments create their own cultural domains where the Islamic tradition and Euro-Christian traditions act as superimpositions that define new complexities under ideological tension and competing sub-thoughts creating contradictory individual and social behaviors in the social decision-choice space, such that the African society as a whole is operating under cultural schizophrenia with confusing rules within individual-community duality.

After critical analysis, Nkrumah came to a realization that the building and development of African societies under democratic decision-choice system have been made extra difficult by the superimposition of a number of negative external elements on the set of conflicts in the ethnic space. The negative external elements include categorial partition of the ethnic space into complex religious divisions by the three segments that are existing under socio-political tension, ideological confusion and cultural disunity. The socio-political tensions, ideological confusion, cultural disunity, lack of understanding of the parameters of African philosophical unity and cultural disunity from the diverse belief systems have created crises in the *African conscious*. The understanding of this crisis in the African conscience requires a toolbox from both the science of complexity and synergetics. At the dawn of the African decolonization and the struggle for Africa's complete emancipation, this socio-political and cultural complexity presented an important question and problem as Nkrumah saw it. The question and problem was how to forge a unity from these historical experiences that will create a reasonable cultural unity and define a new African motion in the qualitative and quantitative spaces at the passage of time taking into account the sankofa-anoma reflection. The first challenge was to organize the African masses within a socio-political unity for the decolonization and the claim of the African continent and its resources of human and non-human elements. The second challenge was the integration of diverse experiences into a common socio-cultural and ideological unity for the protection of the sovereignty and embankment on nation building for the improvement of the welfare of the African masses. It may be noted that Nkrumah accepted the Diopean position of the African cultural unity [R1.84] and the existence of common foundations of the African conceptual system [R1.92] in advancing the African Unity project, and hence it was not necessary to deal with ethnic and tribal divisions as they relate to the tasks at hand. Nkrumah's answer and solution to the question and problem are couched in the following statement:

> A new emergent ideology is therefore required, an ideology which can solidify in a philosophical statement but at the same time an ideology which will not abandon the original humanist principle of Africa.
> Such a philosophical statement will be born out of the crises of African conscience confronted with the three strands of present African society. Such a philosophical statement I propose to name philosophical Consciencism, for it will give the theoretical basis for an ideology whose aim shall be to contain the African experience of Islamic and Euro-Christian presence as well as the experience of the traditional African society, and, by gestation, employ them for harmonious growth and development of that society [R1.203, p. 70].

This statement forms the second organic pillar of the development of the conceptual system of Nkrumah as applied to the African conditions to form the basis of categorial conversion of actual-potential socio-economic states after decolonization. The first organic pillar is the *concept of categorial conversion*. The categorial conversion and the theory that may be developed give substance to the understanding of internal self-dynamics and the nature of qualitative motion [R15.15]. The relevance of categorial conversion to the African conditions lies in the capsule of African internal self-transformation on the basis of the African energy to decolonize herself and set new socio-political conditions for self-development. Having provided the logic of internal self-motion, the next epistemic pillar is to answer the question of how do African people bring about decolonization and solve the problem of socioeconomic transformation over stages from within and by Africa's self-generating energy. The sequential nature of the emancipation problems is first *decolonization* and second *socio-economic development* from within Africa by the African people under their collective consciousness as generated by their cultural confines and unity. It is this sequential nature of the African problems that led Nkrumah to entitle his book *Consciencism: Philosophy and Ideology for decolonization and development*. The problem as identified in the African social space is the existence of conditions of cultural and collectively cognitive disunity that makes progressive social transformation difficult.

These conditions relate to the crises in African conscience creating a zone of cognitive crises and crises in collective and individual rationalities as they relate to the elements in the social decision-choice space. The crises and the zone of crises manifest themselves in the ideological struggles among African humanistic traditions and the superimposed structures of Islamic views and Euro-Christian views of the world and its universe. This African crisis cannot be underestimated. It has the negative aspects of many years of European destructions, thievery of human and non-human resources, colonialism and disruption of African's cultural unity. The crisis forms an important cognitive foundation of behaviors in the African decision-choice space in dealing with African problems and their cost-benefit implications that arise from domestic relations and international relations that may be either bilateral or multilateral. It also forms the foundation of distorted and schizophrenic cultural personalities of a number of individual Africans who have not escaped the zone of the crisis. The intensity of the crisis is amplified by the existing tribal and ethnic cleavages in the traditional African socio-power structures which have tendencies to increase the African confusion and exacerbate Africa's demise.

Within the African crisis zone, a question faced by Nkrumah was simply: What is to be done? Nkrumah has arrived at a fundamental principle of internal self-motion to justify a position that for sustainable freedom, justice and development the African society can be and must be transformed from within through the categorial-conversion process. The working mechanism of the categorial-conversion process requires the creation of conditions of categorial convertibility from within to set the potential against the actual. In other words, the conditions of categorial conversion should lead to the negation of the actual-potential social polarities from within Africa. The convertibility conditions are created by internal forces operating under a complex relational structure of elements of culture which impose conscience on the African people. The culture and the conscience that it holds constitute the foundational vehicle to effect the negations of actual-potential social polarities. To answer the question as to what must be done Nkrumah observed and emphasized that this crisis manifests itself in the African conscience to create an incompatible anti-African personality under the conditions of ideological crisis within the general set of African cognitive characteristics. The anti-African personality must be restructured to give it the characteristics of a true African personality that reflects the African genius to serve the transformative needs of the African people toward the African social vision and organic interest. At this juncture, Nkrumah provided a working definition of African personality required for categorial conversions of African social polarities. He stated:

> When I speak of the African genius, I mean something different from negritude, something not apologetical, but dynamic. Negritude consists in a mere literary affectation and style which piles up word upon word and image upon image with occasional reference to Africa and things African. I do not mean a vague brotherhood based on a criterion of colour, or on the idea that Africans have no reasoning but only a sensitivity. By the African genius I mean something positive, our socialist conception of society, the efficiency and validity of our traditional statecraft, our highly developed code of morals, our hospitality and our purposeful energy [R1.207, p.5].

Toward the restitution of the African personality, which is true to itself, Nkrumah suggested that the conditions of Islam and Euro-Christian traditions must be accommodated as experiences of the traditional African society as the primary cultural category. He advised that *if we fail to do this our society will be racked by the most malignant schizophrenia* [R1.203, p.78]. The philosophical foundations of the African personality and conditions of its effects on the African intellectual space and struggle for freedom and justice are given extensive discussion in [R1.92, pp. 109-138] where it is stated:

> What are the characteristics that must establish and define the new African personality? The characteristics must be such that the new African a) must be dedicated to the creation of a greater Africa, b) must be modest and honest, c) must be willing to devote himself/herself to the service of Africa and her children, d) must detest vanity, e) must abhor greed and embrace humility as a source of our people's strength and integrity abd as a catalyst for African greatness. Furthermore, the new African with these characteristics must be willing to enlist to join the collective to wage war against the worship of pure material interest at the expense of human life. The African personality must promote willingness to embrace justice, truth, freedom and fairness in the management of affairs of the evolving African social order. The enlisted social characteristics must induce willingness to be rightly educated and rightly educate other Africans because the individual is the African society and the African society is the individual [R1.92, p112].

It may be said that these characteristics of African personality constitute the cluster of the basic principles of philosophical humanism that gave directions of African statecraft and practice of decision-choice actions at the level of tradition from antiquity. It is important to note, as it is also understood by Nkrumah, that the African philosophical humanism sees human social formation as a collection of people who has willingly contracted with each other to bring the collective strength in support of the individual rights and collective security without which the individual right and security will be mere mirages. This is the complex relational structure in the formation of the individual-collective duality for freedom and security. This African principle translates to the relational structure of duties and rights of the individual and collective within the individual-collective duality forming a relational foundation of the African social formation. The full discussion of this principle as it relates to tribal, ethnic and state specific affinity and African nationalism is discussed in [R1.91]. This principle underlines communalism with an ideology of each for all and all for each under the rights, duties, justice, freedom and responsibility in relational continuum and unity. The reversal of this social continuum and unity will increase tension in the individual-community duality, enhance the crisis and expand the crisis zone of the African conscience of social existence and decision-choice behavior.

In this context of the need to restructure the African personality and the corresponding conscience, Nkrumah not only identified the African crisis in culture conscience and personality, but define the nature of the crisis zone in which Africans individually and collectively are victims. This crisis zone provides Africans with a dilemma in Africa's decision-choice space due to a distorted personality particularly known as the so-called educated African. He offered a way out of this African crisis zone and the dilemma by specifying the source of the problem and the nature of the solution in terms of an *intellectual revolution* that must support the needed *African social revolution* to change the African terms of domestic existence and relational structure with the external. The task of the intellectual revolution is to decolonize the mind of Africa in support of cultural unity [R1.1]. The intellectual and social revolutions carry with them the needed cultural unity that will ensure unity of thought, practice and social structure. For these dual revolutions to have the intended impact on the African society and its social transformation, every African must ask the following questions: *Who am I?* What are my duties and responsibilities to myself? What is my duty to my state and Africa? What are my rights and obligations? The answers to these questions will depend on the African attitudes to non-traditional experiential information structures and the methods of processing them. In relation to these questions, Nkrumah offered us advice that must help in the unification of the diverse cultural elements of African tradition, Western tradition and Islamic tradition to create a modern intellectual and cultural unity for the guidance of the African social decision-choice processes under progressive African ideology for internal and external relations. He stated

> Our attitude to the Western and the Islamic experience must be purposeful. It must be guided by thought, for practice without thought

is blind. What is called for as a first step is a body of connected thought which will determine the general nature of our action in unifying the society which we have inherited, this unification to take account, at all times, of the elevated ideals underlying the traditional African society. Social revolution must therefore have, standing firmly behind it, an intellectual revolution, a revolution in which our thinking and philosophy are directed towards the redemption of our society. Our philosophy must find its weapons in the environment and living conditions of the African people. It is from those conditions that the intellectual content of our philosophy must be created [R1.203, p.78].

The efficient development of this connected thought will force the people of Africa to state from the onset the problems and questions at issue, the methods of their analytic-synthetic structure, the nature of the solutions and answers, and finally their relevance to Africa's interest and social vision with   supporting goals and objectives for their achievements, where these supporting objectives will collectively constitute a program of action relevant to ideology under Nkrumah's unity of African thought and Diopean cultural unity of Africa..

The structure of the problems and questions and their devastating impact on African society has been extensively discussed in [R1.91]. How these problem-solution and question-answer structures show their faces in the African decision-choice space is also extensively discussed in [R1.92]. The connected thought that will unify the three segments of cultural traditions in Africa Nkrumah called it Philosophical Consciencism which he defined as:

> Consciencism is the map in intellectual terms of the disposition of forces which will enable African society to digest the Western and Islamic and Euro-Christian elements in Africa, and develop them in such a way that they fit into the African personality. The African personality is defined by the cluster of humanistic principles which underlie the traditional African society. Philosophical Consciencism is that philosophical standpoint which, taking its start from the present content of the African conscience, indicates the way in which progress is forged out of the conflict in conscience [R1.203, p.79].

The foundation of this Philosophical Consciencism draws its weapons from the environment and living conditions of the African people as Nkrumah advised. The extensive analytical works in support of the epistemic foundation is developed in *Polyrhythmicity* in which the

Africentric framework is specified, the African personality is conceptually structured, the principles of social behavior in the individual-collective duality are developed for analysis, synthesis and conclusions, and where the methodological *polyrhythmics* as a logic in dealing with the Africentric principles of opposites composed of a system of actual-potential polarities, dualities under negative-positive characteristic sets are provided [R1.92][R9.11].

It is important to understand the practical necessity of Nkrumah's introduction of *Philosophical Consciencism* as a follow up of his introduction of *Categorial Conversion*. The Philosophical Consciencism is required for the practice of categorial conversion as a general mechanism of dynamics of qualitative and quantitative dispositions. The theory of categorial conversion presents a logical system and step-by-step analytical and methodological framework to convert the relational structure of socio-natural actual-potential polarities working through resolutions of conflicts in the system of dualities [R15.15]. The categorial-conversion process shows itself through the joint manifestations of *intra-categorial* and *inter-categorial conversions*. The intra-categorial conversion activities manufacture *categorial moments* in both actual and potential poles. The inter-categorial conversion activities take the categorial moment as input and create *categorial transfer functions* to send the actual to the potential space and send the potential into the actual space to complete the negation.

As has been discussed, the creation of categorial moment in natural polarities is through the natural processes of internal organization and rearrangements of the negative-positive natural characteristic sets. The creation of the transfer function by the inter-categorial conversion activities is also done by natural relational structures of actual-potential natural categories. The process of the internal organization and arrangements that brings about the manufacturing of the *natural* categorial moments and the categorial transfer function is referred to as the *natural technology* that alters the relationship between the actual and potential poles. This natural technology is what is referred to as the *natural Philosophical Consciencism* as has been discussed in Chapter One. There are as many types of Philosophical Consciencism as there are categories in the universal system under the principle of continuum and unity. The creation of categorial moment in social polarities is through the social processes of internal organization and rearrangements of the negative-positive social characteristic sets. The creation of the transfer function by the inter-categorial conversion activities is also by social relational structures of decision-choice activities by people to define policies to act on to affect actual-potential social categories. The social processes of internal organization and arrangements of the negative-positive characteristic sub-sets which define the system of dualities take place through culture, the available social institutions, the system of ideas locked in *ideology* and the guiding philosophy. The culture, institutions, system of ideas, ideology and the guiding philosophy constitute what is being referred to as the *social technology*. The appropriate social technology depends on the relational unity of elements in the social technology which will not generate schizophrenic behavior in the social decision-choice space. Let us reflect on the epistemic foundation in continuum and unity. The actual-potential polarities exist as categories which are the objects of categorial conversion. The actual poles and the potential poles have residing dualities defined by negative and positive characteristic sub-sets. These dualities and residing negative and positive characteristic sub-sets are the instruments for categorial conversion. It is through their relative behavior that forces of resistance and transformations are manufactured to bring about the dynamics in qualitative and quantitative dispositions. The theory of Philosophical Consciencism is about the intentional utilization of the instruments of categorial conversion of socio-natural actual-potential polarities to design sufficient conditions of transformation in order to create new social varieties and categorial varieties. The structure and practice of Philosophical Consciencism is sensitive to culture and experiential information structure.

In the definition and the structure of Consciencism, Nkrumah was particularly careful in recognizing the African philosophical system of the principles of polarity and supporting dualities in relational continuum and unity, and that all socio-natural elements exist in temporary equilibria of qualitative dispositions where such qualitative dispositions are linked to quantitative dispositions. The emphasis is on the intellectual development and transformations to create a dynamic *revolutionary ideology and philosophy* to support African cultural unity which will bring about intra-categorial conversion and inter-categorial conversion of African social actual-potential polarities. Philosophical Consciencism is a map to create both the categorial moments and the categorial transfer functions to alter the relational structure of the actual and potential social poles in Africa as Africans work for decolonization and development. It initializes the categorial conversion dynamics from the present content of the

African conscience as well as indicates the way in which progress is forced out of conflict in African conscience. For this to be successful, the Philosophical Consciencism must restructure the African mind and African personality away from Western, Islamic and Euro-Christian elements that are incompatible with the basic structure of the African humanistic principles which define the basic relation in the individual-community duality for freedom, justice and unity of social purpose. Categorial conversion is impossible without the creation of categorial moment and categorial transfer function. At the level of African historical conditions, Philosophical Consciencism with an African content is the supporting intellectual and ideological map for creating a system of social strategies to deal with categorial conversion of the actual-potential African social polarities.

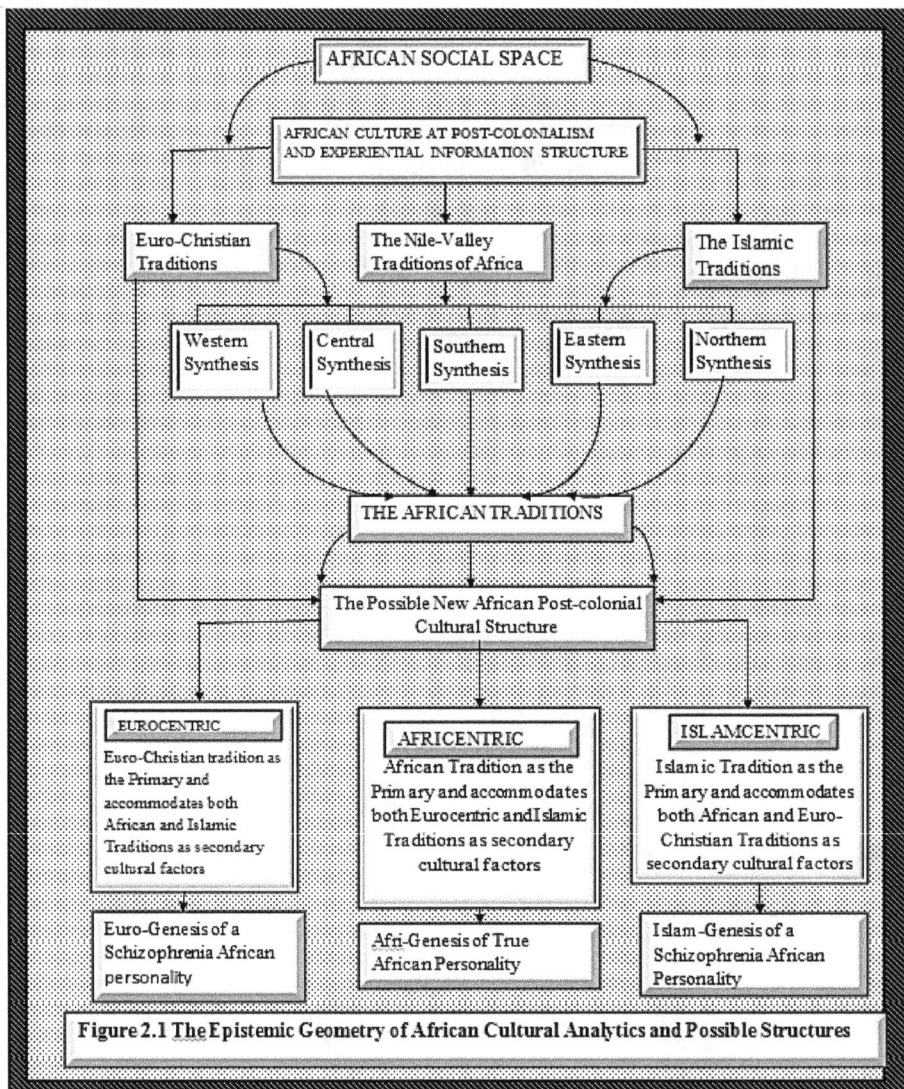

AFRICAN SOCIAL SPACE

AFRICAN CULTURE AT POST-COLONIALISM AND EXPERIENTIAL INFORMATION STRUCTURE

Euro-Christian Traditions

The Nile-Valley Traditions of Africa

The Islamic Traditions

Western Synthesis

Central Synthesis

Southern Synthesis

Eastern Synthesis

Northern Synthesis

THE AFRICAN TRADITIONS

The Possible New African Post-colonial Cultural Structure

EUROCENTRIC
Euro-Christian tradition as the Primary and accommodates both African and Islamic Traditions as secondary cultural factors

AFRICENTRIC
African Tradition as the Primary and accommodates both Eurocentric and Islamic Traditions as secondary cultural factors

ISLAMCENTRIC
Islamic Tradition as the Primary and accommodates both African and Euro-Christian Traditions as secondary cultural factors

Euro-Genesis of a Schizophrenia African personality

Afri-Genesis of True African Personality

Islam-Genesis of a Schizophrenia African Personality

Figure 2.1 The Epistemic Geometry of African Cultural Analytics and Possible Structures

One must always keep in mind that the theory of Philosophical Consciencism with the theory of Categorial Conversion constitutes a general theory of social transformations of varieties. It is applicable to all societies under different cultures and experiential information structures. The emphasis of the contents of the structure and form will vary from society to society due to its sensitivity to culture and experiential information structure.

The intellectual map presented by Philosophical Consciencism requires characteristics-based analytics of the actual pole and the potential pole whenever possible to develop information requirements as inputs into the social decision-making process in order to create the categorial moment and the categorial transfer function to bring about relational transformation between the actual and potential poles. The nature of the information requirement in implementing the logic of Philosophical Consciencism is culturally and ideologically determined in relation to varying information structure. It is discussed and provided in [R1.92] [R3.7][R3.10]. It may be noted that at the level of Africa's cultural setting and its relationship to embrace Philosophical Africanism, Africans are confronted with problems of different concepts of God and the relationship to the people in the definition of relational qualitative disposition in terms of the relational understanding of God to people, and people to God as well as people to people and the guiding principles of freedom and justice, law and order; and love and hate of humanity in resolving problems in individual-community duality in relational continuum and unity. The nature of African cultural analytics in support of an appropriate utilization of Philosophical Consciencism is given as a cognitive geometry in Figure 2.1. This cognitive geometry may be restructure for any social set-up where the particular society is substituted in the place of Africa and the experiential information structure is defined in terms of the particular social history.

AFRICAN SOCIAL SPACE

AFRICAN CULTURE AT POST-COLONIALISM
AND EXPERIENTIAL INFORMATION STRUCTURE

Acquaintance, A_1

Acquaintance, A_2

Acquaintance A_3

Synthesis S_1

Synthesis S_2

Synthesis S_3

Synthesis S_4

Synthesis S_5

THE NATIONAL TRADITIONS

The Possible New Socio-Cultural and
Information Structures

Foreign Dominance Z_1

Z_1-Foreign-Dominated tradition as the Primary and accommodates the national and all other traditions as secondary cultural factors

National Dominance

National Tradition as the Primary, and accommodates all foreign Traditions as secondary cultural factors

Foreign Dominance Z_2

Z_2-Foreign-dominated Tradition as the Primary and accommodates the national and other Traditions as secondary cultural factors

Z_1-Genesis of a Schizophrenic National personality

National-Genesis of True National Personality

Z_2-Genesis of a Schizophrenic National Personality

Figure 2.2 The Epistemic Geometry of National Cultural Analytics and Possible Structures

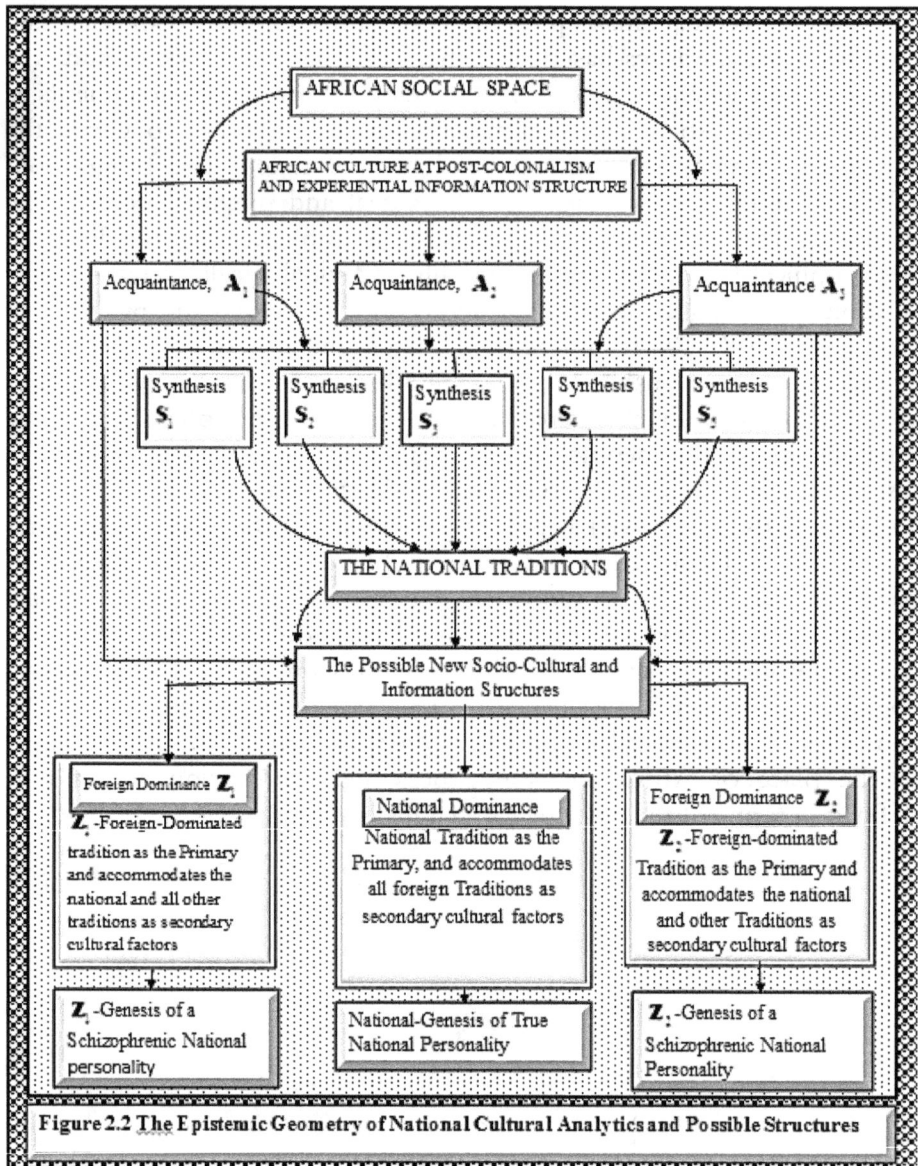

A question arises as to how general is the African cultural analytics when different societies with different cultures and experiential information structures are examined. It may be kept in mind that all cultures and the corresponding social information structures obey the *sankofa-anoma* principle in relation to the past-present-future time trinity.

It is this generality of the past-present-future information-decision-choice interactive processes of the *sankofa-anoma* principle that imposes an explanatory-prescriptive generality on transformative processes of varieties by the theories of Categorial Conversion and Philosophical Consciencism. In this respect, this cognitive geometry of Figure 2.1 may be restructure into Figure 2.2 for general cultural analytics of any social set-up where a particular society is substituted in the place of Africa and the experiential information structure is defined in terms of the particular social history. The set $\{A_i \mid i \in I_A\}$ defines a number of experiential information structure from acquaintances and encounters given the initial culture. The set $\{S_i \mid i \in I_S\}$ specifies the syntheses of the traditional culture with acquaintances and encounter by various geographical areas of a particular society. The set $\{Z_i \mid i \in I_Z\}$ defines foreign domination in cultural developments of personalities in the individual-collective dualities of the social set-ups and the directions of Schizophrenic National personalities. The symbols (I_A, I_S, I_Z) are the corresponding index sets. The nature of the schizophrenic stresses on the individual and collective personalities of the nation may be obtained by the applications of further analytics to the intra-personality and inter-personality characters of the inner structure the national life in time and over time. The important outcomes of the adverse domestic-external interactions of the external control of a national sovereignty may be seen in terms of how the outcomes explicitly and implicitly shape and affect the national spirit and behaviors of the individuals and collective over the national decision-choice space, where adverse distortions are created in the cost-benefit balances of social decision-choice elements. The organic domestic outcome of imperialist control of a national sovereignty is the time-path of the nature of the national behavior as seen through the practices of both the collective and individual.

At the level of practice of Philosophical Consciencism, a number of questions tend to arise in the test of Africa's readiness to escape the development of a schizophrenic personality. What factors for the struggle of Africa's independence and emancipation are dead within you and the African social collectivity? Similarly, what factors of the struggle for Africa's emancipation are awakened in you and the collective for Africa's redemption by defeating neocolonialism and imperialism at the levels of material and spiritual elements? What cultural foundations do you and the collective claim as your thought and reasoning processes in support of your social actions? What vision do you claim or do you think that Africans collectively must claim for Africa and her children? Philosophical Consciencism points to new conditions for Africa with a new African personality, a new African mind and a new African vision to replace conditions of colonial vestiges of slavery and European imperial atrocities where the new African would not accept internal oppression, and where the elected officials will cultivate honesty in the service and management of African affairs. When this new African with a true dynamic and progressive African personality emerges, then the Africa's road to complete emancipation is defined and the true principles of a transformative Philosophical Consciencism are Africentrically entrenched to bring about continual categorial conversions of varieties and categorial varieties in the African social polarities in favor of Africa's progress. The current African conditions of conflicts and wars are completely predicted by the schizophrenic African personality and these maladies will continue until the African personality is restructured in line with Philosophical Consciencism. An extended discussion of this problem in relation to tribalism, ethnicity, tribal affection, ethnic affinity and others is made in African Union [R1.91]. Let us keep in mind that Nkrumah's Philosophical Consciencism does not only contain African content but is constructed from the analytical and epistemic instruments drawn from the African conceptual system the origins of which date back from Africa's antiquity [R1.37] [R1.60b] [R1.83b] [RI.243] [R1.247].

2.4 Nkrumah's Application of Philosophical Consciencism to Colonialism-Decolonization Polarity in System Dynamics

Let us examine how Nkrumah applied Philosophical Consciencism to the conditions that were central to his intellectual and political efforts. During Nkrumah's active political effort, the categorial conversion was to be applied to colonialism-decolonization polarity where colonialism was the actual pole and decolonization was the potential pole. Each of these poles has a residing duality with a relational continuum and unity as has been theoretically developed and discussed in [R15.15] and amplified in [R9.11] and [R1.12]. To decolonize the African territories, the convertibility conditions of categorial moment, categorial transfer function and categorial transversality conditions at each point of time must be created from within each African colonized territory. The creation of the required convertibility conditions must come from the people who constitute matter, energy and information in relational continuum and unity. Such a creation must come from the people's intellects and conscience acting on the African experiential information structure to manufacture positive action to alter their conditions. Philosophical Consciencism, as it was conceived by Nkrumah is to indicate a way in which sustainable African progress is forged out of the conflicts in African conscience within Africa's thought system as seen from the damages of colonialism, Christianization and Islamization where the African is colonized and enslaved for his own good by those who claimed to hold the golden keys to God and civilization. In this respect, what is the nature of African progress that must be constructed by the use of Philosophical Consciencism?

2.4.1 Philosophical Consciencism and the Nature of African Progress

Progress passes through categories from the standpoint of intra-categorial conversions and inter-categorial conversions with a neutrality of time. At the level of society, social progress must always be defined as an actual-potential polarity and in relation to a social vision and collective interests defined in terms of collective social welfare. Every social actual is a stage as well as a category in the categorial enveloping. It is always the actual pole with a residing duality that is under conflict and social action. Every actual pole has a set of corresponding elements, each of which can constitute the potential pole with residing duality in relational continuum and unity. For the categorial conversion to take place, a need is required to develop an inter-categorial moment and an inter-categorial transfer function which are relationally connected and exist in unity as well as abstracted to create a potential pole with a positive net cost-benefit configuration for the collective. The intra-categorial moment alters the relational composition of the negative and positive sets of the dualities in the actual (colonialism) and potential (decolonization) poles. The inter-categorial conversion uses the intra-categorial moment to create a categorial transfer function by altering the characteristics of the social polarities and change the behaviors of the negative and positive poles within the context of African colonial and development conditions. If the potential is actualized according to the will of Africans, then it may be concluded that a progress may have been made. The framework of

Philosophical Consciencism is to restructure the African mindset in order to set social vision and national interest relevant to Africa's progress in addition to creating the convertibility conditions.

There are three organic phenomena that must be understood concerning the associated problems dealt with in the social decision-choice space under culture and experiential information structure. One such phenomenon is the politico-economic institutions that define the decision-choice sovereignty, such as the ones associated with colonized states, decolonized states, neocolonial states, dependent states and others. This phenomenon involves the problem of a sovereignty phenomenon regarding the control of individual and collective sovereignties in the affairs which affect the individual-collective duality in the national political decision-choice space. The second such phenomenon involves the problem of the internal transformations of socio-economic categories within a defined sovereignty. This phenomenon involves the problem of a *socioeconomic development phenomenon* in transforming socioeconomic states over time to change the material and spiritual conditions of the individual-community duality in the national socio-economic space. The first problem of social polarity is about polar conflict of domestic-foreign ownership of the people's sovereignty, requiring the categorial conversion of domestic-foreign polarity within the space of political polarities. This is the theory of political development. The second problem of social polarity is about polar conflict of development-underdevelopment process of a socioeconomic structure within sovereignties requiring the categorial conversion of development-underdevelopment polarity within the space of socio-economic polarities. This is the theory of economic development. The third such phenomenon involves the problem of the internal transformations of legal categories within a defined sovereignty. This phenomenon involves the problem of a *legal development phenomenon* in transforming legal states over time to change the judicial conditions of the individual-community duality regarding freedom, justice and law-and-order process in the national legal space. The third problem of social polarity is about polar conflict of compatible-incompatible judicial process of a socio-legal structure within sovereignties requiring the categorial conversion of socio-legal polarity within the space of legal polarities. This is the theory of legal development within sovereignties.

In the African case, the first problem involves categorial conversions from externally controlled sovereignties to domestically controlled sovereignties for the individual colonized states. The second case is about the problem of internal qualitative and quantitative transformations of socioeconomic states in higher order of preference under domestic ownership of sovereignties of decolonized states, their peoples and institutions. The third case is about the problem of internal qualitative and quantitative transformations of compatible legal structures that account for important balance in the individual-community duality in relation to freedom and justice under the Africentric organizational principle of *each for all and all for each for social progress* in a higher order of preference within the domestic ownership of sovereignties of decolonized states, their peoples and institutions. These were the three problems of *decolonization and development* that formed the subtitle of Nkrumah's book entitled Consciencism and gave rise to the inscription of *freedom and justice* in Ghana coat of arms. Here, the Philosophical Consciencism has two demands. The first demand is to point to the first directional force to create *categorial moments* from within Africans that will alter the relative negative-positive actions to bring about the intra-categorial conversions of the colonized African states in order to set the conditions of decolonization, as the potentials, against the conditions of colonization, as the actual, through the manufacturing of the required *transfer functions* at the level of political sovereignty where domestically controlled sovereignties are actualized. The second demand is to point to the second directional force from within Africa to create new *categorial moments* that will alter the relative negative-positive actions to bring about the intra categorial-conversions of the decolonized states in order to set the conditions of development, as the potentials, against the conditions of underdevelopment, as the actual, through the manufacturing of the required interstate *transfer functions* at the level of domestically controlled sovereignties, where increasing socioeconomic developments, as potential, are actualized in favor of Africa's progress. Let us keep the epistemic focus on the relational continuum and unity of Categorial Conversion and Philosophical Consciencism as defining an important decision-choice framework within the individual-collective duality in dealing with the practice of Nkrumaism in negation of negation, transformation of social varieties within social polarities for Africa's complete emancipation in terms of the role that the *unitary state* and *African unity* play in the institutional casting required for the development of a compatible organizational structure and how the positive and negative actions must be managed in the African social

141

space for decolonization and development. It must be kept in mind that the organizational social structure is made up of sub-structures of politics, economics and law.

2.4.2 Negative and Positive Actions in Categorial Conversion of Social Polarities in System Dynamics under Philosophical Consciencism

From Nkrumah's conceptual system as seen within the theories of categorial conversion and Philosophical Consciencism, every entity like a country, every phenomenon like development, every living thing like a person or every decision-choice action like decolonization is composed of negative-positive duality within actual-potential polarity in action and conflict. The conflicts are between the negative action and positive action in a perpetual struggle for either maintenance of self or a change of the existing. The manner in which the positive and negative actions are associated will depend on the entity and the environment in which the identity of the entity is defined. In the case of t Africa's struggle for freedom, justice and emancipation Nkrumah stated:

> One may say that in a colonial situation positive action and negative action can be discerned. Positive action will represent the sum of these forces seeking social justice in terms of the destruction of oligarchic exploitation and oppression. Negative action will correspondingly represent the sum of those forces tending to prolong colonial subjugation and exploitation. Positive action is revolutionary and negative action is reactionary [R1.203, p.99]

A territory is a colony and remains so because the negative action is greater than the positive action of the occupied people. In this case, the negative duality dominates the positive duality to maintain the colonial social pole as the actual and dominating pole in the actual-potential polarity under relational social continuum and unity. Here, the sovereignty rests in an external hand. To reverse the socio-political structure of colonization-decolonization polarity requires the introduction of conditions of categorial conversion. For a colonial state to be decolonized it is necessary to continually increase the social positive action in the colony to be greater than the social negative action. This necessity of positive action brings about a change in the *direct control* of the domestic sovereignty. To attain independence, the decolonized territory must protect against external indirect control of the domestic sovereignty by increasing the positive action to overwhelm and dwarf the negative action to a minimum and maintain it in the state of insignificant social action. The external control of domestic sovereignty after decolonization is simply what Nkrumah called *neocolonialism*. The structure of neocolonialism and its impact on a decolonized territory with a resemblance of independence is discussed in [R1.204], while the process of creating and maintaining a neocolonial states is discussed in [R1.91].

It is important to keep in mind that sovereignty comes in degrees of ownership ranging from zero-independence to complete independence. The independence is a linguistic quantitative variable whose meaning may be subjectively defined in a *fuzzy space* in terms of degrees ranging from zero to one. Since dependence and independence exist as a duality, what is true of independence relationally exists as opposites. The zero-degree of independence is actually complete dependence with a unit-value degree, and the zero-degree of dependence is actually complete independence with a unit-value degree in a dualistic structure. Let **D** be a fuzzy set of independent states defined by categories of degrees of domestic control of sovereignty as $\mu_\mathbf{D}(d) \in (0,1)$. In this specification, $\mu_\mathbf{D}(d)$ defines a category of independent states with the same degree of domestic control of the sovereignty of the state. In other words, by the use of *fuzzy decomposition*, the collection of all states may be classified into categories of domestic control of national sovereignty. The degree of foreign control may also be specified in terms of fuzzy set **G** with a membership function of the form $\mu_\mathbf{G}(g) \in (0,1)$. We must notice that $(\mathbf{G}, \mu_\mathbf{G}(g))$ is basically the inverse of $(\mathbf{D}, \mu_\mathbf{D}(d))$ such that $\mu_\mathbf{D}(d) + \mu_\mathbf{G}(g) = 1$ and hence the degree of independence is one minus the degree of foreign control of the nation's sovereignty $(1 - \mu_\mathbf{G}(g))$.

The cognitive geometry of this duality in relational continuum and unity is presented in Figure 1.3 in Chapter One. The relational structure in continuum and unity must be linked to the African traditions in concepts and thoughts of epistemological spaces of polarities, dualities with characteristic sets that define their nature, and relative negative and positive characteristic sets that reveal their identities of ontological existences. The conceptual system and its logic of reasoning are such that, given any entity, there is a negative characteristic sub-set with a supporting positive characteristic sub-set without which its ontological existence is undefined. These dualities have built in conflicts, contradictions and categorial struggles induced by the negative and positive characteristic sub-sets seeking dominance, and may conceptually and computationally be presented as an important fuzzy structure. This is the case of colonialism-decolonization polarity with its relevant dualities.

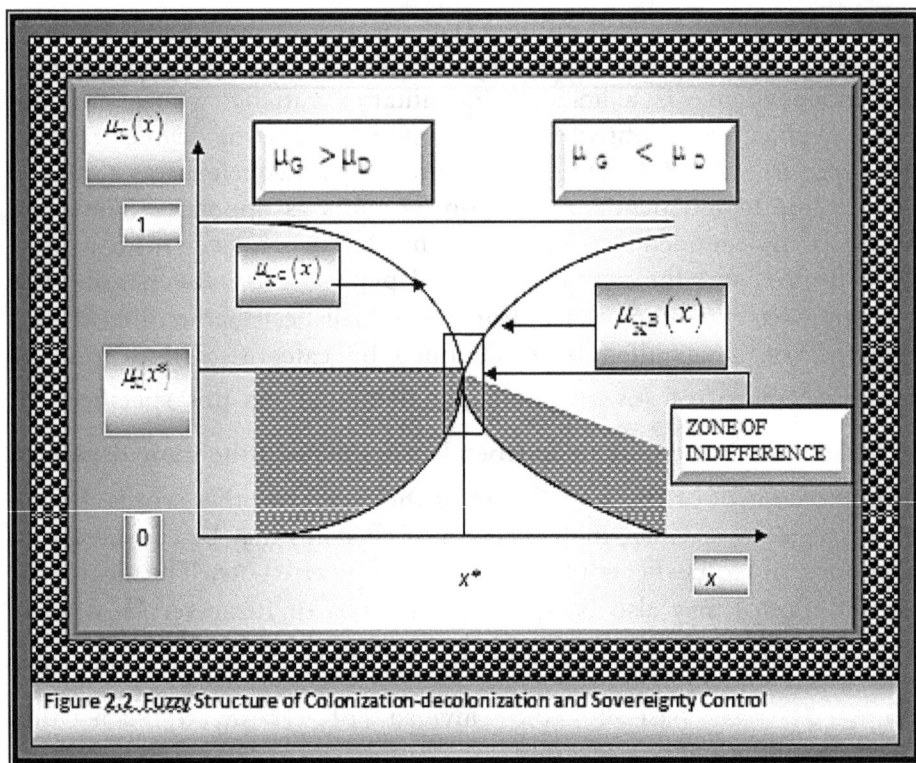

Figure 2.2. Fuzzy Structure of Colonization-decolonization and Sovereignty Control

These degrees of national sovereignty define political stages in the global system of international politico-economic struggles and conflicts of nations in the control of resources, where each sovereignty stage specifies a category of belonging by *fuzzy decomposition*, which contains a set of countries with similar degrees of domestic control of sovereignty. As politico-economic categories, there is an important distinction between a colonial state and a neocolonial state seen from their characteristic sets. In a colonial state the sovereignty of the decision-choice actions are directly controlled by external forces to exploit the domestic human and non-human resources for the benefit of the people and the ruling class of the colonizing country. The whole domestic population is under exploitation and some form of servitude and the natural resources are under active brutal exploitation to serve the interest of the foreign country. Things are, however, different in the neocolonial state. Here, sovereignty is domestically held under the resemblance of independence and indirectly controlled from external source for the benefit of external forces and a small set of domestic political elites in the neocolonial system. In a neocolonial state, the domestic political elites and their supporting socio-economic groups play a negative action to keep the state in a new exploitive stage for rent seeking at the expense of the domestic population. The basic characteristics of a neocolonial state and the modus operandi are discussed in [R1.91]. Nkrumah's characteristics-based analytics put it as:

> In neocolonialism, however, the people are divided from their leaders and, instead of providing true leadership and guidance which is informed at every point by the ideal of the general welfare, leaders come to neglect the very people who put them in power and incautiously become instruments of suppression on behalf of the neo-colonialists [R1.203, p.102]

In neocolonialism, independence of domestic sovereignty is a mirage locked in the illusionary walls of comfort and flag-flying, where the domestic decision-choice system is externally designed and its process externally controlled. In this neocolonial information-decision-control system , the members of the domestic political elite are elevated by the imperialists to positions of neocolonial cronies as useful domestic idiots who perform the role of supervision of the work of a neocolonial subjugation-oppression machine of the domestic population through the developing and marketing of views, propaganda and deception through the use of a neocolonialist mindset by indirectly convincing the domestic population that it is being enslaved for their own progress and good. The domestic sovereignty is indirectly arrested from the people and imprisoned in walls of enticements of different kinds of sweet but dangerous music for the creation of self-inflicting of external domination, exploitation and operation of the domestic population. The self-inflicting external exploitation and societal servitude reflect the basic confusion of what national independence connotes. As has been explained in [R1.197] and further amplified in [R1.91], the people of a nation have not mastered their independence until they have mastered the cost-benefit relation of freedom and its fruits. They have no claim to freedom and security until they have control of the *cost-benefit configuration* of their national sovereignty, government and governance.

If a people of a nation do not collectively master the cost-benefit configuration of their national life and existence through freedom, justice and servitude, they will not understand the glorious path out of the various paths to their social progress, and will stay subservient to foreign dominations. They will be moved to the zone of historic irrelevance with continual natation towards the devastating shores of oppression and suffering where freedom and justice are packaged in a capsule of mirage by the imperialist resource game. The choice of the path within the historic relevance-irrelevance duality is the work of social thought that is packaged in the capsule of national collective conscience. It is through this understanding that the theory of Philosophical Consciencism brings some relieve, and hope, as well as indicates the relevant path of national progress. Here, among all the forces that can bring about social progress of a nation within the dynamics of social actual-potential polarities in creating increasing welfare of time-point varieties within the individual-collective duality, there is no powerful element as national conscience with hope. With national Philosophical Consciencism, the individual and the nation can correctly think and work in the true assessments of relative cost-benefit balances between the set of actual elements and the set of potential elements. If the people of a nation have true national Philosophical Consciencism, they have hope, control their destiny and create the path of their national history through a continual creation of social varieties.

In a neocolonial state, the leadership acquires negative characteristics and amplifies the aggregate negative action in the social set-up to overwhelm the domestic positive action required for the creation of intra-categorial moments and inter-categorial transfer functions required for the maintenance of domestic sovereignty and national progress. In this framework, the domestic leadership becomes a *strategic partner* with the neocolonialist to exploit the domestic masses under complex conditions of corruption that form the incentive of their belonging and social practice. The members of the leadership and their supporting groups have acquired in them either a Euro-genesis schizophrenic or

Islam-genesis schizophrenic African personality or both where they are stripped of their African essence leading them to operate in the zone of cognitive imbecility in the domestic national decision-choice space. This neocolonial strategic partnership and the amplification of the negative action by the domestic leadership for joint enslavement of the domestic population are justified by illusions of an ideology of development, the uselessness of begging and aid solicitation, the disgrace and humiliation of carrying a global beggar's basket, and the mirage of democracy designed with and supported by the intellectual craft of the predators, neocolonialists and imperialists for confusion and deception of the domestic masses to accept slavery as good for them. Implicit in the neocolonial revisionist philosophy are a set of principles of anti-African ideology. The central core of this anti-African ideology is that Africans, whether colonized or decolonized could not be considered as true citizens of the world, and hence they have no legitimacy to control their human and non-human resources and decision-choice actions that affect them. The complementary principle of anti-African ideology is that the Africans have no freedom, liberty and interest that the West is bound to respect. The only freedom and justice they have are what the members of the Western world give them and that they must always work under the principles of law and order as designed by the imperial system of oppression. This anti-African ideology of imperial system must be resisted at all fronts of human existence through the precepts provided by the theory of Philosophical Consciencism.

The current global system of capitalist production, national resource needs and the flow of goods and services requires us to take a look at the nature of colonialism, neocolonialism and national independence within the framework of the global space of imperial aspirations. A colonial state is simply a loss of domestic sovereignty to a direct external control. However a neocolonial state is not a simple loss of national sovereignty but a strategic partnership created between the leadership of bureaucratic capitalists of an imperial state and the leadership of bureaucratic capitalists of a domestic neocolonial state to exploit the neocolonial masses for the benefit of the membership of the capitalist imperial bureaucratic state. The leadership of a neocolonial state has lost sight of its transformative role after decolonization, and has acquired an anti-African personality operating in a zone of anti-African ideology which is supported by anti-African philosophy whose implicit principles have been stated above. Furthermore, it has abandoned its historic mission where the members reduce themselves to useful political idiots, useful intellectual idiots and useful religious idiots under the principle of cognitive imbecility required for the functioning and management of any efficient slave system and obedient slaves on behalf of their imperial masters, obtain the rewards of simple socioeconomic leftovers full of insults and humiliation to themselves and the people that they are supposed to lead. In addition, the leadership finds itself in the creation-begging duality under conditions of conflict zone of categorial conversion where it opts not to create but to beg under the neocolonial mindset.

There are two classes of leadership that are taking place in the current global politico-economic system. The entrepreneurial class in the industrially advanced imperial countries has been overthrown and replaced by the bureaucratic capitalist class whose interests are driven by global resource seeking. The nationalist leadership of the decolonized countries has been overthrown and replaced by a domestic *neocolonialist bureaucratic class* whose interest is personal wealth accumulation for ostentatious living at the expense of the neocolonial masses through external bribery and internal corruption to allow the neocolonial state to solidify and entrench itself. The existence of the neocolonial state is maintained by the existence of a residential class of bureaucratic capitalist and political idiots whose interest is rent-seeking through corruption and exploitation of the domestic population. The leadership of the neocolonial state is in strategic partnership with the imperial bureaucratic capitalist class to work against the creation of true national sovereignty which belongs to the control of the domestic masses. To create an independent state the neocolonial system must be resisted by the national population without being the victim to its sweet music of aid, technical assistance and the offer of all kinds of deceptive elements of strategic partnerships and systems of ideological precepts. Nkrumah put it as:

> In order to be able to carry out this resistance to neo-colonialism at every point, positive action requires to be armed with an ideology, an ideology which, vitalizing it and operating through a mass party shall equip it with a regenerative concept of the world and life, forge for it a strong continuing link with our past and offer to it an assured bond with our future. Under the search light of an ideology, every fact affecting the life of a people can be assessed and judged, and neo-colonialism's detrimental aspirations and sleights of hand will constantly stand exposed [R1.203, p.105].

From this advice, Nkrumah indirectly and wisely observed the twin difficulties facing Africa. The first difficulty was the territorial division of Africa into colonial territories as was done by the European imperialists for resource interest without regard to the African people. Each colonial territory was united and maintained by force of the European colonialist under an ideology of Christianization and civilization. Each ethnic group had its own initial struggle with the colonialist for freedom and justice. Under this historic event, a decolonized territory is a disunited territory from which a unitary nation must be constructed. In other words, immediately after decolonization, each African decolonized territory is confronted with the problem of disunity-unity polarity that must be solved by categorial conversion. Notice here, *disunity* is the actual and *unity* is the potential that must be actualized by the construct of a categorial moment and categorial transfer function to satisfy the categorial transversality conditions in each state. The failure to solve this relationality problem within the disunity-unity polarity will lead to internal conflicts that will be exploited by external neocolonialists for extra-imperial aspiration for resource seeking. The construct of the categorial moment and transfer function can be done by decision-choice actions under the development of appropriate *social technology*. The change of the relational terms in the internal disunity-unity polarity is an important precondition for sustainable social stability, nation building and socio-economic transformation which includes transformations of the political, legal and economic structures.

This problem was critically understood by Nkrumah in thought and action. He recognized that without internal domestic unity of a decolonized territory, the decolonization will produce a useless concept of sovereignty under a mirage and the return to a neo-slavery of neocolonialism. The protection of domestic sovereignty must be linked to other African decolonized territories and mobilization of African positive action to assist in decolonizing the territories under colonialism and imperial bondage. The linkage to other African decolonized territories led to the policy of *African unity*. These two unity questions and their relational foundation to African micro and mega sovereignties are discussed in [R1.91]. Nkrumah's first problem after decolonization is and was the construction of *internal unity* of the decolonized African territories. The *first African Unity problem* to solve in order to achieve a meaningful domestic sovereignty is the problem of consolidation of diverse internal domestic tribal affections and ethnic affinities to create territory specific nationalism, with the intent of surrendering part of this nationalism to the supra-African nationalism and sovereignty in defense of the state sovereignties. By the very nature of the Africentric logic of the working mechanism of polarity, duality, relational continuum and unity under tension, Nkrumah realized that decolonization does not bring about unity of purpose and state. As a categorial conversion of colonization-decolonization polarity, it only moves the system to another actual-potential polarity with its own internal dynamics, where decolonization offers the possibility of constructing a unitary state without which a claim of political sovereignty is and was not possible. Under the logic of Africentic conceptual system, life is in death and death resides in life without which existence and transformations are impossible. Categorial conversion of polarities works in dualities under the principles of opposites with relational continua though the conflicts between the residing negative and positive characteristic sets for negative and positive actions. It is a double-edged sword. Like the cost-benefit process, the negative characteristic sub-set supplies what the positive characteristic sub-set lacks and needs and vice versa creating mutual interdependency and negation.

A colonial territory can be decolonized by the application of increased aggregate positive action by introducing categorial-conversion operators of categorial moments and categorial transfer function to change the relative position of aggregate negative and positive actions, so as to empower the aggregate positive action to overwhelm the negative aggregate action for the destruction of colonialism. The opposite direction works in the sense that a decolonized territory can be moved into a category of neocolonial territories by the application of an increased aggregate negative action by an introduction of categorial operators of a categorial moment and categorial transfer function, to change the relative position of the aggregate negative and positive actions to overwhelm the aggregate positive action in the state to render it a neocolonial state. This analytical structure of system of social polarities, dualities defined by negative and positive duals of characteristic sets under relational tension that generates forces of categorial conversion, was the foundation of the practice of Philosophical Consciencism that guided Nkrumah's policy constructs and his social vision and organic African interests. No other leader in Africa has come close. It is this analytical power based on the African thought system that guided

Nkrumah to investigate the nature and characteristics of neocolonialism, and to examine the decolonization-neocolonialism polarity with residing dualities, as well as the nature and characteristics of African unity under disunity-unity polarity [R1.204] [R1.202].

To resist the development of neo-colonial states in Africa after decolonization, Nkrumah observed that even if the internal forces are consolidated for unity, the domestic conditions required to generate the power and positive action of resistance to neocolonial forces are weak and insufficient to confront the overwhelming force of the imperial incumbents who are now consolidated into an imperial club such as NATO (North Atlantic Treaty Organization) creating temporary monopoly over violence through waging war over freedom and justice. There is, therefore, a power asymmetry in all forms to the disadvantage of the decolonized territory in internal unity and social stability except to take refuge in the blanket of imperial protection under neocolonialism, thus rendering the whole exercise of decolonization a mockery of Africa as well as defeating the whole practice of sovereignty. To change the terms of power asymmetry and secure the domestic sovereignty for domestic control, the decolonized territory after or in the process of constructing the unitary states must form unity with other decolonized African states. This was the *second African unity problem* as seen by Nkrumah. The relevance of the solution to the problem of categorial conversion of the organic African disunity-unity polarity is to correct the power asymmetry that has been established between the imperial incumbents and the African decolonized states. The search for African politico-economic unity is to change the global power distribution and its dynamics over resources, sovereignty and implied cost-benefit distribution in the global space where Africans are no longer living in the zone of irrelevance and powerlessness, and where African power of resistance will be amplified by the Pan-African unity [R1.87] [R1.91]. Under the principle of complete African unity, Africans will not concede to their former colonizers and the consolidated imperial club of the Western Alliance the monopoly of violence without resistance. In this way, Africa will continually remember the past atrocities of the West and avoid the repetition of human right abuses of the West through the intensive and extensive exploitation of human and non-human resources. Under the principle of complete African unity guided by Africentric Philosophical Consciencism, Africans will be equipped with the necessary epistemic instruments to speak the language of the imperial club, whatever the language is, in terms of diplomacy and violence or both.

The solution to the second African unity problem requires the development of a different types of categorial operator of the categorial moment and categorial transfer function to set the potential of African unity against the existing actual of the African disunity among the decolonized African states, in order to consolidate and protect the sovereignty of the federated state and the sovereignties of the individual decolonized states within the federal sovereignty. This analytical solution sequence to the African struggle to freedom and justice in the global system of subjugation and imperial terror has not been clearly understood by the members of the African political elite and their supporting intellectuals and clergy. The solution of the first problem of unity is a necessary condition for freedom of the individual decolonized states. The solution to the second problem of Pan-African unity is sufficient condition to guarantee this freedom within and across the African states and African diaspora. It is in the search for the solution to the problem of sufficient conditions by African thinkers, that the Western influence, intellectual order and justification for the Western continual destabilization of African territories have their most damaging effects. A complete analytical work is provided in [R1.91]. The lack of the creation of unitary states of a number of decolonized African territories generates the necessary conditions for instabilities and categorial reversal for re-colonization under the neocolonial structure. The lack of the creation of Pan-African Unity generates the sufficient conditions of inter-state and intra-state wars, conflicts and the categorial reversal for the creation of neocolonial African states. The creation of both the sufficient and necessary conditions of the categorial reversal is well discussed by Nkrumah in his books [R1.202] [R1.204].

In this respect, every nation in the global system may be placed in a category of one of the structures in Figure 2.3 where a nation is either a victimizer, victim or neutral in terms of resource seeking, rent seeking and transfer of wealth. By the application of negative action to overwhelm the national positive action, a decolonized territory can be moved into a category of neocolonial and neo-slavery states. It is always useful to keep in mind the methodological structure for the Africentic conceptual system in which Nkrumah was working. The methodological structure is under the principles of opposites. The principles of opposites are composed of a set of social polarities that have primary category and derived category, a set of internal polar dualities in relational continuum and unity, and a set of characteristics which is composed of the negative and positive sub-sets that establish the set of negative and positive dualities. The primary category of polarity is the actual-potential polarity which must be defined in every decision-choice time point. The global social information-decision-choice system is involved in controllability and convertibility of socio-political systems under domestic sovereignties and resource endowments where neocolonial states are sought for imperial control.

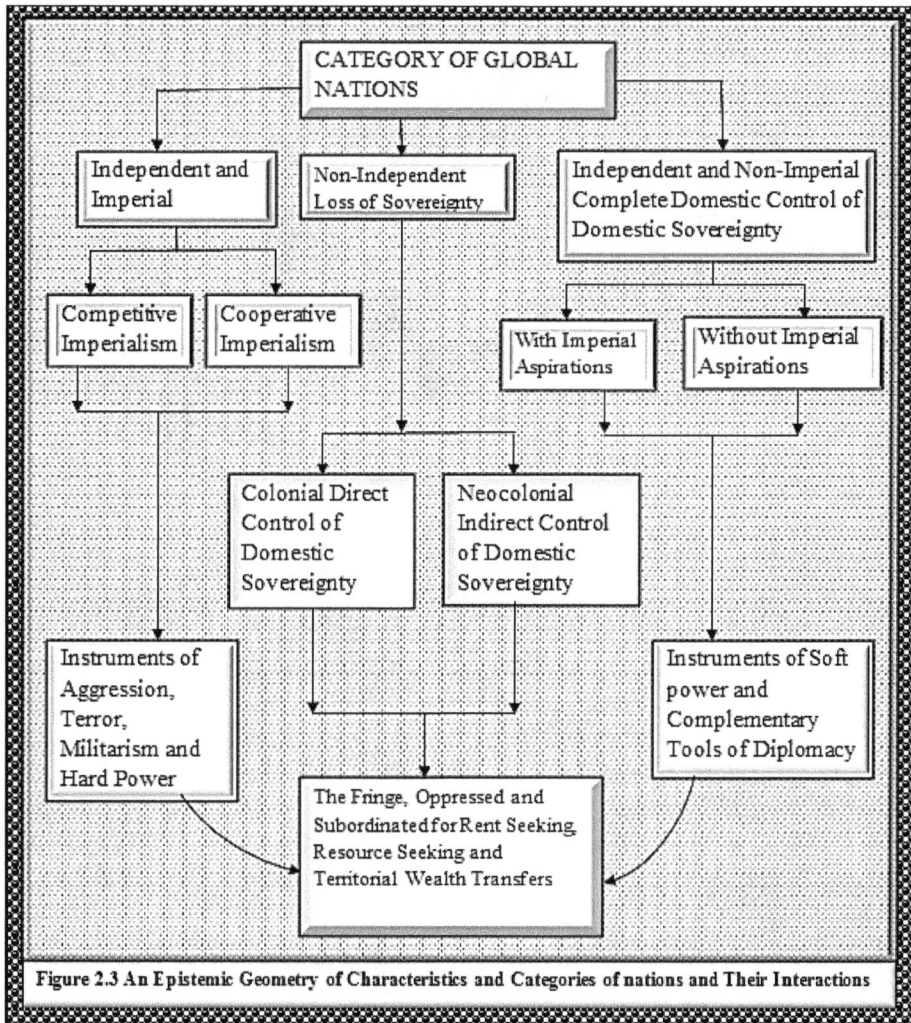

Figure 2.3 An Epistemic Geometry of Characteristics and Categories of nations and Their Interactions

153

Figure 2.3 presents a system of the category of nations from which social polarities, dualities, positive and negative characteristic sub-sets are presented for the basis of which the analytical path of categorial conversion of sovereignties and the control of individual socio-political economies in relation to positive and negative actions are presented in Figure 2.4. The conditions are such that aggregate negative action presents a disadvantage to the domestic socio-political economy of the nation while aggregate positive action presents a progressive advantage for the socio-political economy of the same nation. The aggregate negative action is generated by the negative characteristic sub-set while the aggregate positive action is generated by the positive characteristic sub-set of the nation or territory. The characteristic-based analytics of social action presents three possibilities for any pole of the relevant polarity. They are: 1) the positive aggregate action is greater than the negative aggregate action in the struggle for decision-control of the apparatus of sovereignty $\left[(+ve)\square\ (-ve)\right]$; 2) the negative aggregate action is greater than the positive aggregate action in the struggle for decision-control of the apparatus of the national sovereignty $\left[(-ve)\square\ (+ve)\right]$; and 3) either the positive or negative aggregate action overwhelm one another for the maintenance of the existing conditions of control of the domestic sovereignty: $\left[(+ve)\ggg(-ve)\right]$ or $\left[(-ve)\ggg(+ve)\right]$ in the categorial conversion processes of socio-political polarities. The processes of categorial conversions take place through the changes in the residing respective dualities of the poles of the relevant socio-political polarity. These changes are produced by the struggle between the negative and positive characteristic sub-sets.

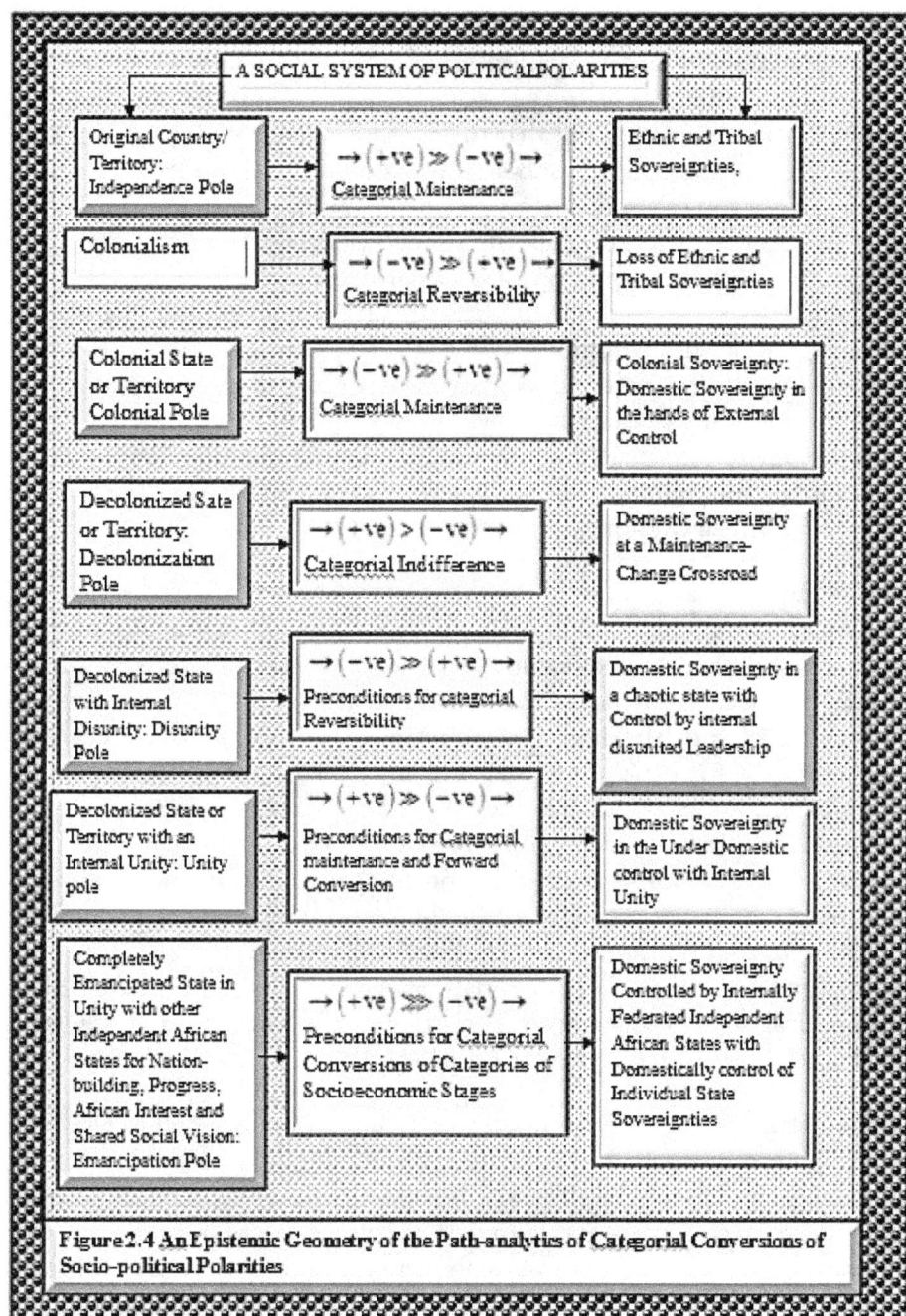

A SOCIAL SYSTEM OF POLITICALPOLARITIES

Original Country/ Territory: Independance Pole	$\rightarrow (+ve) \gg (-ve) \rightarrow$ Categorial Maintenance	Ethnic and Tribal Sovereignties,
Colonialism	$\rightarrow (-ve) \gg (+ve) \rightarrow$ Categorial Reversibility	Loss of Ethnic and Tribal Sovereignties
Colonial State or Territory Colonial Pole	$\rightarrow (-ve) \gg (+ve) \rightarrow$ Categorial Maintenance	Colonial Sovereignty: Domestic Sovereignty in the hands of External Control
Decolonized Sate or Territory: Decolonization Pole	$\rightarrow (+ve) > (-ve) \rightarrow$ Categorial Indifference	Domestic Sovereignty at a Maintenance-Change Crossroad
Decolonized State with Internal Disunity: Disunity Pole	$\rightarrow (-ve) \gg (+ve) \rightarrow$ Preconditions for categorial Reversibility	Domestic Sovereignty in a chaotic state with Control by internal disunited Leadership
Decolonized State or Territory with an Internal Unity: Unity pole	$\rightarrow (+ve) \gg (-ve) \rightarrow$ Preconditions for Categorial maintenance and Forward Conversion	Domestic Sovereignty in the Under Domestic control with Internal Unity
Completely Emancipated State in Unity with other Independant African States for Nation-building, Progress, African Interest and Shared Social Vision: Emancipation Pole	$\rightarrow (+ve) \ggg (-ve) \rightarrow$ Preconditions for Categorial Conversions ofCategories of Socioeconomic Stages	Domestic Sovereignty Controlled by Internally Federated Independant African States with Domestically control of Individual State Sovereignties

Figure 2.4 An Epistemic Geometry of the Path-analytics of Categorial Conversions of Socio-political Polarities

155

2.4.3 A Simple Comparative Analytics of Marx, Schumpeter and Nkrumah in Social Systemicity

Finally, it may be pointed out that the changing behaviors of the negative and positive characteristic sub-sets are due to the available internal institutions through which decision-choice policies are transmitted to generate individual and collective actions to affect socio-political outcomes where the potential is set against the actual, or the actual is maintained to resist the actualization of any potential. It is here, that political and legal institutions come to restrict the behavior of the morphology of the national economy in terms of its institutions of production and distribution. It is within this institutional framework that one must relate Nkrumah's conceptual framework to Marx's conceptual framework and Schumpeter's conceptual framework for qualitative-quantitative social dynamics.

MARX: In Marxian dynamics, the driving force of categorial conversion (categorial moment and categorial transfer function) is the disparity in *income distribution* between capital and labor as factors of production where such disparity is the surplus value creating a conflict between the capitalist class and laboring class. Capitalism as a mode of production-consumption formation is an income distribution duality with contradiction and conflict creating a social class actual-potential polarity which is expressed by *capitalism-socialism polarity* in capital-labor process of the control of political power where profits and costs are socialized. The instruments of battle in the categorial-conversion game are technology and new capital creation on behalf of the capitalist class while the instrument of battle in the categorial-conversion game are mental labor and physical labor in the part of the working class in the same domestic sovereignty and colonized sovereignty [R15.14][R15.34].

SCHUMPETER: In Schumpeterian dynamics, the driving force of categorial conversion (categorial moment and categorial transfer function) is the disparity in *profit distribution* between two capitalist classes of innovators and non-innovators where the class of innovators with innovating investments establishes abnormal profit and the class of non-innovators with non-innovating investment depletes the abnormal profit to zero calling for new innovation investments to establish a new abnormal profit. Capitalism as a mode of production-consumption formation is a profit distribution duality with contradiction and conflict between two capitalist classes creating a social class actual-potential polarity in relation to the social power. The initial instruments of the battle are innovative investments on behalf of the innovating capitalist class and non-innovative investments on behalf of the non-innovating capitalist class in the same domestic sovereignty and colonized sovereignty leading to bureaucratic capitalism, where the new capitalism creates its own internal conflict of government-bureaucratic-capitalism polarity with social power-distribution duality with new instruments of investment of power acquisition on the part of the capitalist class to control the government, while the government invest to control the capitalists where the battle is within the politico-legal control-non-control duality within the dynamics of categorial conversion of *capitalism-socialism polarity* where the power asymmetry is such that profits are privatized for the capitalist class and costs are socialized to the public in the same sovereignty [R15.14][R15.46] [R15.47] [R15.49].

NKRUMAH: In Nkrumaist dynamics, the driving force of categorial conversion (categorial moment and categorial transfer function) is the disparity in power distribution between two countries such as a colonialist or imperialist and the colonized. The colonialist controls the sovereignty of the colonized for resource exploitation. The colonized is forced into a zone of humiliation and conditions of slavery to create colonial poverty and underclass depleting its freedom and justice to nothingness. The instruments of battle are investments in military power supported by ideological propaganda and a deceptive information structure on behalf of the colonialist and imperialist in imperialist sovereignty over the colonized to organize the negative action in the colonized. The imperialist ideological propaganda and the deceptive information structure work in manufacturing colonial concept in a colonized in their favor, and also manufacturing internal domestic descent for social conflict and political destabilization for regime change. The instrument of battle at each round on behalf of the colonized is the investment in resistance through the organization of positive action in the colonized territory for the struggle against the colonialist, neocolonialism and imperialist. It is also a legitimate struggle for freedom, justice and human dignity against a system of arbitrary law and order imposed by imperial forces of inhumanity and terror [R15.14] [R1.203]. The categorial-conversion game is negation of power relations in a sequence of the colonization-decolonization, independence-neocolonialism and underdevelopment-development polarities.

CHAPTER THREE

CHARACTERISTIC-BASED ANALYTICS, A COUNTRY'S IDENTITY, SOCIAL ACTIONS AND PHILISOPHICAL CONSCIENCISM IN SOCIAL SYSTEMICITY

The path analytics, as presented and discussed with an epistemic geometry of Figure 2.4 in Chapter 2, relate to the understanding of the basic characteristics that provide a national identity. The epistemic relevance of the examination of the characteristics of national identity is found in the idea that the general category of Philosophical Consciencism is the primary category from which specific types of Philosophical Consciencism are the derivatives on the basis of specific experiential information and culture that provide the content of the specific social set-up. For an efficient operative process of categorial conversion, it is necessary to know the actual conditions of national sovereignty and the control of the socio-political decision-making process. The category in which a nation in the global system belongs is presented in an epistemic geometry of Figure 2.3. The identity of any nation, big or small is defined by the basic characteristics of the politico-economic structure which places it in the category of belonging. The essential characteristics of category formation are related to sovereignty, and decision-choice control of national interest and social vision that impose behavior in the political, economic and legal structures form the pillars of the society.

3.1 National Identity, Sovereignty and the Power System

The three pillars relate to the power system and its micro and macro-controls of decision-choice activities. When a territory is colonized or militarily occupied there is a categorial reversal of domestic sovereignty and the power system in the nation by the introduction of a categorial moment and categorial transfer function. A territory or a country stays either as colonized or as occupied when the aggregate negative action substantially overwhelms the domestic aggregate positive action. In this respect, the actual pole is the colonized state and the potential pole is the decolonized state whose actualization requires a categorial reversal. To bring about categorial reversal, new institutions must be created to empower and unite the domestic positive actions into a forceful aggregate positive action to overwhelm the aggregate negation action in the colonized state to decolonize the territory in order to bring about a domestic claim of the domestic sovereignty and the control of the national decision-choice system. This decolonization establishes temporary qualitative politico-economic equilibrium where the decolonized state becomes the actual that faces a competition from a set of potentials such as neocolonialism, military occupation and others. All sovereignty control systems are defined within the domestic-foreign duality with relational continuum and unity where the control is established by proportionality distribution in dualistic sense.

To maintain the domestic sovereignty and power system, the people of the decolonized territory must create material, intellectual and relevant culture for their support. They must recognize that a new polarity has been created with a new system of struggles that will generate a categorial-forward motion or categorial-backward motion to actualize an element from the potential space. In this respect, when a territory is decolonized, the categorial conversion requires a new and positive orientation away from the strategies of decolonization toward the creation of a completely emancipated state that has a capacity to protect its sovereignty and power system. Toward this end, a series of increased positive actions is required to secure complete sovereignty through a process of nation building which in turn requires leadership with a *social vision* and a judicious understanding of the *national interest*. This domestic leadership with a social vision is linked to the social characteristic set that leads to social actions and creates forward categorial conversion for higher categorial planes with greater social welfare. Similarly, domestic leadership with no social vision is also linked to the same social characteristic set which leads to negative social actions and creates backward-categorial conversion for categorial reversal into a lower categorial plain with a lower social welfare. The country's or decolonized territory's characteristic set finds expression in the country's cultural foundation. The nature and potential role of culture in generating a categorial moment and categorial transfer function have been discussed in the previous chapter. The domestic control of domestic sovereignty, the power system and the domestic ability to maintain as well as protect them against foreign powers are greatly influenced by the domestic individual and collective preferences. Such preferences, the practice of them and the direction of their aggregate use, have their roots in the domestic culture and its evolution. It is this domestic culture that produces the national personality in either defense or sale of the domestic sovereignty to an external power for enslavement and exploitation of the domestic population. The process of defense or sale of domestic sovereignty is the work of the nature of political leadership and the supporting intellectual system of beliefs and ideology that help to define the country's characteristic set and its division into negative and positive characteristic sub-sets which helps to reveal its identity and category of belonging.

3.2 The Characteristic Set, Negative Actions and Positive Actions

Let us examine how the characteristic set relates to the production of the individual and collective actions for a change or no change in a society. As has been discussed, every element and its identity exist in duality where such a duality is revealed by a characteristic set which is divided into negative and positive sub-sets of characteristics under the principles of opposites with internal tension. Here, an individual is viewed as a unit element just as a nation is viewed as a unit element defined by dualities under relational continuum and unity from the principles of opposites. Let $\mathbb{X}$ be the characteristic set of a nation with a generic element $x \in \mathbb{X}$. The characteristic set is composed of positive sub-set $\mathbb{X}^P$ and negative subset $\mathbb{X}^N$ such that $\mathbb{X} = \left(\mathbb{X}^P \cup \mathbb{X}^N \right)$ with $\left(\mathbb{X}^P \cap \mathbb{X}^N \right) \neq \varnothing$. The characteristic set defines the social character of the nation. Associated with the characteristic set is an *action set* $\mathbb{A}$ with a generic element $\left(a \in \mathbb{A} \right)$ which is composed of *negative action set* $\mathbb{A}^N$ and a *positive action set* $\mathbb{A}^P$ such that $\mathbb{A} = \left(\mathbb{A}^P \cup \mathbb{A}^N \right)$ with $\left(\mathbb{A}^P \cap \mathbb{A}^N \right) \neq \varnothing$. Let $\mathbb{P}$ be the population set of the nation whose generic element is $\left(p \in \mathbb{P} \right)$ with an index set $\mathbb{L}$ and a generic element $\ell \in \mathbb{L}$. Similarly, let the index set of the national characteristic set be $\mathbb{I}$ with a generic $i \in \mathbb{I}$. Let also the index set of $\mathbb{A}$ be $\mathbb{J}$ with a generic element $j \in \mathbb{J}$. The social characteristic set defines the culture of the social setup in individual and collective preferences, behavior, motivation, decision-choice activities, social thinking and net social action. The non-exclusivity condition relates to the principles of opposites where every negative has a supporting positive in relational continuum and unity just like every cost has a benefit support and vice versa and the decision-choice action must obey the *asantrofi-anoma rationality* where every decision is cost-benefit dependent.

The defined social behavior and motivation are generated by social preferences which are a composite aggregate of individual preferences within the cultural confines of the social set-up. It is the structure of social preferences that helps to define that national character, national interest and social vision on the basis of which decision-choice activities are established for social actions in the political, economic and legal spaces. It is also the social actions that create categorial moments to bring about categorial conversions which in turn bring about changes of qualitative dispositions of social states through the social manufacturing of categorial social transfer functions to move the society from one category to another. The social manufacturing of the categorial transfer functions emerges from the struggle between the negative social forces and the positive social forces of the negative characteristic set and the positive characteristic set with their corresponding negative actions and positive actions respectively where the resultant force shapes the direction of change.

It may be kept in mind that the system is a collection of polarities and dualities with relational continuum and unity over time. The polarities and dualities are given contents and structures by the residing negative and positive characteristic subsets. There is thus forward-conversion and backward-conversion processes. Care must be taken in the conceptual meaning and the processes of the forward and backward conversions. Each social state is an actual-potential polarity. The actualization of a potential element moves the existing actual into the potential space. The existing actual is a child of the potential space as well as a derivative of the previous actual which was moved to the potential space by a categorial conversion. Each social state is defined by a set of properties and the backward conversion means that the negative characteristic sub-set has generated negative force to overwhelm the positive force that is generated by positive action. The result is that the social system is moved by categorial conversion to a less favorable social state relative to the current actual as seen by the preferences of the majority. This is *categorial reversibility* not to the previous state, but to a state in the potential space where such a state is less preferred than the currently existing social state. The forward categorial conversion means that the positive characteristic subset has generated positive force to overwhelm the negative force that is generated by negative action. The result is that the social system is moved by categorial conversion to a more favorable social state as seen by the preferences of the majority. The direction of the qualitative motion and its speed will depend on the relative strength of the positive and negative forces which are generated by the positive and negative characteristic subsets respectively. The elements in the negative and positive characteristic subsets act in mutual negation of one another to create a game-theoretic situation requiring the development of strategies and tactics, counter strategies and tactics under conflicts in the negative-positive dualities under relational continuum in order to establish an advantage in shaping the outcome of the resultant social forces. The concepts of categorial reversal and negation of negation must be seen in terms of interchangeability and substitutability of some varieties in the space of actual, and the space of the potential within the space of social preference ordering.

For the forward qualitative motion, the strategies and tactics are concerned with increasing the positive force through the increase of the positive social action. This requires increasing the size of the positive characteristic subset by converting some elements of the negative characteristic sub-set into the positive characteristic sub-set. The intent here is to reduce the size of the negative characteristic sub-set to weaken the negative forces and hence the negative social action. The same process works for the backward motion, where, through the strategies and tactics involving the reversal changes in the positive and negative characteristic sub-sets, the negative forces and negative social action are empowered to be greater than the positive forces and positive social action. These strategies and tactics must emerge from within the social system itself, just as the natural transformations emerge from the natural organizational arrangements of the characteristic sets within elements. The process of rearrangements and qualitative conversions is to either set the positive against the negative for progressive change or the negative against the positive for retrogressive change.

3.2.1 Social Institutions, Decision-Choice Activities and the Characteristic Set

The required strategies and tactics are decision-making elements that are translated into social policies and then transmitted through social institutions. If the necessary social institutions are not in existence then they must be constructed. It is on the basis of this recognition that led Nkrumah to state:

> Progress does not come by itself, neither desire nor time can alone ensure progress. Progress is not a gift, but a victory. To make progress man (person) has to work, strive and toil, tame the elements, combat environment, recast institutions, subdue circumstances, and at all times be ideologically alert and awake [R1.207, p. 113].

The power of this statement of Nkrumah regarding the success of categorial conversion of social states under the conditions of a system of actual-potential polarities and dualities with relational continuum and unity cannot be underestimated. The whole statement involves creating sufficient conditions that will allow the potential to be set against the actual by creating the categorial moment to internally convert negative characteristics in the society into positive characteristics in order to create categorial transfer functions that will move the society through higher levels of progress in the national sovereignty, freedom, justice, reconstruction and nation building for increasing collective well-being. The creation of the sufficient conditions must be guided by social decision-choice action which is social conscience-dependent where such a social conscience must reflect a particular national experiential information structure, social goal-objective set and vision.

In the African case, the organic categorial conversion presents itself in sub-categorial conversions from colonialism to decolonization and to complete emancipation by strict avoidance of neocolonialism. Colonialism and neocolonialism are defined by sovereignties under external control where institutional configuration is cast under the *anti-African guiding principle and thought* (anti-Philosophical Consciencism) to promote categorial conversions of socioeconomic stages, where a movement from one stage to another is induced by externally created negative categorial transfer functions that must overcome the domestic positive resistance function for the benefit of the external power which controls the sovereignty at the expense of the domestic population. The complete emancipation, on the other hand, is defined under domestically controlled sovereignty with categorial conversions of a series of socioeconomic development stages, where a movement from one stage to another is induced by socially created positive categorial transfer functions which must overcome internal and external negative resistance function.

By the inclusion of the phrase *to be ideologically alert and awake*, in Nkrumah's quoted statement, Nkrumah purposefully linked the guiding principle and the thought basis for the casting of the appropriate institutional configuration and the needed policy constructs to *Philosophical Consciencism*, derived on the basis of African traditions where the Euro-Christian and Islamic traditions are mere historic reflections. The use of Philosophical Consciencism with African content is to empower the African positive action from within Africa and set it against the negative action through Africa's own internal dynamics. Nkrumah recognized that the fight to decolonize and liberate for freedom and justice, as well as the fight to develop must proceed under a revolutionary philosophy with a powerful ideology for decolonizing the African mind, and create a directing hand that will point the right direction to complete emancipation where the traps of neocolonialism are avoided.

The negative resistance function is produced by negative social action which is generated from within the activities of the negative characteristic sets. It must be kept in mind that both the social negative and positive characteristic sub-sets reside in the same social set-up under conflict through which energy is generated to produce social force which creates categorial moment and categorial transfer function. The manner in which external factors affect the behavior of the internal characteristic set has been explained under the treaties of the Theory of Categorial Conversion [R15.15] and enhances in [R9.11 [R0.12]. The explanations and discussions relate to the manner in which one views open and closed systems. The simple idea is that for the external factors from any other social environment to affect the internal qualitative dynamics and hence categorial conversion, the internal characteristics must accept and incorporate the external factors as part of the system. An example of this is provided in medical treatments where external intervention is common and this helps to explain medical transplantations. Here, the social categorial moment defines the effectiveness of the social transfer function to move the social system in such a way that the social system is categorically converted from one social state to another in the same sovereignty. Let us keep in mind that different institutional arrangements of the same social set-up with the same qualitative disposition belong to the same category. The word *social* that qualifies the relevant phrases is extremely important. These phrases are limited to the discussions on statics and dynamics of qualitative dispositions of social organisms. The same analytical process will apply at the level of natural organisms where the methods and techniques in relation to matter, energy and information for the creation of categorial moments and categorial transfer functions will be different. In respect to nature, the qualifier, social will be replaced by natural. The qualifier, universal, will be used when we speak of qualitative dispositions of both nature and society.

It must be kept in mind that static states in the evolutionary or revolutionary process are categories for any entity in both ontological and epistemological processes which encompass natural and social organisms. Any social unit is defined by a particular institutional arrangement of its internal characteristics. The result of an institutional arrangement defines the social system's qualitative disposition. Different institutional arrangements may produce different results which may fall in the same social category if they produce the same qualitative disposition otherwise they will fall in different social categories. For example, different institutional arrangements where domestic sovereignty is directly in the hands of external power place the social system in the category of colonial states, while, if the sovereignty is indirectly controlled by an external power it places the social system in the category of neocolonial states. Different institutional arrangements create different conditions for the emergence of categorial moments and categorial transfer functions required for categorial conversions where for example, national independence emerges from colonialism, and development emerges from underdevelopment through the operations of internal domestic forces of Africa. For sovereignty transformation from external control to domestic control, the political institutions must define conditions that promote the development and growth of the needed categorial moment and categorial transfer function. It was this recognition that led Nkrumah to advise that seek ye first the political kingdom and the rest will be added onto it. This advice by Nkrumah has not been understood by scholars, non-scholars and African political elite at the levels of philosophy, mathematics, theory and practice in the space of social transformation. At the level of mathematics, it involves the establishment of socio-political categories and game space. At the level of theory, it involves the understanding of decision-choice theory involving strategies and tactics in the game space. At the level of philosophy, it involves thought and ideological guidance of social principles, establishment of social vision and national interest. At the level of practice, it involves the nature of programs and actions. In the case of a social system, the intra-categorial stages and inter-categorial conversions are decision-choice defined, where the decision-choice structure is culturally determined on the basis of internal dominant philosophy and ideology which cannot and should not be allowed to mimic those of the anti-African philosophy and ideology of colonialism and imperialism on the basis of civilization, democracy and development. Let us keep in mind that colonialism and imperialism have no respect for democracy neither do they have respect for civilization based on humanity and peace. The role of neocolonialism and imperialism is resource seeking through subjugation of non-members. Imperialism has no claim to peace and justice, democracy and maintenance of human rights. Its nature is defined by war, violence and injustice and slavery in the global decision-choice space.

3.2.2 Political, Legal and Economic Institutions under Sovereignties

Given the sovereignty, the legal and economic institutions define the conditions that promote the development and growth of the needed categorial moments for intra-categorial and inter-categorial conversions. Here, the interactions among the political, legal and economic structures and their relationships with the general welfare of the members of the society must be clearly understood. The political structure defines the control and exercise of the domestic sovereignty through the established *political institutions* which create the government and the structure of governance and its inheritance. The political structure also confers the power to govern as well as the power to create the legal structure which is defined by a set of *legal institutions* that establishes constraint sets on general social behavior in the economic political and legal structures as defined by *economic, political and legal institutions* in time and over time. The legal constraints on the economic structure impose material restrictions on the boundaries of all decision-choice actions and on the degrees of success of social categorial conversions within sovereignty, while the legal constraints on the political structure restricts the nature of interactions that can be established among the three structures in terms of freedom, fairness and justice over the national decision-choice space

The possible success of any social action in the categorial conversion depends on the resultant of different social forces that are generated by the outcome of the conflicts between negative and positive actions from the forces produced by the positive and negative characteristic sub-sets in the society. The social positive characteristic sub-set is a *weighted fuzzy aggregation* of the individual positive characteristic subsets. Similarly, the social negative characteristic sub-set is also a *weighted fuzzy aggregation* of the individual negative characteristic sub-sets. The individual decision-choice agent in the social set-up is under positive-negative tensions and the fuzzy aggregation of the individual tensions leads to social tension giving rise to a plenum of social forces from the interactive processes of the social system's characteristic set. The net activities of the elements of the characteristic set produce social action at various degrees of effectiveness in bringing about the categorial conversion and in the direction that is dictated by the dominance of either the negative or positive characteristic subset.

The individual member of the society, like any object in the universal system, is under conditions of duality defined by the negative and positive characteristic sub-sets under social tension. The proportionality of the negative and positive characteristic sub-sets varies over individuals where such proportionality can be manipulated by *philosophy and ideology*. The individual in the social setup is a decision-choice agent. The behavior of the individual decision-choice agent is governed at the personal level by the incentive of self-interest which is mapped into the collective decision-choice space. The activities in the collective decision-choice space are governed by collective self-interest under the African traditional conceptual system of human organisms. In the traditional African conceptual system and governmental organizations, the collective interest overrides the individual interest. The relationship of the individual interest to the collective interest is an important foundation of the cultural complexity of Africa where the ideological principle guiding practice is *each for all and all for each*. Here, the question to the individual is not what the community can do for an individual, but what can an individual do for the community. Similarly, the question to the community is not what the individual can do for the community, but what can the community do for the individual within the individual-community duality in continuum and unity, where the individual and the community draw a continual mutual support for collective existence and progress.

Since the individual as a decision-choice agent exists as a duality, he or she is under positive-negative tension in relational continuum which is established by an enveloping of various degrees of intensity of the individual tensions. The individual tensions are mapped into the social decision-choice space by fuzzy aggregation to produce social tensions that give rise to a plenum of social forces from the country's characteristic set. The plenum of social forces leads to social actions for creating the required categorial moment and categorial transfer function for categorial conversion. The social forces generated by decision-choice actions under the distribution of different individual preferences are composed of a set of negative social forces and a set of positive social forces that exists in duality and under relational continuum and unity. The analytical and operational power of Philosophical Consciencism with its defined African content and methodology of construct is to shape the individual and collective actions to empower the positive social action over the negative social action to bring about progressive change in favor of Africa's progress under the philosophy and ideology of decolonization and development on the basis of Africa's internal effort. In this respect, Nkrumah stated as part of the practice of categorial conversion guided by Philosophical Consciencism that it is better for Africa to do for herself or himself and fail rather than to let others do for Africa. The idea is based on the understanding of the analytical process in a success-failure duality which presents different opportunities of learning under categorial conversion as the positive and negative forces battle themselves to set the path of resultant forces.

The set of negative social forces emerges from the activities of the elements in the negative characteristic set to produce a set of social actions which then produce negative categorial moments for either negative categorial conversion or to prevent positive categorial conversion from taking place. Similarly, the set of positive social forces emerges from the activities of the elements in the positive characteristic set to produce a set of positive social actions which in turn produce a positive categorial moment to bring about either positive categorial conversion or to prevent negative categorial conversion from taking place. The direction of the categorial conversion is always under the resultant forces of the negative and positive forces in mutual negation. The concepts of dualistic continuum and relational unity imply that all the negative and positive elements under mutual negation seek for dominance in the resultant force to define the direction of the qualitative motion at each round of an existence of actual-potential polarity. Let us keep in mind that the concepts of polarity and duality imply that for every negative there is supporting positive without which these concepts make no sense in the Africentric principles of opposites with relational continuum and unity which affirm the conditions of ontological analog and the justification of digital and discrete as human epistemic limitations.

The activities of the negative and positive characteristic sub-sets, the corresponding tensions, social forces and social actions are the results of decision-choice actions. The decision-choice actions are the works of information, methods of reasoning, and elements in the institutional structure through which individual and social policies and decision-choice activities are constructed and transmitted. The information, methods of thinking and institutions are the product and works of the country's cultural characteristic set that produces the social personality of the country under the dominant philosophy and corresponding ideology which must be consistent to the progress and welfare of the nation. As it is argued in this monograph, the dominant philosophy and ideology for the African are defined by Philosophical Consciencism with African contents to produce Africentricity and the African personality required to set the African potential against the actual through categorial conversions in the actual-potential polarities. At this point, it must be clear and noted that the theory of categorial conversion is based on the existence of *actual-potential polarity* where every actual pole has a supporting potential pole to establish the actual-potential polarity. Let us keep in mind that every actual phenomenon is a singleton set as well as a candidate for destruction and change. There are many varieties in the potential space that can be actualized for its replacement. In social set-up, the potential variety that may be actualized requires a judicious application of social decision-choice action guided by national specific Philosophical Consciencism.

Every ontological element belongs to a category in either actual or potential space. Thus every element in the universe is either an actual pole or a potential pole. The universe is conceived to be infinitely closed under continual categorial conversion operating on matter, energy and information. Each pole has a residing duality which is either negative or positive that defines the existence and character of the pole in the actual-potential polarity. The negative duality has a supporting positive duality in relational continuum and unity in the same pole. Each duality has a residing positive characteristic sub-set with a supporting negative characteristic sub-set whose relative balance establishes either the negative or positive duality. Each duality is under relational tension to produce energy and forces. The forces are split in mutual support of negative and positive forces which constitute constraints on each other's action to create a categorial moment and categorial transfer function for categorial conversion in qualitative motion which takes place in the actual-potential polarity. The role of Philosophical Consciencism with African contents, where the weapons of its construction are drawn from the African conceptual system under the African experiential information structure, is to present an Africentic philosophical and ideological system to produce an African personality which will understand Africa. In this respect, such an understanding will relate her history from antiquity to the present in all areas of human endeavors. The Africans with such personality will use and further develop and expand the Philosophical Consciencism to guide the creation of *social technology* composed of relevant social institutions and social policy configuration which will be appropriate in creating the needed categorial moment and categorial transfer function for continual transformation of the socio-political system of Africa. It is on the basis of Philosophical Consciencism that Nkrumah stated: *Only Africa can fight for its destiny. In this struggle we shall not reject the assistance and support of our friends, but we will yield to no enemy, however strong* [R1.207, p. 2]. The conceptual discussions of Africentric foundations of Philosophical Consciencism and African Personality are provided in [R1.92] [R1.91][R1.200].

3.3 Setting the Positive Action against the Negative Action in Political Polarities under Philosophical Consciencism

Every true and lasting social transformation has an underlying thinking system which indicates its philosophy and corresponding ideology crafted from within. In this respect, let us examine the operative framework presented by Philosophical Consciencism in socio-political transformations of Africa in order to set the potential against the actual at each important point of socio-political polarity. The Theory of

Categorial Conversion is used to develop an explanation or prescription to transformations on the basis of internal dynamics of ontological elements as seen from the Aficentric conceptual system of the principle of opposites. The driving force of conversion is the creation of categorial moments and categorial-transfer function from within the elements themselves. The explanatory and the prescriptive conditions are necessary but not sufficient for transformation of varieties. The sufficient conditions must be manufactured. In socio-political systems, such manufacturing is done with social technology which is created by human decision-choice processes acting to produce positive and negative actions. Both positive and negative social actions are defined within a duality and social polarity under the principles of relational continuum and unity. They are the transformative links in the actual-potential polarity through the residing polar dualities. They mutually define their existence and identity. In the social set-up, they are the products of the culture and function through the existing institutions which are also culturally induced in relation to social needs and policies. The social needs and policies, given the national resource constraints, set the required institutions and the direction of the resultant force which will emerge from the positive and negative actions as produced by positive and negative characteristic sub-sets. To appreciate the emergence of the resultant force and the direction of categorial conversion, we need the general and specific collective understanding of the concepts of the positive and negative actions in bringing about categorial conversions that may lead to transformations of socioeconomic structures of the social setup.

Positive and negative actions are only possible in dualities and polarities with relational continuum and unity under the principles of opposites which give rise to conflicts and tensions for the game of mutual negation in actual-potential polarities, where there are changes in the relation of forces under tension and continual power struggles. As has been explained in [R1.90b], there are *natural polarities* and *social polarities* and corresponding to them are natural and social positive and negative forces involved in categorial conversion of natural polarities and social polarities where there is always competition between the actual and the potential for maintenance or replacement. The processes of categorial conversions have the same structures in both nature and society, except in the manner in which the categorial moments and the categorial transfer functions are manufactured for intra-categorial and inter-categorial conversions. The analytical process of the categorial conversion from life to death in the life-death polarity is no different from the analytical process of the categorial conversion of acolonial state to a decolonized state in the colonialism-decolonization polarity. In the former, life is the actual and death is the potential, while in the latter, the colonial state is the actual and the decolonized state is the potential. In fact, they are both governed by the interplay of matter, energy and information under the principles of decision making within the opposites. In a social set up, the objective of power struggles between the negative characteristic sub set with its negative social action and the positive characteristic sub-set with its positive social action is either to maintain the existing socioeconomic relations and the corresponding cost-benefit configuration and its distribution, or, to establish new socioeconomic relations and a corresponding new cost-benefit configuration with a new distribution.

The relational structures may be seen as social antagonistic games where the negative and positive social actions are the instruments to carry on the game. The game is very complex with a changing information structure that produces fuzzy-stochastic uncertainty and risk in the dynamics of the categorial conversion of socio-political polarities under the general principles of opposites with relational continuum and unity. The principle of opposites in relation to negative and positive characteristic subsets and corresponding forces as seen in pyramidal logic is presented as an epistemic geometry in Figure 3.3.1. The production of the social action is to produce categorial moment and categorial transfer functions needed to change the relation in the actual-potential social polarity through the behaviors of the residing dualities.

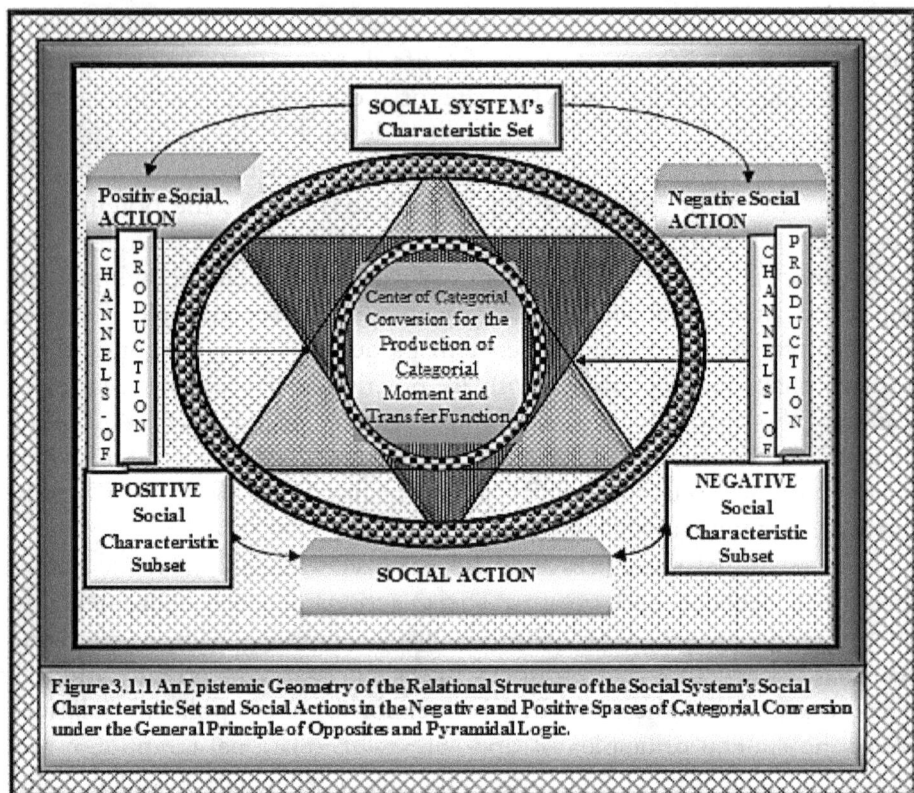

Figure 3.1.1 An Epistemic Geometry of the Relational Structure of the Social System's Social Characteristic Set and Social Actions in the Negative and Positive Spaces of Categorial Conversion under the General Principle of Opposites and Pyramidal Logic.

3.3.1 The Nature of Social Positive Action of Social Categorial Conversion in Systemicity

Let us turn our attention to the nature of both positive and negative actions in categorial conversion. The concept of social positive action is complex in its understanding. There are two structures of positive action in the dynamics of categorial conversions as has been discussed in [R15.15]. They are positive action in nature that characterizes natural positive action in the principle of opposites which relates to natural processes and positive action in society which characterizes social positive action in the principle of opposites which relates to social processes. At the level of nature, positive action relates to progressive natural transformations. Here, positive action must be carefully interpreted in relation to the judgment of nature as well as cognitive agents and in relation to both open and closed systems and their relationship to the costs and benefits of transformations. The cost-

174

benefit assessments are different in nature and in the framework of society as cognitive agents interpret categorial-conversion costs and benefits. Every categorial conversion involves destruction-construction duality that is mapped into the space of cost-benefit dualities. Let us examine the concept of *social positive action.*

3.3.1.1 *The Nature of Social Positive Actions in Social Categorial Conversions of Social Actual-Potential Polarities under Philosophical Consciencism*

Social positive action has special characteristics that define its distinction from the natural positive action. Such a distinction may be cast in a definition.

Definition 3.3.1: Social Positive Action

Social positive action is the product of internal activities of the social positive characteristic sub-set of an element where such a positive action seeks to negate the nature of the actual existence of an element. It is a fuzzy aggregation of the effectiveness of all the forces working to establish social justice, freedom and fairness in order to free the masses from the yoke of the dominant class in terms of domination and oppression.

An example of social positive action is that which is designed to abolish slavery, colonialism, imperialism, or wars of aggression, racism and other such entities. The social positive actions are generated by collective decision-choice actions which may or may not be allowed by the politico-legal structures of the social set-up. The social positive action seeks to change the existing socioeconomic relation through the alteration of the existing institutional arrangement by revolutionary activities the intellectual background of which is revolutionary philosophy and ideology. The nature, attributes and culture of the revolutionary philosophy and ideology have been discussed in [R15.15] [R1.92][R3.7] in relation to the conceptual system of Nkrumah to show its African-centered roots and its relevance to African social transformation. Their connections to the general theory of knowledge and its foundation have been provided in [R3.10] [R3.13]. Since the social positive action is a weighted fuzzy aggregation of individual positive actions in line with the positive characteristic sub-set, its social effectiveness depends on the degree of individual commitment and contribution as constrained by opposite actions.

The individual, like all other ontological objects, is under the principle of opposites with relational continuum and unity in characteristic sets and actions in positive and negative spaces. Meanwhile, the individual positive action is derived from the structure of behavior of individual positive characteristic sub-set with positional reference to the structure of collective behavior. The structures of such individual and collective behaviors depend on the cultural foundation and the guidance of *revolutionary philosophy and ideology*. In this respect, the effectiveness of the individual contribution to the social positive action will depend on the individual ability to understand and accept the basic core of the content of the revolutionary philosophy and ideology that guides the manufacturing of the social positive action intended for the socio-political transformation required to bring in a new order of socioeconomic relations. The individual positive action is to strengthen the effort of the positive social characteristic set toward efficient processing of the available social information. It is here that the interplays of defective and deceptive information structures affect social action. It is also here that explanations can be logically abstracted for the creation of deceptive information structure and propaganda for the social system's controllability. It is also here that a decision-choice explanation can be found for governmental secrecy, misconduct, and institutional corruption and information classifications for social control in favor of selected classes.

Let us keep in mind that the material inputs of categorial conversions are matter, energy and information as have already been discussed in the theory of categorial conversion [R15.15]. In this respect, the people in the social setting are the material backbone of the development of social positive actions for categorial conversions. The revolutionary philosophy and ideology are the conceptual foundations that provide the integrated positive decision-choice social actions for the creation of the categorial moment and categorial transfer function required for positive categorial conversion to bring about desired results and for the new order inherent in the potential to emerge. In terms of categorial conversion of socio-political actual-potential polarities, the people constitute the matter, energy and information whose relational structure is continually changing as well as embedded in the cultural confines of philosophy, ideology and knowledge. The positive action in the social action must be provided by a judicious development of socioeconomic strategies. The needed socioeconomic strategies are produced through the available and potential social institutions which will allow a coordination and integration of the people and ideology within a given revolutionary philosophy. In other words, every social positive action is armed with a positive revolutionary philosophy that gives content to the social ideology on the basis of which social decision-choice activities are undertaken to create the *social technology* required for the manufacturing of the social categorial moments and transfer functions.

3.3.1.2 *The Nature of Social Negative Actions in Social Categorial Conversion of Social Actual-Potential Polarities under Philosophical Consciencism*

Just as the concept of social positive action in categorial conversion is complex, so also is the concept of social negative action. Here again, there are two types of negative action. They are social negative action and natural negative action. At the level of nature, the negative action relates to either the maintenance or reversal of the order of existing arrangements of natural relations among the natural characteristics in an ontological object, and factors that maintain it and prevent unwanted transformation of such relations such as life to death. This requires that negative action must have an interpretation that connects judgments of nature to the judgments of cognitive agents. The concept of negative action has been discussed in a companion volume [R15.15]. It must be kept in mind that the creative destruction process in nature and society and its relationship to substitution-transformation structure which is mapped into the cost-benefit space are different for society and nature. Categorial conversion is a creative destruction process of varieties and categorial varieties. It involves also substitution-transformation dynamic activities within the space of cost-benefit dualities in both nature and society under the principles of opposites. The effects of substitution-transformation activities are revealed in actual-potential polarities and dualities with relational continuum and unity with a destruction of an actual elemental variety and creation of a vacuum. The vacuum is the filled by an actualization of a potential element no matter what it is. It is these conceptual properties of elements as varieties, seen from within and under forces of tension that provide the claim that The Theory of

Categorial Conversion is a general theory that provides the necessary conditions of qualitative and quantitative transformations.

The characteristics which define and establish social negative action are different from those which establish natural negative action. The difference resides in the role that they play in categorial conversions in nature and society. Let us define social negative action.

Definition 3.3.2: Social Negative Action

Social negative action is the product of internal activities of the social negative characteristic subset of an object where such negative action seeks to maintain the nature of the actual existence of the object. It is a fuzzy aggregation of the effectiveness of all the forces working to maintain or impose attributes of subjugation, exploitation, injustice, unfairness and other real costs required to oppress the masses under the yoke of the dominant class for domination, rent-seeking, resource-abstraction and wealth-transfers to enhance the welfare of the dominating class.

The social negative actions are produced by collective decision-choice actions that are legally allowed or overlooked under unfair practices in the established politico-legal structure of the social set-up which is composed of the political and legal social structures given the economic structure. The social negative action seeks to maintain the existing order of the socioeconomic relation and the institutions that keep it. The negative action has an intellectual backing rooted in reactionary philosophy and ideology. This reactionary philosophy and ideology is connected to the development of the general theory of knowledge. The individual negative actions are rooted and guided by the degrees of understanding and acceptance of the reactionary philosophy, and the manner in which this understanding with the corresponding social ideology is integrated into the creation of the individual negative actions that must individually contribute to the overall social negative action in the social set-up. Given the conceptual and content understanding, the effectiveness of the contribution of an individual negative action will depend on the degrees of individual commitments and the perceived individual real cost-benefit calculations as related to the positive changes and net real cost-benefit structure.

The individual negative action is an important derivative from the structure of the behavior of the individual negative characteristic set. The boundaries of the individual behavior are established by the culture of the society and the degrees of the understanding and belief in the reactionary philosophy and ideology for example, those associated with neocolonialism and imperialism that produce neocolonial and intellectual dependency mindsets which restrict thinking in the social decision-choice space and indirectly control social action. Generally speaking, the effectiveness of individual contributions to the social negative action in the social setup will be driven by the individual ability to understand and accept the basic conceptual framework underlying the reactionary philosophy and ideology that must guide the individual degree of participation in the social negative action intended to maintain the existing social order or the possibility of negative reversal of the existing social order to a more primitive oppressive state such as from colonialism to either military occupation or neocolonialism which is basically colonialism in disguise. The individual negative action is to strengthen the effectiveness of the activities of the negative characteristic sub-set toward an efficient processing of the available social information. It is here that education driven by the teaching of thinking under the principle of the practice of doubt becomes transformative as well as revolutionary in support of individual action. It is also here that the creation of a deceptive social information structure acquires an important role in the social games within the social opposites.

From the viewpoint of active social transformation and change, or maintenance of the existing order of socioeconomic relation, the people in the social set-up are the material and spiritual backbone for the development of the negative action required to maintain or transform the existing social arrangements. The reactionary philosophy and the corresponding ideology are the conceptual foundations that allow subjective assessments of the real social cost-benefit implications for decision-choice actions needed for the development of resistance against the negative forces required to maintain the existing social actual. The negative action in the social set-up must also be produced in the same social set-up. It is easier to produce it by the use of methods and strategies which appeal to the comfort zone of the familiar where the people may be imprisoned in the ideological box of the current cost-benefit deceptions.

It must be kept in mind that the needed socioeconomic strategies are produced through the available institutions which coordinate and integrate the peoples' decision-choice actions within the logical framework of the reactionary philosophy and ideology that provide the cognitive mechanism for the social information-processing. These institutions are under the control of the sovereignty of the existing politico-legal arrangements which seek to maintain the existing order of socioeconomic relations on the basis of real social cost-benefit configuration for those who control the politico-legal structure. The concept of imprisonment of an individual in the dark walls of the familiar implies that every negative social action is armed with negative reactionary philosophy which gives content to the social ideology of the politico-legal controllers to deny the cognitive practice of the principle of doubt through propaganda and non-thinking conditions.

An example of social negative action that which is needed to maintain institutions of slavery, colonialism, imperialism, racism, wars of aggression, terror and others. The social negative actions are produced by collective decision-choice actions which may be allowed by unfair applications of the politico-legal structures of the social set-up. The social negative action seeks either to change or maintain the existing outmoded socioeconomic relation into a worse one through the alteration of the existing institutional arrangement by oppressive social activities the intellectual background of which is reactionary philosophy and ideology of evil-doing. The nature, attributes and culture of the reactionary philosophy and ideology have been discussed in [R1.91] [R1.92] [R3.7] [R3.8][R7] [R7.4][R7.7] [R7.12] in relation to the conceptual system of Nkrumah to show its anti-African-centered roots and its relevance to the battle of neocolonialism and racism and the work toward the complete African emancipation in African social transformation. It must be emphasized that philosophy and ideology define mindsets in the decision-choice space and space of social actions. Neocolonialism, imperialism and racism draw their mindset from reactionary Philosophical Consciencism with anti-African contents based on the foreign experiential information structures. Complete African emancipation from imperialism, neocolonialism and racism draw its mindset from progressive Philosophical Consciencism with African contents based on African experiential information structure.

3.3.1.3 *Comparative Analytics of Social Positive and Negative Actions in Social Categorial Conversion in Social Actual-Potential Polarities in Systemicity*

To appreciate the interdependent logical frameworks as defined by the theories of *Categorial Conversion* and *Philosophical Consciencism* in relation to the dynamics of society, we need to understand the relative structure of the positive and negative actions in social polarities and dualities and how the dynamics of society are embedded in the social collective decision-choice space The individual and the collective behaviors in the social decision-choice space are molded by production-consumption knowledge processes which produce self-excitement and self-correction of socio-political transformations. Every pole of a social polarity has a residing duality. Every dual of the social duality has a residing negative-positive characteristic set that ensures its identity, internal conflicts and force-production capacity. The social negative characteristic sub-set produces a social negative action which follows from the existing reactionary philosophy and the corresponding social ideology. The social positive action is produced by the activities of the positive characteristic sub-set as guided by a revolutionary philosophy with the corresponding social ideology. The existence of the negative and positive characteristic sub-sets in relational continuum and unity under the general principle of opposites produces internal struggles for dominance; where the social internal struggles that generate tensions for categorial conversion are expressed through the interactions of the negative and positive actions. The direction of the conversion will depend on the resultant force which may produce positive conversion, negative conversion or simple maintenance of the existing order of the socioeconomic relation, and the structure of politico-legal institutions which maintain them as well as the social power distribution for development of social policies, programs and actions.

The negative and positive actions are increasing or decreasing in accord with the activities of the respective negative and positive characteristic sub-sets in addition to the behavior of the intra-negative-positive negation in the duality. In the negation-of-negation process and the conflicts of the negative-positive sets, positive social action may exceed the negative social action. Alternatively, negative social action may exceed the positive social action to dictate the direction of conversion. The former involves the dominance of the activities of the positive social characteristic sub-set over the negative social characteristic subset, while the latter involves the dominance of the activities of the negative characteristic sub-set over the positive characteristic sub-set. The dominance relates to the degree of effectiveness measured in terms of zero to one as a fuzzy number where pure dominance is one and complete lack of dominance is zero. In other words, dominance is viewed as a linguistic number that appears in the degree of action effectiveness [R4] [R4.17] [R4.37] [R4.38][R4.47].

The social categorial conversion process is such that neither positive nor negative characteristic sub-sets requires complete dominance in the social action, and hence they operate with conditions of dominance and not conditions of complete dominance on the part of negative and positive characteristic sub-sets. We must keep in mind that under the principles of opposites, the actual and potential poles exist in relational unity, the negative and positive dualities are in relational continuum and inseparable unity, the positive dual and negative dual are also in relational continuum and unity, for every negative characteristic subset there is a supporting positive characteristic sub-set, and for every cost there is a supporting benefit which helps to define the meaning of substitution-transformation process in categorial conversions. The appropriate computational reference condition is categorically overwhelming. The space of categorial overwhelming in any dual is expressed in a zero-one overlapping of the social positive-negative characteristic sets to define the terms of relative negative-positive action. The path of enveloping between each pole is defined by a pair of negative and positive effectiveness as $(\alpha, \beta) \ni \alpha < \beta \in (0,1]$. In order to bring into the focus of the effectiveness of positive categorial conversion, it is necessary for positive action to overwhelm the negative action through the dominance of the action of the positive characteristic set. The sufficiency requires the development of politico-legal institutions that enhance education of the understanding and thinking of the philosophical basis of change as well as increasing the degree of individual and collective consciousness through the revolutionary philosophy and supporting ideology. The process is to prevent the reversal of the categorial conversion to take place as well as hold on to an effective check on the negative action as generated by the negative characteristic subset in the social set-up.

In this respect, the social categorial conversion by the positive action involves two stages of a) the conversion of the category or actualization of a potential and b) the maintenance of the new category or the maintenance of the new actual. The first state involves the development of strategies to overcome the activities of the power structure of existing institutions and sovereignty. The second stage involves the development of institutions that will help to maintain the new order as well as restrain and prevent the growth and development of the power of the previous institution in which the negative characteristics function to generate negative action as a contestant to the new reality. The social negative characteristic set and the corresponding negative action do not completely go away. They are, through a limiting process, simply reduced to a manageable minimum in order for the positive characteristic set to function and provide positive action. The new institutions that emerge from the triumph of the positive action must be used to develop clear strategies of containment against the behavior of the negative characteristic set. After the success of the positive action and an overthrow of the actual by a potential, a new actual-potential polarity is established to define a new antagonistic relation for a new process of categorial conversion. The new strategy will define the degree of sufficiency that will be required to maintain the new category of the actual and its forward conversion through the construct of strategies, counter-strategies and social policies which must be transmitted through the restructured existing or created new social institutions. The categorial-conversion process of social actual-potential polarities is a never ending one. This is the meaning of continual creation without end. This is also the meaning that every problem has a solution and that every solution to a problem generates a new problem that affirms the idea that knowledge production is a never-ending process and that human life takes meaning is the infinite problem-solution space.

3.3.2 Social Polarity, Duality, Games, Positive Action and Negative Action in Systemicity

The general concepts of socio-natural category, polarity, duality, and the residing positive and negative characteristic sets have been discussed in the works in [R1.92] [R3.7] [R3.8] [R3.13]. These concepts constitute logical foundations of reasoning about states, transformations and processes involving qualitative and quantitative dispositions. Their internal extensions to the development of strategies and counter-strategies for categorial conversions have been discussed in [R1.92]

[RI5.15] in terms of socio-natural games in polarity and duality around the basic objects of matter, energy and information. In social polarity, the game is played between two social poles whose strategies are developed from the game between the two social duals of the duality. The social duals depend on the activities of the social negative and positive characteristic sub-sets which generate negative and positive forces for negative and positive actions respectively. There is a clear linkage between the poles of polarity to the negative and positive characteristic sets through the respective dualities. The poles of any polarity function as the cognitive center that provides instructions for the positive and negative actions while the positive and negative characteristic sets act as information collection modules to send the information to the central processing unit and receive instructions

In the social categorial conversion, there are two types of strategies which are developed by the social poles of any actual-potential polarity. One is the *containment strategy* and the other is the *transformative strategy*. If the social negative pole has the dominance, it uses its power of dominance to develops strategies of containment to restrict the social positive pole to the state of insignificance as well as the ability of the social positive pole to develop strategies that will convert the existing relations of the social category in terms of the will of the social positive pole. In fact, the social negative pole may develop strategies for social categorial conversion to an object of a potential that is deemed more favorable to the negative pole and less favorable to the positive pole. Within the containment and transformational strategies of the negative action, the positive action is still in existence but in an ineffective state even though producing some level of aggravation and challenge. For example, the negative pole may be associated with a neocolonial state while the positive pole may be associated with an independent state to produce the existence of a social actual-potential polarity

In the situation of dominance of anegative pole, such as colonialism, neocolonialism or any form of foreign rule with a maintenance and backward transformative strategy to bring about say some form of slavery as it happened in Africa, the positive pole has a number of strategies available to it through its residing duality and constituent negative-positive characteristic sets. The development of these strategies cannot draw guidance from the *reactionary philosophy and ideology* that justify the unjust activities of the negative social pole. The social positive (potential) pole must have or develop a *revolutionary philosophy* with a corresponding ideology of liberation for socio-political transformation to alter the relational terms of cost-benefit balances of the social set-up. The revolutionary philosophy and ideology must be developed from within the positive pole with its own weapons of thinking and on the basis of its experiential information structure. The social positive pole must develop strategies to empower the social positive characteristic set for a continual increase in the positive social action. The necessary requirement for the successful strategy for social categorial conversion is a policy of continual education on the framework of the revolutionary philosophy whose weapons and contents are based on as well as derived from the poles' experiential information tradition of the oppressed. Such education should define a framework for the elements of the social positive characteristic set to develop and intellectually operate on the *principle of doubt* in the activities of the negative characteristic set in order to reduce its relative set quantity while increasing the relative set quantity of the social positive characteristic set. This is to reduce the social power of the negative pole while increasing the social power of the positive pole. The sufficient conditions require the development of appropriate institutions for the education and policy transmission. The educational process must be such that the basic structure of the core ideas of the revolutionary philosophy and the corresponding ideology must discredit the foundations and ideas of the opposing reactionary social philosophy with the corresponding ideology on the basis of which the negative action draws social strength, dominance and force of motion. The required institutions in support of positive social action must find expressions in the political, legal and economic structures of the social set-up as well as the organizational structure of the people who have the right of ownership to the social set-up, its history and progress. In this respect, the degree of the people's conscience must be raised to intensify the social positive actions of the individual and the collective. It is here that institutions of learning and transmission of ideas must be transformative in the development of thought in every segment of knowing and teaching with integration of the culture of thinking. The institutions of learning, creation of new thought and transmission of ideas must avoid the culture of mimicry since this moves the population onto the oppressive and subservient zone of cognitive imbecility, and destroys the transformative forces of collective creativity, learning and knowing.

The direction of categorial conversion is always a socio-political game whose cost-benefit relations are drawn from the economic

185

structure. The game is played between the poles of the social polarity through the behaviors and activities of the respective dualities and the residing negative and positive characteristic sub-sets that are given identities by the people under the epistemic guidance of a philosophy and an ideology whose content is called Philosophical Consciencism. Philosophical Consciencism defines the social mindset which guides the individual and the collective decision-choice actions in the social decision-choice space. Philosophical Consciencism in the case of Africa must contain African content where the weapons for its development are drawn from the African conceptual traditions and the information input is the African experiential information structure as pointed out by Nkrumah [R1.203].

CHAPTER FOUR

PHILOSOPHICAL CONSCIENCISM AND GAMES IN THE CATEGORIAL CONVERSIONS OF SOCIAL POLARITIES WITHIN SOCIAL SYSTEMICITY

The nature of socio-natural actual-potential polarities in relational differences and similarities has been discussed. The similarities are found in the unified logical unity of categorial conversions where every categorial conversion is a game between opposites and the understanding is analytically game theoretic. The unified logical unity requires the manufacturing of *categorial moments* in intra-categorial conversion and *categorial transfer functions* for inter-categorial conversions from the internal qualitative-quantitative dynamics of   all socio-natural categories. The differences are found in the manner in which the categorial moments and categorial transfer functions are internally manufactured to bring about the respective categorial conversions.   In natural polarities, the competitive game is naturally created where the technology for the activities of manufacturing is the work of nature or *natural technology* from within the ontological objects. In social polarities, the manufacturing is done by the use of *social technology* which must be created by the people under various negative and positive characteristics to generate negative and positive actions respectively.

4.1 The People, Positive Action and Negative Action

The social polarity, duality, negative and positive characteristic sets have been abstractly presented as social games in the social categorial-conversion processes. Whenever there are social actual-potential polarities with residing polar dualities, conflicts arise to present game situations between the actual and the potential. The game situation requires the development of strategies and counter-strategies for their resolutions to establish a new actual-potential polarity. Here, strategies and counter-strategies must be viewed as enveloping of tactical actions, and counter-enveloping of counter actions from the action space [R15.15]. These strategies and counter-strategies are the works of the people in the social set-up who are acting in either individual or community interests or both. From the perception of the individual and the community, the strategic social actions appear as opposites of positive or negative. In this respect, the social categorial conversion must at any time be seen in terms of game struggles between the negative and positive characteristic sets, where such game struggles are expressed in terms of strategic actions in actual social duality of the actual social pole of the social actual-potential polarity. The conflicts and the game are always within the actual pole to either maintain it or to change it by actualizing a potential.

Stated directly, the game of categorial conversion is about the game of social transformation and the change from the conditions of social relations expressed in terms of the cost-benefit distribution of the existing social system to a new social system with new social relations that will define a new cost-benefit distribution. In this respect, every categorial conversion is a conversion of social cost-benefit distribution in the social actual-potential polarity. The existing system is the actual, the possible substitution-replacement is the potential, and the two define the social actual-potential polarity. The nature and direction of the categorial conversion depend on the sovereignty and social power distribution. The people in the social set-up are the sovereign owners of their existing conditions. They are also the sovereign owners of the tools to destroy the existing conditions and create new ones. The people, therefore, own the conditions of the actual as well as the conditions of the potential. It is by the effort and actions of the people that the social actual is defined. It is also by the effort and actions of the people that a desired social potential is actualized through the collective activities in the decision-choice space. This ownership is the sovereignty that defines the social freedom, justice and happiness of the people. The people lose their personal and collective freedom for the social decision-choice actions if they fail collectively and individually to grasp their sovereignty ownership of their conditions of yesterday, today and tomorrow in the *Sankofa-anoma* tradition where the past is connected to the present in which future resides [R1.92].

4.1.1 The People, Social Conscience and Social Action

The understanding of the concept of the people's ownership of sovereignty is enshrined in the *social conscience* backed by a well-structured positive revolutionary philosophy the conceptual foundation of which provides an ideological and intellectual framework for integrating diverse social conditions of experiential information for the creation of continual positive action to set the preferred potential against the existing actual, in order to change social relations which establish cost-benefit distributions in the social actual-potential polarities. The positive revolutionary philosophy establishes a set of rules to be followed to create united social positive action, as well as defines a revolutionary ideology to justify the necessity and usefulness for establishing the social conscience in the individual and community, to motivate them to follow the rules in carrying them out to either defend the existing actual or to bring out a new actual from the potential pole. The relationship among philosophy, ideology and social consciousness is what Nkrumah called *Philosophical Consciencism*. An analytical definition will be useful in understanding its general meaning and its general application to socio-natural transformation and specific application to the African conditions.

Definition 4.1: Philosophical Consciencism

Philosophical Consciencism is the intellectual path and logical framework, made up by a relational unity and continuum of philosophy, ideology and social consciousness, with a supporting ideology that is developed from the experiential information structure of a given social set-up with its mode of reasoning for integrating competing social forces in a society, and develops them in a manner that allows for the digestion and reconciliation of the social contradictions in such a way that these contradictions fit into a progressive social personality for progressive social transformation of the given society, to set the potential against the actual or to defend the actual against an emerging potential.

The definition of Philosophical Consciencism is general and applicable to any social setup as viewed in terms of distribution of actual-potential social polarities. It unifies the cognitive social activities in the information space, decision-choice space and social action space of the social system and its organizational parts. It seeks to establish the purposeful foundation of social decision-choice action and the framework of social intentionality for the management of the social system in both static and dynamic states. The nature of the philosophy, ideology and social conscience that it engenders will vary from society to society in the human social set-up, and from generation to generation for any given society depending on the historical circumstances, experiential information and their interpretive structures and the condition of social

189

welfare. In each society and for each generation, the conditions of experience must historically be identified and analyzed for epistemic integration to develop the required individual and collective personalities and transformative leadership for the right decision-choice activities in the social space. In terms of the African context, for which Nkrumah was concerned, the historical circumstances are defined by the conditions of Western Euro-Christian and Islamic traditions as impositions on the African traditions. The experiential information structure, given these traditions, is expressed in terms of colonialism, slavery, brutality, inhumanity, terror, racism and complete resource exploitation without shame and without mercy by European and other imperial powers or neocolonial exploiters. . The interpretive structures find expression in human degradation, racism, suffering, mass killings, and disruption of African societies with social development reversal, poverty and global humiliation [R1.4][1.43] [1.69] [R1.70b][R1.239][R1.02][R1.87][R1.43]. All these historic facts combine to define the African experiential information structure and conditions for the construct of Philosophical Consciencism with African contents. The relational structure is presented as a cognitive geometry in Figure 4.1.1.

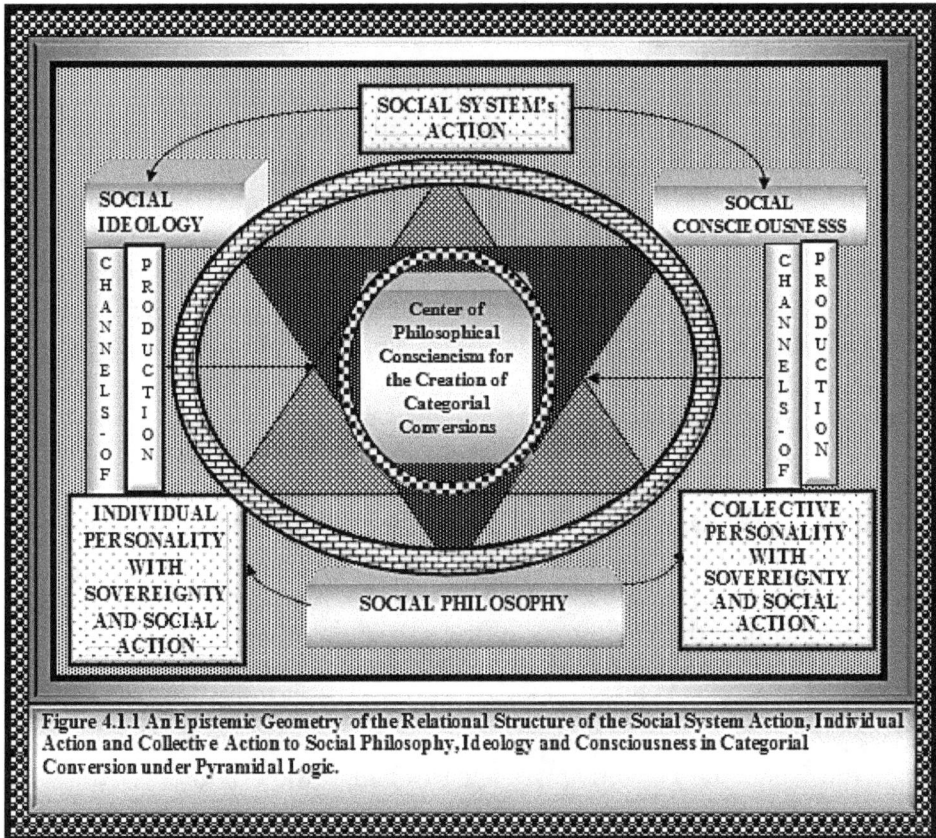

Figure 4.1.1 An Epistemic Geometry of the Relational Structure of the Social System Action, Individual Action and Collective Action to Social Philosophy, Ideology and Consciousness in Categorial Conversion under Pyramidal Logic.

In terms of the African case, and in the context of African freedom, justice, social transformation and progress, the general definition of Philosophical Consciencism was restricted by Nkrumah to give it content that fits into African conditions, needs and social revolution. With regard to the needed revolutionary philosophy and ideology to point to the right direction in support of an African revolution of decolonization and development, Nkrumah offered an appropriate definition of Philosophical Consciencism with African contents. He stated:

> The philosophy that must stand behind this [African] social revolution is that which I have once referred to as philosophical Consciencism; Consciencism is the map in intellectual terms of the disposition of forces which will enable African society to digest the Western, Islamic and Euro-Christian elements in Africa, and develop them in such a way that they fit into the African personality [R1.203, p.79].

191

4.1.2 Philosophical Consciencism and African Personality in Social Systemicity

The concept of social personality in the general definition of the Philosophical Consciencism, just like the African personality in Nkrumah's definition, must be defined and explicated to fit into the social characteristics and particularities of any social set-up. Every social set-up will have its own social characteristics and corresponding individual and collective personalities. These personalities are molded by a social philosophy and ideology that have taken hold of the society to shape the prevailing dominant culture. These personalities can be changed by changing the social philosophy and ideology that guide cognition in the interpretational space of the experiential information structure. In colonial Africa, the ruling personality is a colonial personality that was developed and shaped by anti-African philosophy and ideology which created and maintains a schizophrenic type of behavior in the African decision-choice space. The goal of Philosophical Consciencism in the African case of experiential information structure is to negate this process by presenting an African-centered philosophy and ideology to replace this anti-African philosophy and ideology which will shape, develop and mold a true African personality at the level of an individual and the collective needed to change the colonial relations by decolonization and development. For the social personality to be progressive and revolutionary, it must incorporate into its framework the element of Philosophical Consciencism with African contents. In the needs of Africa's revolutionary progress to set the African potential against the actual, Nkrumah defined the concept of African personality as: *The African personality is itself defined by the cluster of humanist principles which underlie the traditional African society* [R1.203, p.79]. Further discussion and clarification of the concept's analytical value and applications that may be required of it in the African struggle are discussed in [R1.200] [R1.214]. A long dissertation is also available in [R1.92]. Within the definition of the concept of African personality, Nkrumah sharpened the concept of Philosophical Consciencism that is relevant to the African colonial and post-colonial conditions, liberation and required revolution which must be supported by a revolutionary philosophy and ideology. He stated:

> Philosophical Consciencism is that philosophical standpoint which taking its start from the present content of the African conscience

indicates the way in which progress is forged out of conflict in that conscience [R1.203, p.79].

The conflict in the content of social conscience has two parts of negative and positive characteristic subsets. The negative part of the conflict in the social conscience seeks to destroy social cohesion and prevents a positive social transformation which will serve the increasing welfare needed to improve the quality of life for the masses. The positive part of the social conscience acts in the opposite direction. It seeks for progressive social transformation where there is a fair and just production and distribution of the cost-benefit configuration of social productive activities. In this respect, every conflict in the content of social conscience exists as a duality in relational continuum and unity. The content of existing conscience and the content of new possible conscience exist as social polarity whose categorial transformation and the nature of such transformation depend on the ruling philosophy and ideology. The nature of this philosophy and ideology will depend on experiential conditions and the needed social personality for decision-choice actions in the conflict resolutions within the socioeconomic space. For harmony and progress of every social set-up, it is important to assess the conflict, analyze and understand its social significance in order to allow the creation of reasonable degrees of social cohesion, and define an appropriate range of social conformity. It may be argued, as it has been done in [R1.91], that the contemporary difficulties of the individual decolonized African territories and the African Union may be traced to the failure of the *leaderships* to understand the nature of this conflict in the African conscience, which has been further amplified by lack of understanding of the relational structure of philosophy, ideology, personality and categorial conversion to advance the African interest and social vision.

Without social stability and unified social conscience, social progress and improvement of quality of life are impossible, and become mere mirages in linguistic monuments in living structures of deception of the masses. The nature and structure of the conflict in the content of the African conscience are identified and analytically explained by Nkrumah. He stated:

African society has one segment which comprises our traditional way of life; it has a second segment which is filled by the presence of the

193

Islamic tradition in Africa; it has a final segment which represents the infiltration of the Christian tradition and culture of Western Europe into Africa, using colonialism and neocolonialism as its primary vehicles. These different segments are animated by competing ideologies. But since society implies a certain dynamic unity, there needs to emerge an ideology which, genuinely catering for the needs of all, will take the place of the competing ideologies, and so reflect the dynamic unity of society, and be the guide to society's continual progress [R1.203, p.68].

4.2 Cultural Analytics and Philosophical Consciencism in Social Systemicity

From Nkrumah's analytical system of socio-cultural dynamics of Africa, the needed philosophy and ideology to point to African transformation must *be born out of the crisis of the African conscience confronted with the three strands of present African society* [R1.203, p.70] Each of the three strands has its own ideology and supporting philosophy that place a cultural distinction and competitive struggle in an antagonistic mode. Additionally, each one sits on a duality defined by negative and positive characteristic sub-sets in a relational proportion which presents its identity as well as projects such identity into the social cost-benefit space. In Africa's social transformation, the triple social elements may consolidate into one of the three polarities in philosophical coalitions for ideological game of social control through wars of ideas in the African social intellectual space. The epistemic analytics of the African intellectual space present the following philosophical and ideological coalitions in the possibility space:

1. A coalition of Islamic and African traditions with African synthesis $(\mathbf{Z}_1)$

2. A coalition of Islamic and African traditions with Islamic synthesis $(\mathbf{Z}_2)$

3. A coalition of Islamic and Euro-Christian traditions with Islamic synthesis $(\mathbf{Z}_3)$

4. A coalition of Islamic and Euro-Christian traditions with Euro-Christian synthesis $(\mathbf{Z}_4)$

5. A coalition of African and Euro-Christian traditions with African synthesis (Z_5)

6. A coalition of African and Euro-Christian traditions with Euro-Christian synthesis (Z_6)

7. A synthetic coalition of African, Islamic and Euro-Christian traditions with a complete African synthesis (Z_7)

8. A synthetic coalition of African, Islamic and Euro-Christian traditions with a complete Islamic synthesis (Z_8)

9. A synthetic coalition of African, Islamic and Euro-Christian traditions with a complete Euro-Christian synthesis

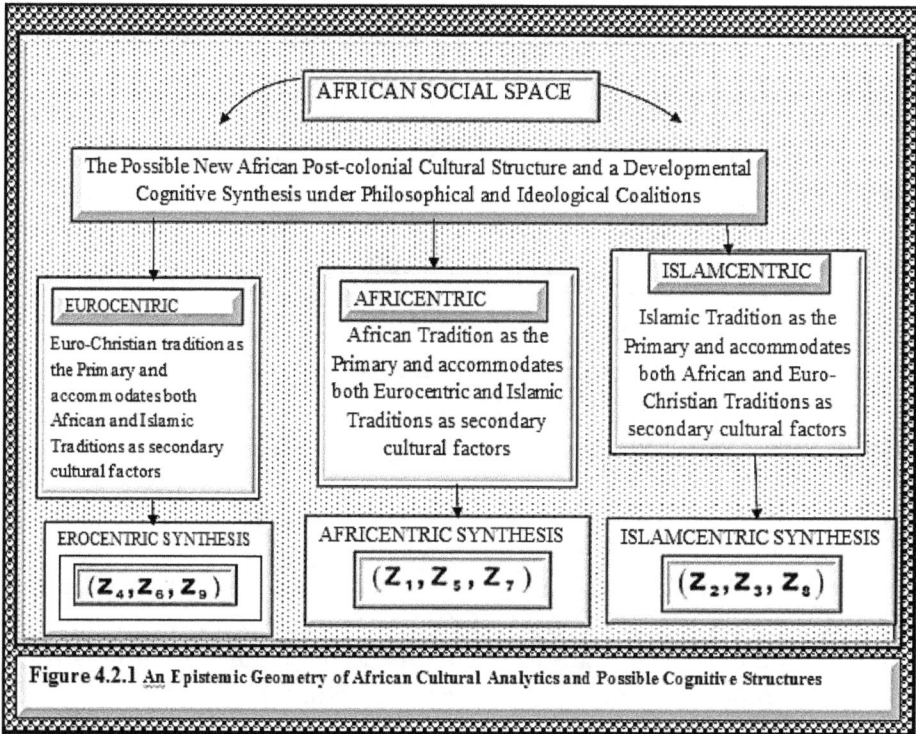

Figure 4.2.1 An Epistemic Geometry of African Cultural Analytics and Possible Cognitive Structures

Figure 4.1 is an analytical continuation of the structure of Figure 2.1 that shows the cultural analytics. This African cultural analytics is combined with epistemic analytics to understand the possible development of philosophy and ideology that will be consistent with Africentric Philosophical Consciencism.

Under the socio-cultural analytics of the African conditions, the African social system is confronted with a possibility set of size nine as an abstraction that presents intellectual paths in the potential space. The decision-choice actions to select and develop any of these intellectual paths are the responsibility of an awakened African intelligentsia as was argued in [R1.91] [R1.92]. Nkrumah put it this way: *The history of human achievement illustrates that when an awakened intelligentsia emerges from a subject people it becomes the vanguard of the struggle against alien rule* [R1.202, p. 43] [R1.207, p. 81]. The duty of the African intelligentsia is not to seek vain glories in the intellectual space of Africa's oppressors, colonizers and imperialists but to construct and develop a philosophical system that will support an ideology to liberate the African masses from the shackles of colonialism, neocolonialism and imperialism as well as from the vestiges of slavery, terrorism, human and non-human resource thievery including intellectual lifting by the Europeans from antiquity to the present. This philosophy and ideology must also seek a new African personality whose duty is to help to prevent a repetition of the past African human suffering. Additionally, the needed philosophy and the corresponding ideology must point to the path of political reconstruction, social institution casting and transformation to create a continuity of the African past, present and future in an uncompromising African tradition of the time-trinity as represented by the philosophical concept of the *Sankofa-anoma* which expresses the relational connectivity of the past-present-future phenomena in transformations and decision-information-interactive processes, where the present is the birth of the past and the present is pregnant with future possibilities conditional on an experiential information structure. This phenomenon of time trinity gives meaning to the scientific idea of forecasting which seems modern to the present but it is actually derived from Africa's ancient.

The nine possibilities are elements in the potential while the current conditions represent the social actual. The combination of this actual with any of the elements in the possibility constitutes an actual-potential polarity the categorial conversion of which requires the intellectual creation of categorial moments and categorial transfer functions. The nine possibilities present important complexities in the search space for an intellectual way that will allow Africa's preferable positions in colonialism-independence and neo-colonialism-emancipation polarities.

The choice of one of the nine possibilities must be guided by Africa's socioeconomic rationality where such rationality must be rooted in Africa's history in order to avoid falling into a gulf of intellectual schizophrenia where the African masses lose value in social stability and roam in cognitive confusion and a mirage of education without a clear understanding of who they are, what they are, and in fact, regard each other as enemies, and foreign powers as friends and strategic partners for missions that are intended to re-enslave the African people operating in the space of cognitive imbecility. The selected intellectual path should keep in mind and in focus the modern African history of colonialism, slavery, racism, and loss of sovereignty. The choice of a possible path to follow requires an analysis of the possibilities available. This is the analytics of African conditions and social decision-choice actions for the construct of a revolutionary philosophy and the corresponding revolutionary ideology that will be derived from the experiential conditions of the African people and culture of traditions. It is this Africentric revolutionary philosophy and the corresponding ideology that must guide African social decision-choice actions, African relations within Africa and outside Africa.

The battle against the actual in order to bring in the needed potential in the intellectual space is a difficult one. In fact, it is a fundamental problem whose solution must be searched and developed. This is because

> Colonialism and its attitude die hard, like the attitudes of slavery, whose hangover still dominates behaviour in certain parts of the Western hemisphere. The social effects of colonialism are more insidious than the political and economic. This is because they go deep into the minds of the people and therefore take longer to eradicate. The Europeans relegated us to the position of inferiors in every aspect of our everyday life. Many of our people came to accept the view that we were an inferior people [R1.207, pp.36-37]

The task of the revolutionary African-centered philosophy is to create a thinking system from the structures of African experiential information and traditions that will decolonize the mind of colonial Africa, empower the Africa's self-confidence and *sankofarize* the true African personality and the supporting conceptual system required to create the sufficient conditions needed towin in playing the games in categorial conversion to further the interest of African social transformations.

4.2.1 The Analytics of the Decision-Choice Action to the Intellectual Pathway

The cognitive pathway of possible coalitions of cognitive development from one to six $(\mathbf{Z_1} - \mathbf{Z_6})$ create a framework of conflicts to promote continual uncontrollable instabilities in the African social set-up. Some hope, however, seems to be presented by possibilities seven to nine. Possibility $(\mathbf{Z_8})$ sees African society and its people in terms of external history where the African experiential information structure is accommodated as experiences of Islamic tradition to the complete neglect of the historic fact that the African traditions were in existence before Islam. Similarly, possibility nine $(\mathbf{Z_9})$ also sees the African society and its people in terms of external history where the African experiential information structure is accommodated as experiences of Euro-Christian tradition and history to the complete neglect of the historic fact that the African traditions predate the Euro-Christian traditions. The possibilities $\mathbf{Z_8}$ and $\mathbf{Z_9}$, thus present no useful choice. The choice of either $\mathbf{Z_8}$ or $\mathbf{Z_9}$ defeats the search for a useful path in constructing a revolutionary philosophy in support of revolutionary ideology. It is these historical realities and Euro-Christian and Islamic distortions that brought about the important works of Ashby [R1.24][R1.25] [R1.26], Ben-Jocannan [R1.35] [R1.37][R1.38], Blyden [R1.42], Diop [R1.83b][R1.85][R1.86] and Massey [R1.179] on the historical realities which also lead Nkrumah to write:

> Africa cannot be validly treated merely as the space in which Europe swelled up. If African history is interpreted in terms of the interest of European merchandise and capital, missionaries and administrators, it is no wonder that African nationalism is in the forms it takes regarded as a perversion and neocolonialism as a virtue.
>
> In the new African renaissance, we place great emphasis on the presentation of history. Our history needs to be written as the history of our society, not as the story of European adventures. African society must be treated as enjoying its own integrity; its history must be a mirror of that society, and the European contact must find its place in this history only as an African experience, even if as a crucial one. That is to say, the European contact needs to be assessed and judged from

the point of view of the principles animating African society, and from the point of view of the harmony and progress of this society.

When history is presented in this way, it can become not an account of how those African students referred to in the introduction became more Europeanized than others; it can become a map of the growing tragedy and final triumph of our society. In this way, African history can thus become a pointer at the ideology which should guide and direct African reconstruction [R1. 203, p. 63].

The Euro-centric methodological path to history and conceptual system must be changed because it is not consistent with the African experiential information structure from human antiquity if such an information structure is to be captured by either an empirical information structure or an axiomatic information structure or both for the constructs of explanatory or prescriptive processes. The task of Philosophical Consciencism, as a revolutionary philosophy that draws its roots from the African conceptual system and true history, is to falsify the core propositions of this Euro-centric method and conclusions of interpreting African history and the oppressive philosophy that it endangers, in order to emancipate the contemporary African thought from the shackles of European imperial oppression, injustices, racism and unjust order so as to strengthen the basic foundations of freedom, justice and quality of human essence.

All the paths that we have discussed for the development of the needed African philosophical construct in support of the required African revolutionary ideology present deep traps of cognitive confusion, analytical instabilities in the zones of thought and social madness in practice of freedom and justice. We are left with possibility seven $(\mathbf{Z_7})$ that relates to Nkrumah's question and answer.

What is to be done then? I have stressed that the two other segments (Islamic and Euro-Christian traditions) in order to be rightly seen, must be accommodated only as experiences of the traditional African society. If we fail to do this our society will be racked by the most malignant schizophrenia.

Our attitude to the Western and Islamic experience must be purposeful. It must also be guided by thought, for practice without thought is blind. What is called for as a first step is a body of connected thought which will determine the general nature of our action in unifying the society which we have inherited, this unification to take account at all times, of the elevated ideas underlying the traditional African society. Social revolution must therefore have, standing firmly behind it, an intellectual revolution, a revolution in which our thinking and philosophy are directed towards the redemption of our society. [R1.203, p. 78]

In the development of the needed African-centered thought and philosophy that will guide the needed revolutionary ideology to direct African social actions in the individual-collective duality towards the redemption of African society, Nkrumah, after deep reflections on African experiential information structure and general history of philosophy, suggested, and rightly so that: *Our philosophy must find its weapons in the environment and living conditions of the African people. It is from these conditions that the intellectual content of our philosophy must be created* [R1.203, p.78]. The required philosophical basis and foundation are what Nkrumah called *Philosophical Consciencism* whose social dynamic foundation is what he called *categorial conversion*. The theoretical system of the categorial conversion is fully developed in its philosophical and mathematical structures on the basis of the African conceptual system from the antiquity to the present in [R15.15] as *The Theory of Categorial Conversion*. Its philosophical foundation is presented as polyrhythmicity and its logic of reasoning as polyrhythmics **[R1.92]**

It has been pointed out that the construct of the Africentic system of thought is an epistemic complexity. It requires an important devoice from the colonial mindset where a substantial part of the African intelligentsia operate in a comfort colonial zone of thought which forces them to play cooperative games in the categorial conversion of social polarities to the disadvantage of true African emancipation. It requires the practice of personal sacrifice and discipline such as the practice of such African scholars and others like Diop [R1.83B][R1.84] [R1.85], Carter G. Woodson [R1.294][R1.295], Ben-Jochannan [R1.35][R1.37], and de Lubicz Schwaller [R1.243] [R1.244][R1.247]. Without intellectual discipline the required philosophy in support of the needed African revolutionary ideology cannot be constructed and developed. It is this philosophy that can ensure Africa's freedom and justice in the global space. It is also this philosophy that can help to create the needed African personality in the African thinking and decision-choice space to create African unitary nations and a united Africa to defend the African masses in support of the African diaspora from territorial confusion left over by imperialists and colonialists now being exploited to the maximum by neocolonialists of the imperial club supported by domestic neocolonial cronies.

Africa has failed to make much progress against neocolonialism and imperialism because the greatest portion of her intelligentsia has been transformed into intellectual zombies living and practicing their thought and craft in the comfort zone of familiar with epistemic colonialism and *askarized* in the rent-seeking zone of neocolonialism with illusions of being rewarded with something from the leftover bones of the intellectual dining table of the neocolonial masters. In fact, the members of this portion of the African intelligentsia have acquired the character of askari to terrorize the members of their decolonized state for personal riches. For example, the current sweeping of Africa with Christian and Islamic fundamentalists attests to the failure to decolonize the mind of Africa which is placing it into a zone of malignant schizophrenia in the global and domestic decision-choice spaces.

The members of this portion of the African intelligentsia have their African traditional roots systematically deprived of sustenance and living essence by *intellectual neocolonialism*. From their education, they have been encouraged to accept and treat *Western* history and its intellectual system as the only worthwhile portion of human history that commands an epistemic universality in support of human behavior in the global decision-choice space affecting all aspects of human life and its socio-natural environment. Questions about the African intellectual foundations of this Western epistemic universality are not asked [R1.86]][R1.1.35] [R1.243]. In this epistemic process in the epistemological space, some members of the African intelligentsia have become so colonized and further neo-colonized that they become alienated from Africa, her intellectual history and her monumental achievements as foundations of human progress and its history. They also fail to understand that knowledge production is a never-ending success-failure process guided by the African principles of opposites in polarity, duality, relational continuum and unity and not from one source but many [R3.13][R3.10], [R8.44]. The members of this intellectually colonized and neo-colonized portion of the African intelligentsia have no knowledge of the great logical system of Africa regarding the principle of opposites which is composed of polarity, duality, relational continuum and unity, and that exactness in human knowledge is claimed by the actions of human decisions and choices in the knowledge-ignorance duality within social actual-potential polarities where each stage of the knowledge acquisition process is a category defined by actual-potential polarity. They criticize Africa and her people in the same frame as the African imperial oppressors and colonizers and encourage neocolonialism in Africa without realizing that they have alienated themselves from their roots and are, in fact, an important part of Africa's conditions. The African leadership is philosophically and ideologically blinded by intellectual colonialism and neocolonialism. As a result of the intellectual colonialism and neocolonialism in education, the personality of the African leadership, as developed from this education, learns to practice in the neo-colonial comfort zone of least resistance rather than to work in the battle zones of independence and emancipation to overcome odds for what the African experiential information structure has revealed to be in Africa's best interest. In this way, the neocolonial education and intellectual practices turn large members of the African intelligentsia into useful idiots who then collectively become an effective apparatus for controlling the African human and non-human resources from without [R1.91][R1.92]. Their education has failed to teach them how to think to overcome odds. It has only taught then the humiliating art of mimicry.

The members of this portion of the African intelligentsia seem not to understand that the pathway to African progress cannot be found with the ideological searchlight of an imperialist's constructed logic and philosophy that generate an anti-African personality in the African people where the Africans abandon their cultural roots and seek integration into cultures alien to them. They fight to be what they are not, and condemn everything that encompasses what they are that of African essence. The fact remains that the pathway to Africa's freedom, justice and progress can only be found with the searchlight of African-centered ideology constructed from an African revolutionary philosophy which is developed on the basis of Africa's experiential information structure and conceptual system. In this respect we can, by complementing Nkrumah's thought on colonialism and independence [R1.203] [R1.207] state that it is by the effort of the African intelligentsia that intellectual colonialism, neo-colonialism and cognitive imbecility are rooted. It is also by the sweat of the brow of the African intelligentsia that the national thinking system can be constructed to decolonize and bring intellectual liberation. The African intelligentsia is the backbone and reality of Africa. However, it is the lack of philosophically and ideologically enlightened leadership and intellectual awareness of Africa in relation to the African experience and problems why the African people are locked in the suffering zone of indignities, depredation, and vestiges of slavery under the yoke of intellectual neo-colonialism. This unpleasant situation within Africa can be internally corrected by applying the core ideas of the theory of *categorial conversion* in understanding the working mechanism of categorial moments and categorial transfer functions of quality-quantity dynamics of socio-natural categories, and the use of *Philosophical Consciencism* to create the relevant categorial moments and categorial transfer functions required for socioeconomic and political transformation of Africa.

4.2.2 The Relationship between Philosophical Consciencism and Categorial Conversion

The theory of categorial conversion presents a logical and philosophical basis of general internal self-motion that alters the quality characteristics from one category to another. It presents a general mechanism of continual change of socio-natural elemental categories. This logical and philosophical foundation provides the reasons for the African internal self-motion for the transformation of the socio-political category of colonialism to a category of decolonization, from the socio-economic category of less developed to a category of more developed, from the socio-economic category of lack of nation building induced by African preferences to a category of nation building according to Africa's own preferences, and from a general category of dependency to a category of complete African emancipation. Here, it must be understood in the strongest term that the conditions of categorial conversion are different for natural categories and social categories. Nature transforms itself from within by creating the Nkrumah's Delta composed of categorial moments and categorial transfer functions by means of which an element qualitatively moves from one natural category to another through changes in the characteristics. An example of this is the categorial conversion involving the life-death polarity as seen from development and decay. This Nkrumah's Delta (δ) in natural systems is internally manufactured by nature, in a given environment, through a complex system of strategic struggles between the negative and positive dualities working with the residing negative and positive characteristic sub-sets to create the needed energy and force required for qualitative motion.

Things are different, however, in categorial conversions of social categories. Here, each socio-political state or socio-economic state over time must be seen as a category with a different identity defined by a differential combination of negative and positive characteristic sets. The energy required to create the force for movement from one social category to another, or to maintain the same social category must be abstracted from the positive and negative characteristic sets of the members in the social set-up. The negative and positive characteristic sets aggregately produce the social negative and positive forces from which social negative and positive actions are manufactured to produce Nkrumah's Delta for either the retention of the old actual, or for a change to a new actual from the potential. In other words, Nkrumah's Delta must be manufactured from the technology of social motion where the inputs are the collective efforts of the people, output is either retention or change, and the change is either negative or positive. The Nkrumah's Delta must be manufactured from the action with consciousness, individual self-consciousness and social self-consciousness as an important part of the required social technology to bring about categorial conversion of social states defined within a social category. Social action without self-consciousness and intentionality is self-defeating and social self-consciousness and intentionality without action is a mirage. This self-consciousness and intentionality must flow from revolutionary philosophy and ideology that give self-confidence to the people to empower the positive forces. It is here that *Philosophical Consciencism* finds revolutionary meaning and applicative power in the context of the African struggle against imperialism and neocolonialism. It is also here that the true understanding of forces and the power of constructionism-reductionism methodological duality in relational continuum and unity of category formation and categorial conversion brings social creativity to the use of Philosophical Consciencism for a meaningful social categorial conversion in line with Africa's preferences and not in line with neocolonial preferences.

Categorial conversions of social categories are motivated by the social consciousness of the existing social actual and the corresponding cost-benefit distribution that defines the individual and collective welfares. The nature of the cost-benefit distribution forms the social incentive to produce action which will bring about a change or no change. In other words, social transformations are due to joint relational operations of social consciousness, the nature of social cost-benefit distribution and categorial conversion. After analyzing the logical structure of categorial conversion, Nkrumah was confronted with the problem of how to internally manufacture categorial moments and categorial transfer functions from within Africa itself. It is here that Nkrumah connected social action and the conscience required to operationalize the categorial conversion process in general and in African society in particular. The social conscience is connected to the knowledge of categorial conversion and to both the negative and positive actions. As has been explained above and analytically developed in [R15.15], the categorial moment and categorial transfer function may be negative, positive or neutral. The negative categorial moment and transfer function bring about negative categorial conversion. Meanwhile, the positive categorial moment and categorial transfer function bring about positive categorial conversion of progressive social transformation. It is useful to note that socio-political stability of any social system is guided by a set of rules and principles and the consciousness in the individuals that must obey and implement them. It is here that the legal structure finds usefulness where reward and punishment in the social order take root. It is also here that a social system may operate in the zone of injustice and oppression to deny humanity to others through distortions in the social cost-benefit distribution.

The social consciousness is a creation of ideology and a supporting philosophy which together Nkrumah called Philosophical Consciencism. Philosophical Consciencism is social philosophy and ideology created to deal with qualitative and quantitative dispositions of social dynamics with practical significance to individual and collective lives of the social setup. Its content is social system dependent on the basis of a social experiential information structure, history and generation. It is on the basis of implied content that a set of rules and principles are created and enshrined in the dynamics of the legal and moral structures of society. The set of rules and principles for any given society is dynamically opened for all generations since the set itself obeys the principle of categorial conversion. Consciousness and social action exist in a relational continuum and unity which in turn affect the construct of philosophy and ideology. The appropriateness of the philosophy and corresponding ideology depends on critical awareness of history and experiential information that give substance and meaning to their content as observed by Nkrumah. It is on this experiential information structure and social history that the content of Philosophical Consciencism is specified for the case of African society by Nkrumah and further developed in [R1.92].

In this respect, a question arises as to the similarities and differences between rules and principles, ideology and principles and ideology and philosophy and programs and actions in a social set-up. The answers to some of these questions have been discussed in Chapter One. In relation to rules and principles Nkrumah suggested:

> It is necessary to understand correctly the relationship between rules and principles. This relationship is similar to that between ideals and institutions and also to that between statutes and by-laws. Statutes, of course, state general principles, they do not make explicit those procedures by means of which they may be carried out and fulfilled. By-laws are an application of such principles. It is obvious that when the conditions in which by-laws operate alter seriously, it could be necessary to amend the by-laws in order that the same statute should continue to be fulfilled. Statutes are not on the same level as by-laws, nor do they imply any "particular" by-laws. It is because they carry no specific implication of particular by-laws, but can be subserved by one of whole spectrum of such, that it is possible to amend by-laws, while the statute which they are meant to fulfill suffers no change [R1.203, p.93-94].

The necessity in having correct understanding of the relationship between rules and regulations finds usefulness when there is categorial conversion of a social polarity in which rules and principles serve their respective behaviorally social stability where consciousness becomes a product of rules and regulations either in the old actual or the new actual after conversion. For example, how different are the rules and regulations in colonial education from the rules and regulations in neo-colonial education and how different are the corresponding personalities? To what extent does the anti-African philosophy and ideology help Africa to set the potential against the actual in Africa's struggle to emancipate? Philosophical Consciencism provides an intellectual framework for defining the social vision, national interest and supporting social goal-objective set within the environment of the African experiential information structure and global situation. These may be defined in terms of ethical principles and statutes on the basis of which by-laws are created. The ideological component of Philosophical Consciencism provides the justification and rationalization of the by-laws and the rules for the implementation of the social vision, the national interest and the social goal-objective set. The philosophical component of Philosophical Consciencism provides the conscience in the individual which must incentivize him or her to accept and implement the rules and regulations for bringing about continual progressive transformation. To understand the incentive structure that allows the individual to accept or not to accept and implement or not implement the set of principles by following a set of rules, the framework of Philosophical Consciencism must lead the African people to understand the human nature in which (s)he responds to different socio-political conditions in which (s)he functions.

Philosophical Consciencism acknowledges the critical role of human internal self-motion in setting the direction and path of social history, progress and suffering. It also acknowledges the power of curiosity and thinking as important pillars of categorial conversions of social polarities that provide true essence of human creativity and the forces of self-organizing as the greatest asset of humankind. The conceptual framework of Philosophical Consciencism also sees human society as both an open and closed system that is self-organizing, self-propelling, self-correcting and self-destructing on the basis of internal energy for changes in the qualitative and quantitative dispositions within itself. The energy for existence or change of social categories is internally derived from the people who constitute matter, where the people as matter and energy from it are relationally linked by information in an inseparable continuum and unity.

Since any social system is simultaneously open and closed, it is being acted on by internal and external forces. The question then arises as to how the external and internal forces are related in terms of the internal dynamics of the social set-up. In terms of the theory of categorial conversion, the effect of any external force on the system is temporary, while sustainable progress is due to internal forces that receive direction from the creative destruction of human ingenuity. For any external force to produce sustainability of behavior in a society and hence in any category, it must be internalized by either the negative or the positive characteristic sub-set to augment the strength of either the negative or positive action respectively. Such an external force enters the social set-up to enhance the effectiveness of negative action viewed relative to social progress. It enters the social characteristic set to infest, corrupt, pervert and dwarf the true aspiration of the people for true and lasting social progress. It is under this understanding that Nkrumah stated:

> The people are the backbone of positive action. It is by the people's effort that colonialism (oppression) is rooted, it is by the sweat of the people's brow that nations are built. The people are the reality of national greatness. It is the people who suffer the depredations and indignities of colonialism (oppression), and the people must not be insulted by dangerous flirtation with neo-colonialism (new form of oppression) [R1.203, p.103].

The idea that the people are the backbone of national greatness points to several notions about people as the creative force producing the categorial moments and categorial transfer functions in intra-state and inter-state conversions. The notion that the people's future is defined in their aspirations is in relation to the time trinity of past, present and future in the philosophical tradition of the *Sankofa-anoma* which links the past experiential information to the present one to define the future by the internal creative essence of the people. These aspirations are related to the establishment of social vision, national interest and a social goal-objective set in support of the social vision and national interest related to internal social creativity, hard work and commitment under an ideological guidance [R13.8] [R13.9]. The analytical and foundational basis for the establishment of social curiosity, creativity and aspiration is social Philosophical Consciencism. In terms of the African conditions of freedom and nation building Nkrumah stated:

> The cardinal ethical principle of Philosophical Consciencism is to treat each man (woman) as an end in himself (herself) and not merely as a means. This is fundamental to all socialist or humanist conception of man (person) [R1.203, p.95].

This cardinal ethical principle may be viewed as the African social vision of a property of a good society, where a system of economic activities serves the people as it was in African tradition instead of the people serving a system of economic activities to the enjoyment of the few, as the performance of the social structure is examined in the individual-community duality in relational continuum and unity. On the basis of this cardinal principle is derived a sub-principle of each for all and all for each, in that the identity of each person has meaning only within the social set-up and the community of the social set-up is defined only by its members. The analytical and philosophical foundation of this African thinking is provided in [R1.91] [R1.92].

It is here that the philosophical power of an integrated conceptual structure of *Sankofa-anoma, Anoma-kokone-kone* and *Asantrofi-Anoma* finds an epistemic beauty in defining social problems and abstracting solutions in human organization, where such human organization is seen in terms of the principle of opposite composed of polarity and duality in relational continuum and unity, as it has been argued that these are the African traditions which guided the Nkrumah's construct of categorial conversion, Philosophical Consciencism and his claim of scientific socialism as a way to organize African societies. For the readers not familiar to these philosophical concepts, Sankofa-anoma defines the concept of the time trinity of past-present-future connectivity in human decision-choice actions. The phenomenon of Anoma-kokone-kone defines individual-community duality under tension where the individual has identity only within the community and the individuals provide the identity of the community, in the sense that the individual is both good and bad (good-bad duality) for the community and the community is also good and bad (good-bad duality) in human social formation. The phenomena of Asantrofi-anoma finds definition and meaning in the cost-benefit space regarding any decision-choice element in that every decision-choice element is a cost-benefit duality in such a way that one cannot select the benefit and leave the cost, and one cannot be forced to take the costs without the corresponding benefits. It is this Asantrofi-anoma problem that gives scientific legitimacy to the development and study of cost-benefit analysis [R4.17] [R4.18.]. The Asantrofi-anoma rationality is not a simple acknowledgement of the existence of costs and benefits in decision-choice element but that these costs and benefits are relationally connected and inter-supportive in dualistic sense. The relational structure defining continuity and unity in problem-solution dualities in social formation is presented as an epistemic geometry in Figure 4.2.2.1.

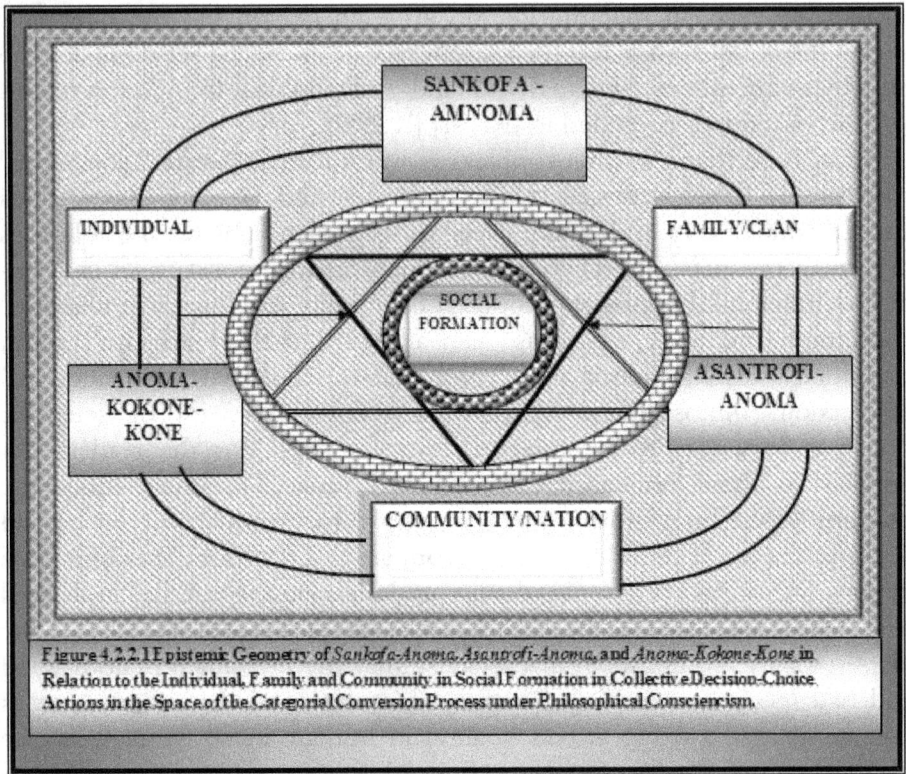

Figure 4.2.2.1 Epistemic Geometry of *Sankofa-Anoma, Asantrofi-Anoma*, and *Anoma-Kokone-Kone* in Relation to the Individual, Family and Community in Social Formation in Collective Decision-Choice Actions in the Space of the Categorial Conversion Process under Philosophical Consciencism.

Through his knowledge of the African principles of opposites, Nkrumah came to understand that the community is a support of the existence and identity of an individual and the individual's existence defines the identity and the meaning of the community. In terms of organization, production and consumption, the relationship of the individual and community is a duality where such duality finds meaning and expression in the cost-benefit duality as projected by *Anoma-kokone-kone* and *Asantrofi-anoma* concepts. The individual is both cost and benefit to society and the society is also both cost and benefit to the individual where the cost-benefit duality is in relational continuum and unity. To create stability and harmony in the social set-up, the distribution of the cost-benefit configuration in the cost-benefit duality must be fair. It is also here that the right-wrong duality is defined to give meaning and define a decision-choice action in the justice-injustice duality. The existence of the *Anoma-kokone-kone* and *Asantrofi-anoma* problems is a problem in duality where there is a cost support for a benefit and a benefit support for a cost in an inseparable relation in all decision-choice activities in never-ending transformations in destructive-constructive duality [R1.244]. A similar problem is defined by Catullus of Rome and cited by March in his essay on bounded rationality [R15.33] with full and extensive discussion in [R3.7]. These principles are the fundamental basis of the African conceptual structure of individual-community duality of the foundations of the African social set-up. The modern version of this form of African social formation where there is a relational balance between an individual and community on the principles of fairness and justice in the cost-benefit distribution is what Nkrumah called scientific socialism. Nkrumah's scientific socialism appeals to the principles of African traditions under changing scientific and technological progress. This Nkrumah's concept of scientific socialism is never understood as it is confused with other conceptual types of socialism such as Schumpeterian socialism and Marxian communism [R13.9] [R15.34][R15.46]. Furthermore, the differences and the comparative logics of social transformation of these powerful thinkers are not clearly distinguished.

In the social set-up, the destruction of the existing conditions (actual) and the replacement of them with new conditions (potential) are the work of the people by the people and for the people. In terms of the African conditions, Nkrumah put the struggle in the actual-potential polarity as:

> Independence is of the people; it is won by the people for the people. That independence is of the people is admitted by every enlightened theory of sovereignty. That it is won by the people is to be seen in the successes of mass movements everywhere. That it is won for the people follows from their ownership of sovereignty. The people have not mastered their independence until it has been given a national and social content and purpose that will generate their well-being and uplift [R1.203, pp.105-106].

The actual-potential polarity represents the colonialism-independence polarity in Nkrumah's statement. When a political element, in general, has been actualized by the application of positive action, the positive action requires a new orientation away from a mere destruction of the old conditions and must be oriented to the defense of the new actual or to actualize a new and higher level of a potential. Nkrumah's understanding of the dynamics of actual-potential polarity, its relation to the residing dualities and the works of the negative and positive characteristic sets in generating tensions in the negative and positive actions for categorial conversions is not only powerful but logically beautiful. He understood the continual process of categorial conversion that allows a new polarity to be traced to the previous polarity and also to an emerging potential. Both the negative and positive actions never quit. They are rearranged in dominance to define new qualitative disposition in all socio-natural categories. It is on this understanding and as applied to social actual-potential polarity that Nkrumah stated:

> When independence has been gained, positive action requires a new orientation away from the sheer destruction of colonialism and towards national reconstruction.
>
> It is indeed in this address to national reconstruction that positive action faces its greatest dangers. The cajolement, the wheedlings, the seductions and Trojan horses of neo-colonialism must be strongly resisted, for neo-colonialism is a latter-day harpy, a monster which entices its victims with sweet music [R1.203. p.105].

Nkrumah had great insight into the vestiges of the brutish nature of colonialism and how it worked to colonize and neo-colonize the African mind to make it an appendage and catholic to the imperial system of thought. This statement must be related to the discussion on statutes and by-laws. Decolonizing the mind is very difficult when the members of a greatest portion of African leadership in politics, education, religion and others are operating in the comfort zone of a familiar colonial and neocolonial intellectual order, trying to use the same colonial thinking system that colonized and enslaved them for social transformation. The African in the decolonized territory continues to be a colonial child where (s)he is restricted by neocolonialism to have joy with things of small and insignificant values. He claims independence and carries a beggar's basket over the globe. His/her colonial education has not taught him/her the meaning and conditions of independence but simply transform him/her to accept the lifestyle that allows him/her to live in the comfort zone of neocolonial intellectualism. His/her neocolonial education has stripped off his/her true *African personality*, separated him/her from his roots, and given him/her a distorted logic to justify his/her corruption and acts of stealing from his people as profit from capitalist production under the principle of each for himself/herself and God for us all. Having been separated from the basic foundation of the

African tradition of organization, hard work and its conceptual system, and having been transformed into an object of charity, the content of the neocolonial African beggar's basket contains increasing subservience to neocolonialism where he/she lives in the comfort of indignities and humiliation. This colonial child has matriculated from colonial slavery into neocolonial slavery. As a graduate from direct slavery to indirect slavery, this child is a paradox. He wants to simultaneously carry both an independence basket and a beggar's basket. (S)he lives in a comfort zone of acute schizophrenia without the true knowledge of his/her identity. (S)he does not ask questions regarding the relationship between national independence and national begging and what is the relationship between a beggar's basket and neocolonialism. (S)he fails by the use of his/her neocolonial education which is devoid of an African experiential information structure, to understand that the significant problems in nation building after independence cannot be solved by the use of colonial and neocolonial thinking systems. A new thinking system is required. This new thinking system is what Nkrumah called Philosophical Consciencism derived from African experiential information structure.

One thing that must be clearly kept in mind is that Nkrumah, as an African had a critical and insightful understanding of the role of tension in all transformations as perceived and projected from Africentric principles of opposites as developed from African philosophical antiquity [R1.35] [R1.37][R1.24], [R1.27] [R1.56][R1.57] [R1.124] [R1.219] [R1.251a] [R1.1.86] [R1.92][R1.243] [R1.246][R1.247]. The principles of opposites are philosophically, logically and mathematically defined in the spaces of polarity and duality with relational continuum and unity which are established by relative structures of the negative and positive characteristic sets as they relate to qualitative and quantitative dispositions of categories of socio-natural objects. Let us analytically notice that Nkrumah, working with Africentric principles of opposites, first established the colonialism-independence polarity, linked it to negative-positive-action duality, and then worked on the strategies required to create the sufficient conditions of convertibility from the colonialism pole to the independence pole.

Here, it is useful to notice that colonialism is the actual pole and independence is the potential pole for the politico-economic set-up. When the pole of colonialism as the actual is internally transformed by categorial conversion into a potential, and the pole of decolonization as a surrogate of independence as the potential is also transformed into the actual, the struggle and conflict between the actual and the potential do not stop. Keep in mind the relational continuum of never-ending struggles and transformations in the actual-potential polarity whose outcome merely changes relationality but retains the actual-potential polarity of different categories of existence defined in terms social varieties. They take on different form to define a new social actual-potential polarity which may be specified in the Nkrumah's conceptual framework as independence-neocolonialism polarity. The independence-neocolonialism polarity works through conditions of new sovereignty control that may be domestic or external. The struggle for domestic sovereignty control is in relation to nation building, composed of the casting of relevant institutions in the political, legal and economic structures which will promote the welfare of the domestic people. The struggle for external sovereignty control is in relation to resource development, composed of the casting of relevant institutions in the political, legal and economic structures which will serve external interest at the expense of the domestic population. In general, as seen from the Nkrumah's conceptual system, there is a continual categorial conversion from one social polarity to another social polarity with a continuum and unity in a never-ending process of change induced by categorial conversions. It is here that the orientation of positive action to the defense of independence and the construction of nation building from within become a challenging task for the people of the new social actual of independence.

The restructuring of the positive action requires the organization of the people to revote against entrenched opposite values that control the collective mind and deprive them of creativity in the nation building, and to escape from new forms of imperialist aspirations. It is on this condition that Tekyi writes: *Let there be sane thinking, sane organisation, sane methods and the future is with us to command* [R1.264, p.404]. Here, Nkrumah's conceptual system saw each stage in politics, law and economics as social categories of actual and potential that are organized in an increasing order of social preferences, where each stage is transformed by the application of conditions of categorial conversion that require the creations of Nkrumah's Delta composed of categorial moments and categorial transfer functions. In the African case, the conditions of creating the needed categorial moments and transfer functions must be guided by Philosophical Consciencism which projects a philosophy and ideology on the basis of both the African experiential information structure and the African conceptual system from antiquity.

4.3 Philosophical Conscienciecism, Categories of Social Action and Methods of their Creation

In social polarities, the matter, energy and information required to bring about categorial-conversion dynamics are the people who are the agents of change to destroy the conditions of the actual and the creation of new conditions for the emergence of a potential. The people constitute the matter, energy and information in the universal system of things. By their activities as the primary category of social actions in the collective decision-choice space, and guided by an appropriate philosophy and ideology, the information structure is created and processed into a knowledge structure to guide the implementation of categories of social decision-choice actions as the first categorial derivatives from the primary. The appropriate philosophy and ideology in the African case is Philosophical Consciencism with an African content. In other words, the Philosophical Consciencism is derived on the basis of African conceptual tools and from the African experiential information structure. The implemented decision-choice actions become the mechanisms to combine matter and energy to create a second derived category of social actions. The second derived category of social actions becomes the mechanism for the creation of *categorial moments* and *categorial transfer functions* of categories of social polarities. In the last analysis, the people are the backbone of categorial conversion. The persistence of conditions that maintain the current actual is rooted in the people's effort. It is also on the basis of the people's ideological awareness rooted in their experiential information structure, and decision-choice activities with hard work, that the conditions of the actual are destroyed and the conditions of their preferred potential are manufactured for social progress. The people, therefore, constitute the transformative link in any actual-potential social polarity through their decision-choice actions guided by a set of philosophy and ideology which is relevant to their information structure and social vision. Each element is a variety in the sequence of the categories of the decision-choice actions that must be carefully selected and judiciously implemented if the desired result is to be achieved in the social actual-potential polarity.

215

The carefulness and judicious implementation derive their guidance from social philosophy and ideology that provide the thinking framework of individual and social actions. The enveloping of social decision-choice actions constitutes the pathways to the creation of categorial moments and categorial transfer functions. The epistemic guidance of this decision-choice process in general is philosophy and ideology where philosophy molds the preferences over the space elemental categories and ideology molds social actions under the cultural confines of the social set-up. It may be useful to refer to Figure 4.1.1. For clarity, it is useful to emphasize the two interdependent conceptual sub-structures that are the foundations of Nkrumaism and Nkrumah's policy practice. The epistemic interdependency as it has been developed in the volume entitled the theory of Categorial Conversion and in this volume devoted to the theory of Philosophical Consciencism may be made explicit. The theory of categorial conversion establishes the necessary conditions of perpetual transformations of socio-natural objects (varieties) from within and how such transformations take place where the transformations depend on the creation of categorial moments and categorial transfer functions. The categorial moments and the transfer functions constitute the necessary conditions of transformability of varieties as identified by the theory of Categorial Conversion. The theory of Philosophical Consciencism on the other hand establishes the conditions that allow the manufacturing of the categorial moments and the categorial transfer functions to induce categorial conversions to be utilized in game activities of social actual-potential polarities, and where the content of the Philosophical Consciencism is dependent on particular cultures and experiential information structures. The theory of Philosophical Consciencism establishes the sufficient conditions required for perpetual transformations of socio-natural objects (varieties) from within and how the categorial moments and categorial transfer functions are decision-choice manufactured to move the objects in order to satisfy the categorial transversality conditions for conversion. The conditional processes for the management of command-control structure of the socio-natural decision-choice systems of the categorial moments and categorial transfer functions constitute the sufficient conditions of transformability of varieties as indicated by the theory of Philosophical Consciencism, the contents of which are dependent on culture and experiential information structure. The contents of Philosophical Consciencism in the African case depend on African culture and the

African experiential information structure. The implication for Africa is simply embedded in Nkrumah's statement: *It is clear that we must find an African solution to our problems, and that this can only be found in African unity* [R1.200, p. x].

4.3.1 Philosophical Consciencism, the People and the Social Decision-Choice Space

The people constitute the vehicle through which categorial conversion of social actual-potential polarity is manifested through their social action. Each manifestation is realized through an internal antagonistic organization which ensures the bringing together of the individual action at both negative and positive sub-spaces of the general social space. The first general analytical action is to examine the structure of the people in relation to the conditions of the existing social actual-potential polarity. There are two categories of polarity that are of concern under the application of Philosophical Consciencism to the African conditions of modern experiential information. They are the categories of sovereignty and the categories of socio-economic development. The categories of socio-economic development may be realized within any element of the categories of sovereignty. Any category of sovereignty may be expressed as domestic-foreign control polarity waiting for categorial conversion. The domestic-foreign control polarity of sovereignty is such that either the foreign control is the actual and the domestic control is the potential, or the domestic control is the actual and the foreign control is the potential. Given a control of sovereignty, any socio-economic stage within a particular sovereignty (social decision-choice system) resides as underdevelopment-development polarity waiting for categorial conversion. The categorial conversion of the categories of sovereignty control and socio-economic transformation are brought by the people and their activities in the social decision-choice space under the principle of some collective rationality guided by social preferences induced by a democratic organizational structure or a dictatorial organizational structure under a social philosophy and a corresponding ideology.

4.3.1.1 *The General Population Analytics and Different Sovereignties*

Since the people are the backbone of categorial conversion of social polarities, it is necessary to understand its structure through the use of population analytics. This is exactly what Nkrumah did with his work in Ghana under the application and thinking system of Philosophical Consciencism. He also recognized that the collective personality of the population under colonialism with foreign control of domestic sovereignty is different from the needed collective personality of the same population under the decolonized territory with domestic control of sovereignty. The population under colonialism must be organized for decolonization, a claim of domestic control of sovereignty and political power without which nation building in accord with the preferences of the domestic population is impossible. It is here that Nkrumah asserted that *Political power is the inescapable prerequisite to economic and social power* [R1.207,p.78]. For extensive theoretical work on this statement see [R13.8], [R13.9]. Here the goal is the domestic control of national sovereignty. The population after decolonization with domestic control of sovereignty and political power, must also be reorganized for national reconstruction and nation building. The goals here are reconstruction and nation building under the guidance of Philosophical Consciencism with an African content. Nkrumah linked the two. He stated: *Every movement for independence in a colonial situation contains two elements; the demand for political freedom and revolt against poverty and exploitation....Political independence is only a means to an end. Its value lies in its being used to create new economic, social and cultural conditions which colonialism and imperialism have denied us for so long* [R1.207, p.80]. Let us take a look at the population under a colonized, occupied, decolonized or independent state schematized in a geometry of population analytics provided in Figure 4.3.1.

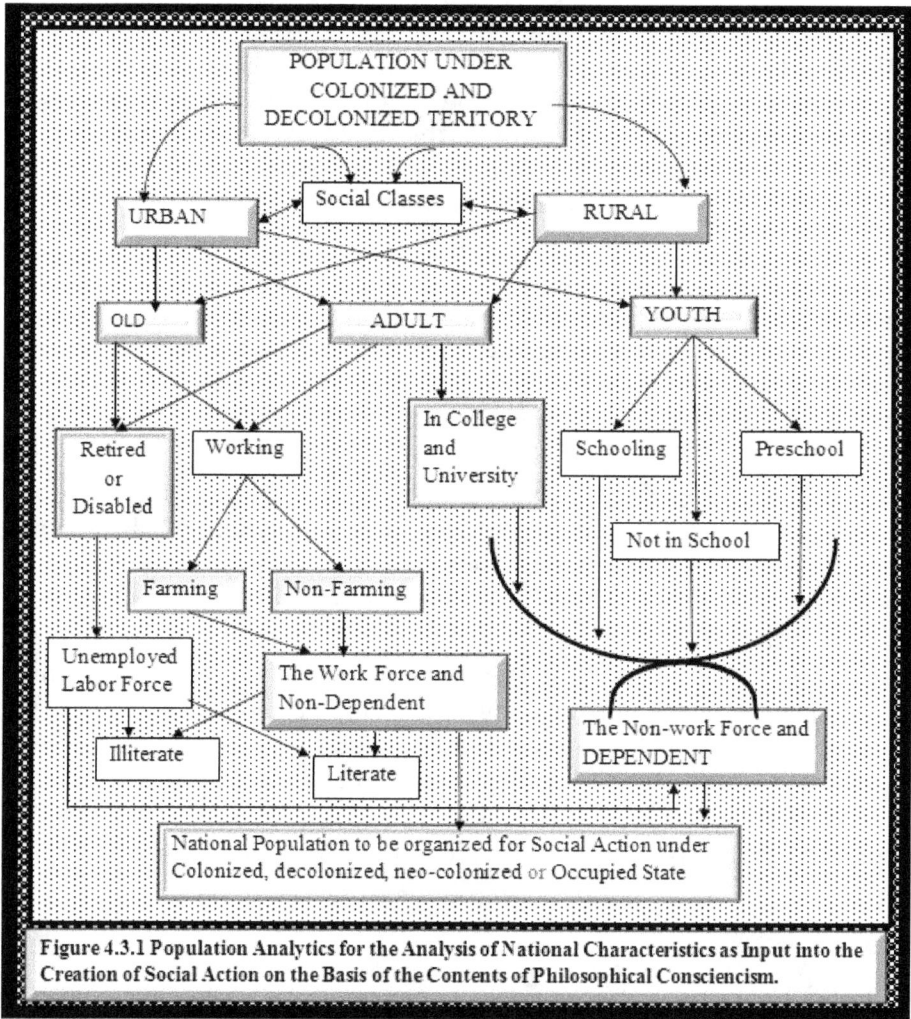

Figure 4.3.1 Population Analytics for the Analysis of National Characteristics as Input into the Creation of Social Action on the Basis of the Contents of Philosophical Consciencism.

4.3.1.2 *The Population Structure and Characteristic Analytics*

From the population analytics emerges another important analytical work in identifying the social and cultural characteristics of the population in relation to each of the two tasks of decolonization and development. This important work is the *characteristics analytics* which will allow the identification of the negative and positive characteristics and then place them in useful analytical sub-sets. The negative and positive analytical subsets are then mapped into the action space where the negative subset is mapped into the space of negative action and the positive sub-set is mapped into the space of positive action in the struggle to determine the direction of the categorial conversion. The resulting direction depends on the manufacturing of a categorial moment and a categorial transfer function. The process of finding the methods and techniques for manufacturing the categorial moment and the transfer function is action *analytics* which separate into violence and nonviolence and combinations of the two which constitute the set of *regime of social action*. The study and use of the connecting paths is the *path analytics* which is presented as a cognitive geometry in Figure 4.3.2.

There are three categories of methods and techniques for manufacturing the categorial moment and categorial transfer function. Each of these categories has social costs and benefits. They also have risks of failure and success which may be computed in the social cost-benefit space. The success or failure of each of these methods and techniques is social-system dependent and in relation to the culture, organizational style, ideological alertness of the society, information-knowledge structure and the commitment of the leadership that may exist or emerge. The choice of the category of social techniques and methods for manufacturing the categorial moment and transfer function will depend on the judicious understanding of the conditions of the social actual, its power distribution, and the social cost-benefit distribution relative to the social potential when actualized. The choice in all circumstances must be under the guidance of Philosophical Consciencism with an appropriate African-derived content in the African case.

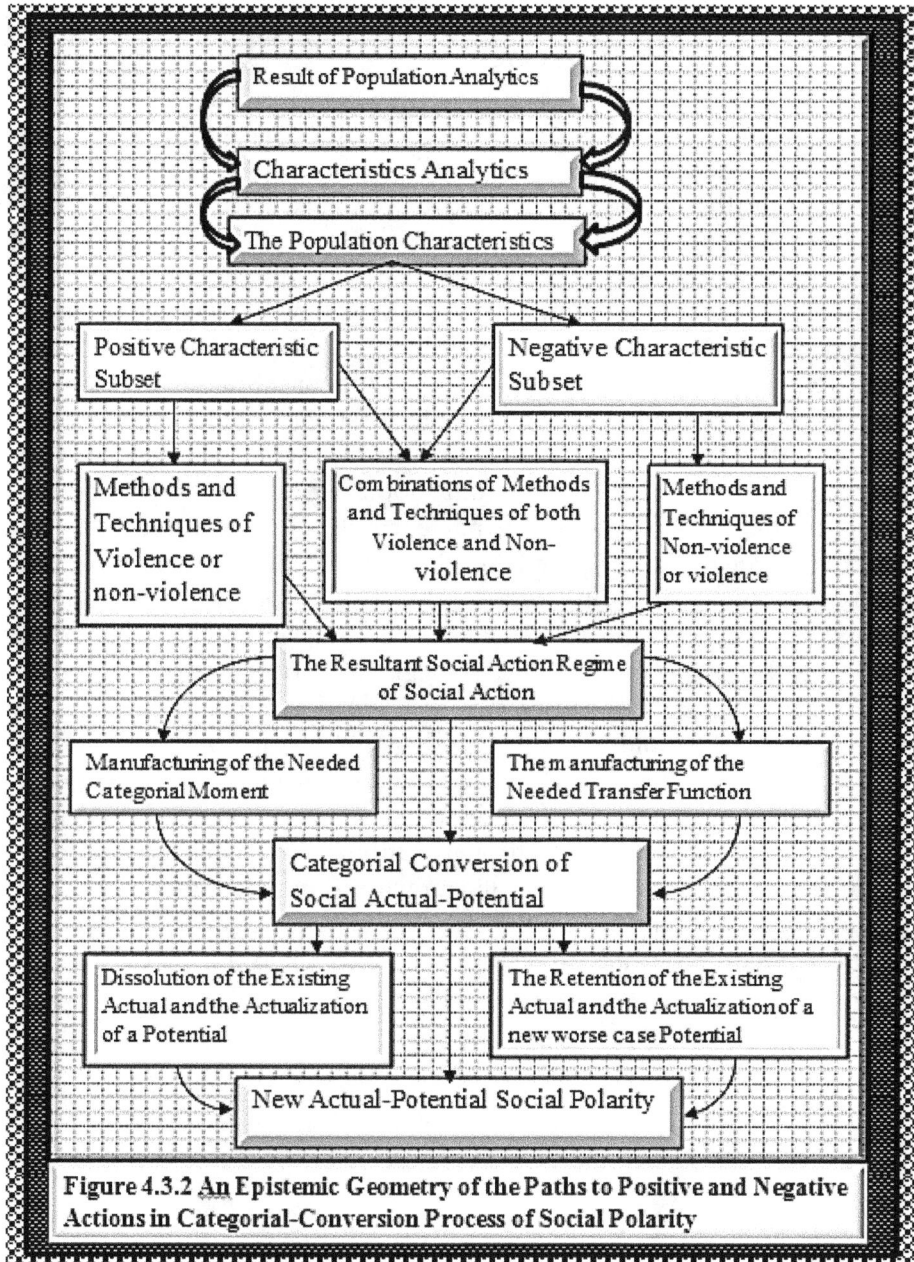

Figure 4.3.2 An Epistemic Geometry of the Paths to Positive and Negative Actions in Categorial-Conversion Process of Social Polarity

4.3.1.3 *The Mobilization of the Population under Positive and Negative Characteristics*

The conflicting roles of positive and negative characteristic sets, the negative and positive social actions, and negative and positive dualities will vary from the nature of social actual-potential polarity. For the case of sovereignty of foreign-domestic polarity, the mobilization and organization of the population will be different from after the struggle for domestic control of sovereignty has been won. The struggle for domestic control of sovereignty involves a decolonization movement which requires different strategies for the organization of the domestic population as compared to the struggle for nation building and socio-economic development within a domestic control of sovereignty. The epistemic geometry of the population mobilization for positive action is shown in Figure 4.3.3. The struggle to decolonize a territory is a collective struggle to acquire the domestic control of the national decision-making process in order to define the destiny of the territory rather than to depend on the generosity of a foreign power to define the historic path of the nation. It a struggle for collective confidence in establishing freedom and justice that are denied through the seizure and occupation of a people and their land by a foreign power in order to exploit human and non-human resources through force, violence, lawlessness and disrespect for democracy under civilized codes of conduct. Here, positive action through national positive characteristic set of the people is called upon to liberate the domestic population from rude conduct of inhumane foreign terror and aggression. The success of the collective decolonization places a new challenge of positive action.

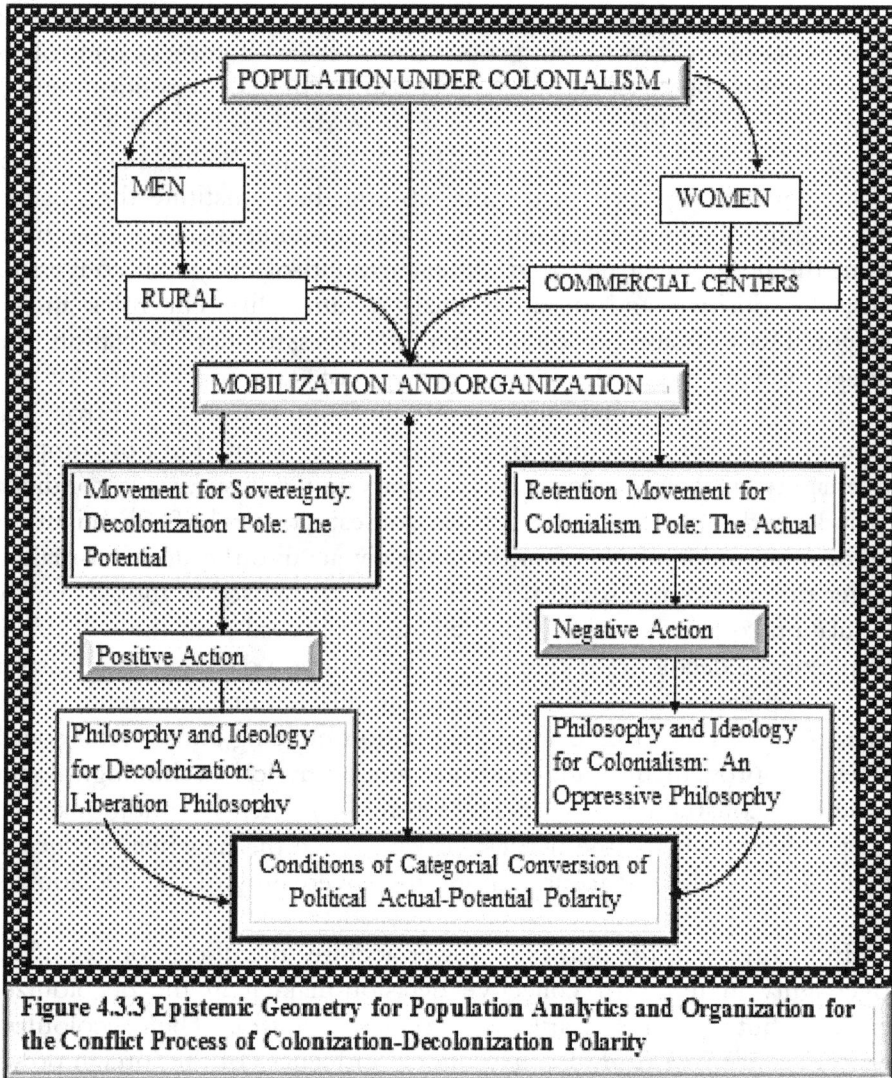

Figure 4.3.3 Epistemic Geometry for Population Analytics and Organization for the Conflict Process of Colonization-Decolonization Polarity

The nation building and socio-economic development involve a social-improvement movement for freedom and justice under domestic control of sovereignty. Both the seeds of complete emancipation and neocolonialism are within the decolonized territory creating an independence-neocolonialism polarity for categorial conversion. This independence-neocolonialism polarity presents more difficult challenges for domestic positive action than during the struggle in the colonialism-decolonization polarity. It must be kept in mind that the political polarity involves the struggle to control sovereignty, and hence, the mobilization of the people for positive action is to claim domestic control of sovereignty through categorial conversion of the socio-political actual-potential polarity. Let us remember that in the logic of the Theory of Categorial Conversion, every pole has a residing duality made up of positive and negative characteristic sub-sets that constitute the positive and negative duals respectively. It is this residing dualities that bring about the negation of negation. At each round of categorial conversion both the positive and negative duals assume different roles in the categorial conversion process. After a successful decolonization, the foreign power loses the direct control of the sovereignty of the decolonized territory but not necessarily loses complete control. The domestic population acquires ownership of sovereignty but not complete sovereignty. Decolonization is a necessary but not sufficient condition for independence. The essence of the domestic control of sovereignty of the decolonized territory is to domestically acquire the decision-making power that allows the social positive action to be used in mobilizing the population to create social improvements in accord with the will of the people of the decolonized territory. This requires a different mobilization of the population in addition to the natural resources to create a new orientation of the positive action for nation building and general social-welfare improvement under various socio-physical technological and resource constraints to manufacture sufficient conditions for independence. This is the hardest process since the negative vestiges of imperialism work hard to undermine the sovereignty of the decolonized territory.

In this respect, the domestic power acquires the direct control of the sovereignty in terms of collective decision-making in the decolonized territory but not necessarily a complete control. The decolonized territory is not free from the institutional influences of the foreign power and its cultural reminisces, and hence cannot claim the status of independence. Decolonization is a necessary condition to do away with direct foreign control of power to decide, but not sufficient condition for independence. The domestic independence on sovereignty must be fought and won at every step in the struggle. Initially, the domestic control of the sovereignty of a decolonized territory is temporally and highly contestable and unstable due to the use of the inherited foreign institutions by the foreign power to destabilize the political and social structure of the decolonized territory and cause chaos and sometimes untold suffering. This foreign destabilization must be fought through dismantling or reducing the foreign power reminisce to ineffective minimum by restricting the strength of the negative action by the negative dual. The process can be done through the development of domestically relevant institutions to replace the colonial institutions, and use these new institutions to develop the material support for the permanency of the domestic ownership of the country's sovereignty. The development of new domestic institution based on different orientation of philosophical Consciencism is to enhance the power of the positive action and reduce the power of negative action in the mobilization of the domestic population. The epistemic geometry of the population mobilization for such a positive action and a reduction of the power of negative is shown in Figure 4.3.4.

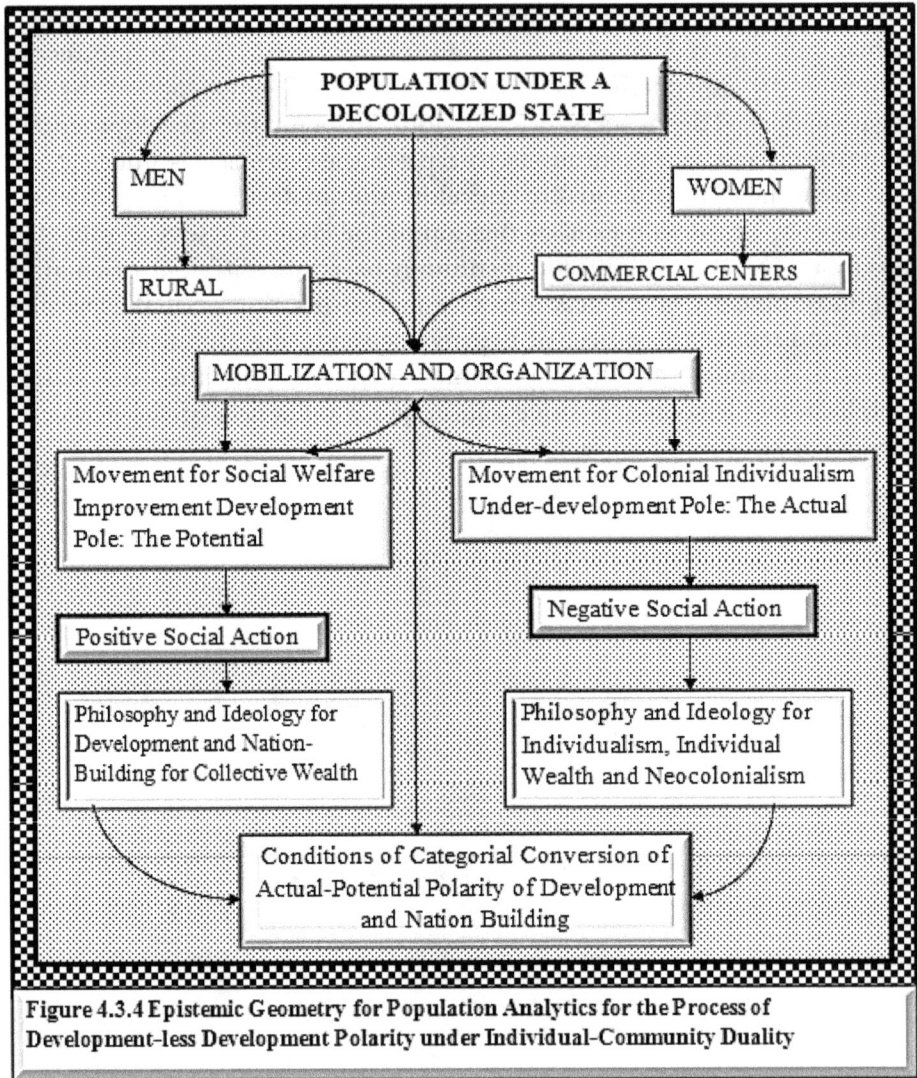

Figure 4.3.4 Epistemic Geometry for Population Analytics for the Process of Development-less Development Polarity under Individual-Community Duality

The essence of the initial and continual domestic control of sovereignty of the decolonized territory is to use the political power invested in it at all times to mobilize the population in order to create social improvements at all levels in accord with the will of the people. This requires a different mobilization of the population in addition to the natural resources to create a new orientation of the positive action for nation building and general social-welfare improvement under various socio-physical technological and resource constraints to manufacture sufficient conditions that will ensure the permanency of domestic control of the sovereignty required for political and economic independence. This is the hardest process since the negative vestiges of imperialism work hard to undermine the creation of political and economic independence of the decolonized territory. The failure to enhance the positive action through the increase of the positive dual to create the material support for the domestic sovereignty will reveal that independence is a failure. The result is a new form of dependency that the negative dual through the negative action intend to produce. The dependency is a politico-economic negation of the sovereignty negation.

The strategies for categorial conversion of socio-economic polarities to create sufficient conditions for independence and complete emancipation become constrained by the strategies of the neocolonialist to create negative action to convert the direct domestic control of sovereignty into indirect foreign control of the sovereignty of the decolonized territory, leading to a client state or a neocolonial state managed by the domestic crony political elite with a neocolonial mindset with little thinking if any. The categorial conversions of socio-economic polarities toward independence in all stages are transformed into a continual categorial conversion of independence-neocolonialism polarity creating a continual uncertainty with risk and danger of losing domestic control to foreign control of sovereignty of the decolonized territory at all future historic points as the categorial-conversion games take place in the actual-potential polarity in the socio-political space for resource and economic control. The greatest problem and transformation difficulties faced by a decolonized territory are the behaviors of the internal cronies who become collaborators of the imperial and neocolonial structures. It is here that an enlightened domestic leadership, equipped with transformative personality developed from the Philosophical Consciencism, becomes a great liberating asset to the country at all levels of the country's life including nation building, institutional casting and material development. It is also, here that culture and relevant education, guided by Philosophical Consciencism equipped with an appropriate content, become effective tools of resistance to the negation and enhance the national reconstruction process. The culture gives the nation the collective confidence. The relevant education decolonizes the mind and molds non-subservient personality encapsulated in freedom and justice.

The risk and danger of loss of domestic control of sovereignty must be seen in terms of loss of domestic resource endowment and the emergence of brutal foreign resource-seeking activities to enhance foreign economies with a continual pauperization of the domestic economy as it was explained in [R1.91]. The conditions of the process of transformation through the active action of the categorial conversion of the socio-economic polarity into categorial conversion of the independence-neocolonialism polarity are found in the existing comfort zone of begging by the decolonized territory and the comfort zone of enticement on the part of the imperialist and neocolonialist to have access to cheap resources. It is the risk and danger in the relational structure of sovereignty and neocolonialism over an enveloping of socioeconomic polarities that drove Nkrumah to the study of neocolonialism and its effect on the domestic control of national sovereignties, especially Africa. After decolonization, the strategies of creating independence require continually solving the problems of categorial conversion of socioeconomic polarities from within on the basis of the internal effort to create positive action from the decolonized territory in order to create sustainable categorial conversions of socio-economic actual-potential polarities. From within the decolonized territory Nkrumah observed:

> When independence has been gained, positive action requires a new orientation away from the sheer destruction of colonialism and towards national reconstruction [R1.203, p105]. He further observed: Progress does not come by itself, neither desire nor time can alone ensure progress. Progress is not a gift, but a victory. To make progress, man [woman] has to work, strive and toil, tame the elements, combat environment, recast institutions, subdue circumstances, and at all times be ideologically alert and awake.

The domestic internal strategies are discussed by Nkrumah in terms of socioeconomic development planning, institutional constructs and mobilization of resources. This further reflects his policy creation and resource mobilization in political, economic and legal structures in support of independence in Ghana. Ghana may be seen as simply a test case for the application of Philosophical Consciencism to categorial conversion of political actual-potential polarities. The mechanisms of the imperialists and neocolonialists to create negative action to arrest foreign control of the domestic sovereignty are outlined by Nkrumah as:

1. Economic control in the form of "aid", "loans", trade and banking.
2. The stranglehold of indigenous economics through vast interlocking multinational corporations with their subsidiaries and affiliates.
3. Political direction through puppet governments;
4. The cultivation of an indigenous bourgeoisie closely linked with the international bourgeoisie to create crony imperialist slave drivers and managers of a-neocolonial state;
5. The imposition of "defence" agreements, and the setting up of military, naval and air basis to control national defense;
6. Ideological propaganda through the mass communications media of press, radio and television – the emphasis being anti-communism;
7. The fomenting of discord between countries and tribes to create internal domestic conflicts for wars and terror that allow justification of humanitarian bombing and interventions;
8. Collective imperialism, such as the politico/military cooperation of the racist minority regimes of central and southern Africa;
9. Activities of intelligence and espionage organizations and international agencies.
10. The sending of "advisers" and "experts" (international bureaucrats) evangelists, Peace Corps etc.

The mechanisms of manufacturing the negative action to create negative categorial moment and negative categorial transfer function by the imperialists and neocolonialists have so far been successful in all African countries. To understand how and why neocolonialists have been successful, it is useful to visit the discussions in [R1.204] [R1.91] with further discussions in [R13.8][R13.9]. At this moment all the decolonized African territories are neocolonial states under the complete control of the Western cooperative imperialism with shared neocolonialist's strategies. The working mechanism of this collective imperialism has been discussed [R1.91], where the cooperative imperialist strategy is the creation of confusion, territorial fires with the slogan of humanitarian militarism, compassionate imperialism, humanitarian bombing and compassionate enslavement under civilization where the victimized neo-colonial countries are blamed for the imperialist actions. These neocolonial strategies have been successful because the African leadership is operating in the intellectual zone of neocolonial philosophy and ideology rather than in the intellectual zone of Philosophical Consciencism with African content that will allow them to rightly conceptualize the complex relationships among independence, sovereignty, collective decision-choice activities, cost-benefit distribution and collective welfare of the national progress. In this this categorial-conversion enveloping of politico-economic polarities, Nkrumah presented a zone of cognition.

> Independence is of the people; it is won by the people for the people. That independence is of the people is admitted by every enlightened theory of sovereignty. That it won by the people is to be seen in the successes of mass movements everywhere. That it is won for the people follows from their ownership of sovereign. The people have not mastered their independence until it has been given a national and social content and purpose that will generate their well-being and uplift [R1.203, pp. 105-106].

The implied essence of Africentric Philosophical Consciencism is the continual creation of cognitive instruments that will be used to intellectually and ideologically seek and destroy the dominating interests created by foreign control on all aspects of the African life by instilling a subservient collective psychology in the people. In this respect, Philosophical Consciencism is an intellectual map to reclaim the psychology of the African people, destroy the colonial mind-set that promotes psychology of dependency, non-thinking personality and reduce the peoples independence into a desert mirage. The basic epistemic result of Philosophical Consciencism is to create a culture of true independence, collective confidence and security of the people where any decolonized African territory will work on the traditional African principle of *each for all and all for each* within the individual-community duality, where there is a continual progressive understanding of necessity for collective actions to set the potential against the actual for individual and collective progress in freedom.

4.3.1.4 *The Sanctions of Actions within Nonviolence-Violence Duality in Categorial Conversion of Socio-political Actual-Potential Polarity of Sovereignty Control*

Most of the African struggles for independence, emancipation, justice and freedom have been against the Western countries and their imperial philosophy of racism and oppression with greed. The Western philosophy of individual imperialism under nationalism has given way to a philosophy of cooperative imperialism under the Western imperial club named NATO [R1.91] This Western imperial club of NATO is nothing but a huge bureaucratic and military system with a machine of injustice and violence that acts as a fearful parasite whose gaming is the exploitation of the militarily weak and resource-rich states, and the containment of existing and emerging rivals. It works with a neocolonial and imperial mindset under the philosophy of violence as political action to justify the terror under the amorphous concept that Western civilization is here to bring democracy and compassionate imperialism only to enslave others for their own good and welfare [R1.69b][1.70b][R1.169a][R1.239][R1.240].

The methods and techniques of the game of categorial conversion under Philosophical Consciencism are of two categorial sanctions as is shown in Figure 4.3.2. They are nonviolence and violence action within a duality in a relational continuum and unity. The sanction of violence is under a negative-positive duality of social action in a relational continuum and unity. Thus, there is a violent positive change or resistance and a nonviolent negative change or resistance. Similarly, the sanction of nonviolence is under a negative-positive duality of social action in relational continuum and unity. There is, therefore, a nonviolent positive change or resistance and a nonviolent negative change or resistance. The energy as the primary category to generate the required power may thus fall under conditions of nonviolence and violence while the power as a derivative of energy may fall under soft and hard categories. Soft power promotes nonviolent sanctions while hard power promotes violent sanctions under violence-nonviolence duality with relational continuum and unity of socio-political action. The most preferred methods and techniques of the Western imperial club are hard power and violence expressed in violent regime changes, military invasion with seizure of political power and coupe d'états. The justification of the choice of violent sanction by the Western imperial club lies in the notion of the principle of military superiority over victims of neocolonialism where defensive military resistance cannot produce severe counter effect. In other words, the benefits of military success far outweigh the costs of success due to power asymmetry. This principle of military superiority under cost-benefit rationality may be used to explain the actions of NATO which is a club of countries with imperial and neocolonial aspirations in the global resource space.

Given the possible existence of power asymmetry between the imperialists and neocolonial states under conditions of categorial conversion of socio-political actual-potential polarities, what methods and techniques must a neocolonial state use in the domestic-foreign control of domestic sovereignty if complete emancipation is sought? This question is extremely important and it is at the center of our contemporary games within socio-political actual potential polarities. This question is implicit and the answer to it is explicit in Nkrumah's Philosophical Consciencism. It is also reflected in the battle to win minds through the methods and techniques of propaganda wars, media control and the creation of a deceptive information machine. Part of the answer to this problem of power asymmetry is to be found in Nkrumah's principle of African unity as a mode of organizing Africa's resistance [R1.202]. The analytical extensions of this African Unity principle and its relationship to the preservation of individual and collective African sovereignties and the global imperial power system are provided in [R1.91]. The relational structure of African Unity, its urgency and global power distribution and structure are yet to be understood by a number of African leaderships. This relational structure is well understood by the Western imperialists and neocolonialists.

The decision-choice actions to settle within a point in the violence-nonviolence duality with continuum and unity in order to bring about desired categorial conversions require   judicious insights into the structural requirements of the game of competing power systems. The game is similar to a pursuit-avoidance dynamic game in an optimal control space. This game problem is complicated by fuzzy-stochastic elements that may lead to unintended circumstances, especially when sanctions of intense violence are used as tools for categorial conversions of socio-political actual-potential polarities. Negative foreign action within the category of either violent or nonviolent sanctions in the activities of social polarities is directed to devolution of domestic power and elimination of the governance by negating the control of domestic sovereignty and placing it in foreign control or in the hands of control of foreign cronies. The process is to establish the capacity to either directly wield effective control of the domestic sovereignty like colonialism, protectorate and politico-military occupation, or indirectly wield effective control of domestic sovereignty like neocolonialism and governance by a selected domestic proxy. The intent of foreign negative action is to control the domestic economic and resource endowment through the control of the legal structure to benefit the foreign country and her people. Here the foreign action is in pursuit of sovereignty control. Positive domestic action within the category of either violent or nonviolent sanction in the activities of socio-political actual-potential polarities is directed toward resisting foreign aggression toward the domestic sovereignty by strengthening the domestic power base. The process is to organize the domestic population and resources to create a capacity to resist foreign devolution of domestic power that will allow them to capture the control of domestic sovereignty. The intent of domestic positive action is to control the domestic economic and resource endowment and her legal structure through the resistance of foreign aggression intended to place the country into the sphere of colonialism, neocolonialism or politico-military occupation.

In the international politico-economic arrangements under some amorphous international legal structure, there is always the presence of the distribution of power asymmetry at the level of sanction of violence of differential order over the distribution of global sovereignties in the game of categorial conversions of socio-political polarities. The maintenance of the Western imperial club is organized under the principle of a bureaucratic and military machinery to maintain superiority of power of violence in the game of categorial conversion of control of sovereignties of militarily weak states. The modus operandi of this imperial club with the power of violence in the game of categorial conversion of politico-economic actual-potential polarities has been discussed in [R1.91]. Given the presence of the power asymmetry in the use of violence sanctions in the game categorial conversion of actual-potential polarities in sovereignty control, it is useful for those going to be involved in national resistance to reflect on the following questions.

1. How can the people of a colonized state initiate categorial conversion of colonialism-decolonization polarity in a way that deals with the problem of national sovereignty-control

and facilitate long-term domestic control of the national sovereignty and socio-economic development through continual protection of the domestic political power and its social decision-choice system, and what kind of sanction of social action must be selected for implementation?

2. Given the success of decolonization, how can a people organize a decolonized society in ways that improve and preserve the national capacity to preserve national sovereignty, remain free and resist aggressions from imperialists and neocolonialists, and what kind of sanction of social action must be selected for implementation?

3. How can the people of decolonized states with weak economic, technical and military power deal with their former colonizers, particularly, when these former colonizers have formed an imperialist club called the Western Nations with military organization called NATO, in a manner that will preserve their sovereignties with development of further resistance to imperial aspirations of their former colonizers and other possible powers, and what kind of sanction of social actions must be selected for implementation?

4. How can the people of a neocolonial state initiate categorial conversion of neocolonialism-independence polarity in a way that deals with the problem of indirect control of their national sovereignty and facilitate a long-term direct domestic control of the national sovereignty, and at the same time bring about socio-economic development through a continual protection of the domestic political power and its social decision-choice system, and what kind of sanction of socialaction must be selected for implementation?

These questions and the corresponding answers are both implicitly and explicitly implied in Philosophical Consciencism, where in the African case, the Philosophical Consciencism must be developed with weapons from the African conceptual system on the basis of African experiential information. The awareness of these questions and possible answers moved Nkrumah to deal with the conditions and meaning of direct positive social action. The development of cost-benefit analytics on the basis of characteristics of violence and nonviolence sanctions will be useful in the decision-choice activities of the methods and techniques.

Because of differential power distribution over sovereignty distribution, different choices may be required depending on individual social conceptual systems and experiential information structures.

4.3.2 Philosophical Consciencism and the Information-Knowledge Requirement for Categorial Conversion

The use of Philosophical Consciencism as a guide in mobilizing the population to implement two categories of categorial conversion of foreign-domestic polarity of sovereignty control and Categorial Conversions of socio-economic polarities have been discussed. Also, as previously discussed in [R3.13], any social information-knowledge system is self-organizing, self-correcting and self-exiting. This information-knowledge system is a production of the social system through the activities in the decision-choice space by decision-choice agents. It is also the support, as well as a subset of the social decision-choice system which provides the power and energy of self-organization and self-correction. In other words, the social information-knowledge system is in relational unity with the social system through interactions of the activities of individuals and the collective in the general decision-choice space in a feedback dynamic process in a relational continuum and unity. It is these relational interactions and the property of the feedback dynamic process that endow the social system, composed of an information-knowledge sub-system and a decision-choice sub-system with the power of error-making, self-correction and self-destruction. It is here that the audacity of curiosity and imagination in thinking, combined with hard work persistence with vision create a space of success in relation to freedom and justice, broadly defined.

Given the social decision-choice structure, every social system has a social information structure from which a social knowledge structure is constructed by the use of some laws of thought derived from the ruling social philosophy and ideology, the Philosophical Consciencism, to support the activities in the decision-knowledge structure. The general quality of the social knowledge structure, given the laws of thought, depends on the quality and quantity of For critical understanding of social information requirements and the role of social information in the categorial-conversion process of social polarities, it is useful to divide the social information structure into a *defective information structure* and a *deceptive information structure* which are related to quantity and quality of information. information and its sources. The philosophical aspect of Philosophical Consciencism guides the knowledge-production process to create relevant personality. When the content of the Philosophical Consciencism is African centered, in the sense of being constructed from African experiential information, the individual and the collective acquire the *African personality* which will give rise to an enlightened intelligentsia. The ideological aspect of Philosophical Consciencism guides the people with African personality to navigate the decision-choice process in order to enhance the positive action over the negative action in the categorial conversion of social polarity. In the last analysis, therefore, the nature and guiding the creation of categorial moments and the implementation of the categorial transfer content of Philosophical Consciencism induce categorial conversions of social polarities by guiding the creation of categorial moments and the implementation of the categorial transfare functions.

4.3.2.1 Categories of Social Information Structure, Propaganda and Categorial Conversion and Directional Control of the Categorial-Conversion Process.

The social information structure for processing in support of decision-choice actions in categorial conversion of social polarities is composed of two important substructures of defective and deceptive information substructures. The defective information substructure is made up of a *limited information sub-structure* and a *vague (fuzzy) information substructure*. The deceptive information sub-structure is made up of a *disinformation sub-structure* and a *misinformation sub-structure*. The vague, disinformation and misinformation sub structures relate to the quality of information and the credibility of the sources. The limited information substructure relates to the quantity (volume) of information. The sum of these information sub-structures creates complexities in the production of social knowledge in support of decision-choice actions to manufacture either positive or negative categorial moments with corresponding transfer functions.

In the decision-choice system under the guide of Philosophical Consciencism for categorial conversions of social actual-potential polarities, the *defective information structures* may be complicated by a second postulate of *deceptive social information structures*. The disinformation and misinformation characteristics enhance the vagueness and the spectrum of the penumbral regions of the social decision-choice activities to bring about a change in the relational structure of the social actual and the social potential. This principle of information structure used in this analysis is consistent with the games of polar conflicts under the principle of uncertainty and opposites with relational continuum and unity, where the resolutions of the polar conflicts are guided by Philosophical Consciencism with different content. It is opposed to either the *principle of the perfect information structure,* or the *principle of the exact probabilistic information structure* used in other decision theories. The perfect information structure is exact and full, while the exact probabilistic information structure is exact and incomplete. These two information structures are the foundations for the *classical paradigm of thought* in explanatory and prescriptive science and knowledge for the study of social dynamics. The defective and deceptive information structures are defined in the *fuzzy spaces* and are handled with fuzzy logic, mathematics and soft computing, where every claimed truth has a supporting falsity to establish duality and conflict with some doubt and a fuzzy measure of doubt. In general, the social information structures available for processing as knowledge inputs into the decision-choice actions for categorial conversion either by the individual or by the collective may be represented in terms of a cognitive geometry as in Figure 4.3.5. The concepts of the defective and deceptive information structures in information-knowledge-decision-choice processes are discussed in [R3.7][R3.10][R3.13][R13.8][R13.9].

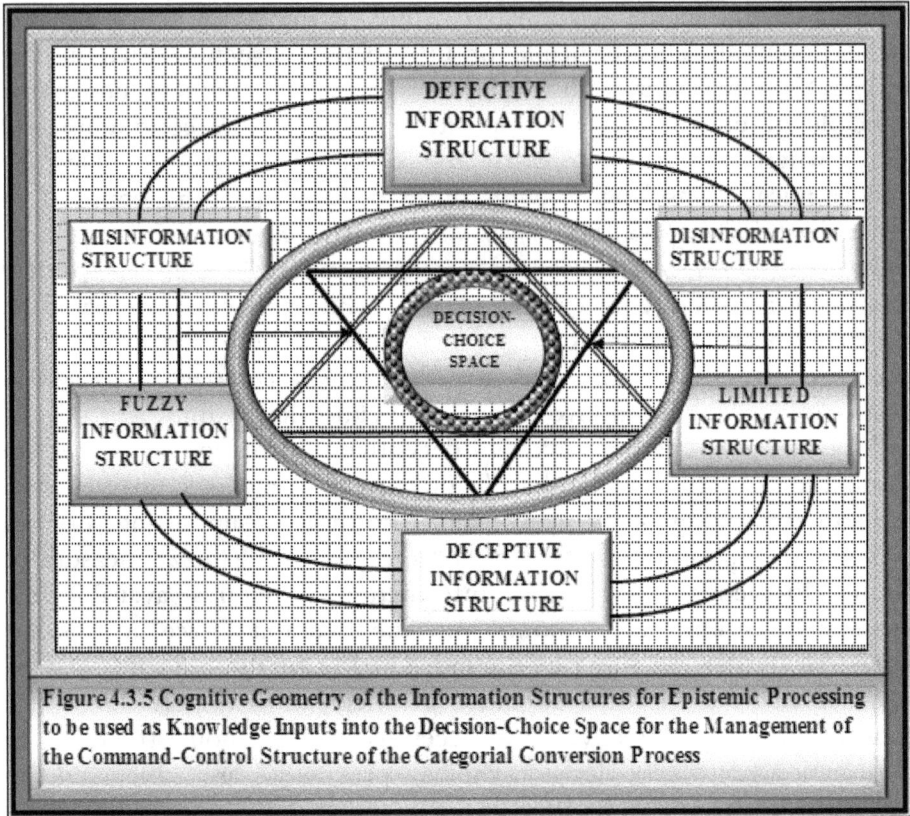

Figure 4.3.5 Cognitive Geometry of the Information Structures for Epistemic Processing to be used as Knowledge Inputs into the Decision-Choice Space for the Management of the Command-Control Structure of the Categorial Conversion Process

Let us keep in mind that the quantitative and qualitative dispositions of the general social information structure give rise to general uncertainty composed of *possibilistic uncertainty* associated with quality of information, and *probabilistic uncertainty* associated with quantity of information in the social knowledge production and the outcomes of the decision-choice actions which will affect the direction of the categorial conversion process of the social actual-potential polarity. For example, these uncertainties will affect whether domestic control of sovereignty will be actualized or foreign control of sovereignty will persist in maintaining a neocolonial state. The general uncertainty gives rise to the general *fuzzy-stochastic risk* which is composed of *fuzzy risk* due to the quality of information and *stochastic risk* due to the quantity of information available in the categorial conversion process. The component of the fuzzy risk is amplified by the presence of the deceptive information structure whether it manifests itself as misinformation, disinformation or both to give rise to *propaganda* of all kinds [R3.7] [R3.13] [R13.8] [R13.9]

Our ability to conceptualize these two types of uncertainties, develop analytical frameworks to capture their essence and translate them into risk computations is the basic and fundamental challenge of Philosophical Consciencism developed from the African experiential information which is intended to assist cognitive computing in categorial conversion to set the positive action against the negative action. These information sub-structures and the relationships that they have to one another, to the social knowledge production and decision-choice systems generate the phenomena of extreme complexity and synergetics with regard to application of Philosophical Consciencism to the categorial-conversion processes in political polarities and socio-economic polarities. The understanding of the relational structures, therefore, will offer an increasing opportunity to understand complex systems of categorial conversion of social actual-potential polarities which must be related to social revolutions and evolutions through the methods and techniques of either violence or non-violence or a combination of both within the violence-non-violence duality in relational continuum and unity. These methods and techniques constitute regimes of social action that may be employed by the negative duality or the positive duality. The prevention of non-violent transformation may always lead to violent transformation. It must be remembered that the mechanism to change the relational structure of the actual-potential polarity is through a system of dualities whose existence is defined by negative and positive characteristic sets while the nature of the relative combination of the negative and positive characteristic sets defines their identities and the category of belonging.

4.3.2.2 *Social Information Structure, Collective Decision-Choice Systems and Command-Control Structure of Social Actions*

Philosophical Consciencism will always influence as well as determine the social information-knowledge structure and the set of individual cost-benefit preferences in any collective decision-choice system in relation to categorial conversion of social actual-potential polarity. The individual preferences are influenced by the cost-benefit calculus of the collective decision-choice system in terms of the relative degree of domestic and foreign control of national sovereignty with cost-benefit flows that are associated with the elements of the decision-choice set under either domestic control or foreign control of the national social decision-choice system. The general information on the cost-benefit characteristics is crucial for correct individual decision-choice actions to participate in creating either positive or negative action to establish the collective outcomes by the application of the principles of categorial conversion. Such participation entails cost-benefit trade-offs. Let us keep in mind that costs and benefits exist as duality in continuum for any social decision-choice action. The outcomes of the categorial-conversion decision-choice system may be distorted completely to produce unintended social decision-choice actions by information manipulation through propaganda holding the social preference set constant. Any information manipulation has the power to change the calculations of individual cost-benefit imputations in favor of the unintended decision-choice actions. The success of information manipulation will depend on the philosophy and ideological system that guide the individual and social cost-benefit calculations. It is here that an adherence to Philosophical Consciencism with African content as proposed by Nkrumah is powerfully useful in all African struggles against neocolonialism and underdevelopment. Information manipulation through the negative pole has the tendency to influence the outcomes of the categorial conversions of social polarities by changing the individual preferences which have been culturally shaped from the beginning in the social environment. Philosophical Consciencism with African content provides the intellectual and cultural power of resistance by providing an alternative logic of information validation and processing to create a useful input to ensure the maintenance of individual preferences in relation to the collective and individual will of African people.

CHAPTER FIVE

PHILOSOPHICAL CONSCIENCISM, LEADERSHIP AND INSTITUTIONS IN CATEGORIAL CONVERSION OF SOCIAL POLARITIES: THE AFRICAN PERSONALITY IN SOCIAL SYSTEMICITY

In Chapter 4, discussions were undertaken on a number of relevant general instruments that help the categorial conversions of social polarities. Among them are the people's actions, information, knowledge and decision. The social information-collection process, the social information-processing capacity, the knowledge input and the decision-choice activities are all influenced and determined by Philosophical Consciencism and its content which together present a justification for a belief system that supports individual convictions to act.

5.1 Belief System and Philosophical Consciencism in Decision-Choice Preocesses

Within this belief system, Philosophical Consciencism holds a position and conviction that great, lasting and sustainable revolutions and the resulting social edifices that may be constructed are the creative works of honest principles of philosophy and ideology, and those who implement them in the social decision-choice space. Such principles acting through cognition are achieved through the transformation of the mental and spiritual personality of the individual and the collective as guided by Philosophical Consciencism. In this respect, Philosophical Consciencism, by establishing firm principles of honesty in action, moral values in conviction and dedication to social service, becomes the moving force for revolutionary change to set the potential against the actual leading to the creation of a new social order which will replace the old. Every social actual has a particular form of collective personality and distribution of individual personalities where such individual personalities find meaning and identity from the collective. Analytically, every revolution is a struggle against the old order and also a contestant for new ones within the potential. It is important to always maintain perseverance of the

241

struggle against the old order and to maintain and increase the social momentum of the positive action after every successful revolt. The problem is that every actual is in competitive struggle against a potential element in categorial conversions in the sense that every problem has a solution and every solution has a new problem in the never-ending process of ontological existence. The successful revolutions are the crafts and creative works of forms of collective personalities and thinking as cultivated from the Philosophical Consciencism.

This thinking system as established by Philosophical Consciencism on the basis of the African experiential information structure while drawing its intellectual weapons from Africa's conceptual system provides a universal justification of African socio-natural rights, responsibilities and duties to self and the African society. The realization of these rights, responsibilities and duties must spring from profound and firm principles that are rooted in the belief system of the major core of Africans from Africa's antiquity where each is responsible for all and all is responsible for each in a relational continuum and unity. The core is constituted by the members of the social decision-making class who must move the African masses on the right path of emancipation through its decisions, choices and actions. This social decision-making core constitutes Africa's leadership to whom we shall now turn our attention. The social decision-choice activities and implemented actions are the products of the personality which the members of the leadership collectively share. The profound and firm principles are those that are being advanced by the basic social logic of Africentricity or African-centeredness with the supporting foundation of Philosophical Consciencism. The philosophical content establishes the rules of reasoning and creates capacity of methods and techniques of social information-processing to create knowledge input for decision-choice actions. The ideological content establishes the justification of unity, African nationalism, a belief system that is self-assuring and actions that are socially self-empowering. Together, these structures are required to empower African leadership and the masses to successfully establish a categorial-conversion process to set the potential against the actual after decolonization and bring the Africa's complete emancipation and African unity into being. The logical structure of the African nationalist agenda is presented in [R1.91].

The objective of this chapter is to relate how the principles held in Philosophical Consciencism with an African content can help to mold the minds within the African leadership to acquire the African personality that will be useful for sustainable categorial conversions as well as struggle against neocolonialism and imperial aspirations of cooperative imperialism of the Western imperialist club and other imperialist advances. It is also to provide an intellectual map to answer the questions posed in Chapter 4 regarding the establishment of social vision, a national goal-objective set and the decision-choice on methods and techniques from the nonviolence-violence duality with a relational continuum and unity. The practice of the principles projected by Philosophical Consciencism should lead to the disintegration of cognitive chains that have held Africans in mental servitude from colonial times to the present, and awaken the sleeping minds of Africans to the painful realities of the global order in which we find ourselves. This sleeping mind is what Nkrumah called the African genius. This African genius resides in the true African personality which expresses itself in a multiplicity of context from the African experiential information [R1.35] [R1.37] [R1.83b] [R1.84]. The principles will equip Africans with a unified system of social reasoning that is internally consistent with the goal and mission of complete emancipation of Africa and her children irrespective of their place of residence. Philosophical Consciencism is intended to redefine the African collective personality since in all social setting the collective personality spins the vector of social changes and hence categorial conversion of social actual-potential polarities. The set of principles of Philosophical Consciencism relates to a broad and general system of ideas whose structures may not be directly obvious to someone who has not followed nor intensely studied Africa's originality, its creative essence, and the conditions of colonial suffering and problems of modern socioeconomic transformation in the global competition for resources.

The objective of Philosophical Consciencism developed on the basis of African experiential information is to take the colonial African and create a new African who thinks as an independent African, acts as an independent African and spread this knowledge to the masses. In this respect, the African leadership that immerges from the masses imbued with the principles of African Philosophical Consciencism becomes the moving force of positive action and hence positive African history. It thus becomes active, not reactive. It becomes independent and a creator instead of subservient and a beggar. It acquires the dynamism of evolution of a kind that sweeps unwanted properties out of its way and replaces the old stubborn mental attitudes of colonialism, slavery and neocolonialism (basic characteristics of current African social settings) with true African essence. It becomes a force of categorial conversion and a model of personality transformation of the neo-colonial masses for sustainable categorial conversion of socio-economic polarities in Africa. The transformation dynamics will bring positive elements for the creation of socio-economic development as well as for Federated and Greater Africa. In such an environment, Philosophical Consciencism will define and establish the path of the transient process creating *categorial transverality* conditions through the stages of Africa's modern history reflecting nation building on African's terms and not otherwise. The practice of the principles embodied in the Philosophical Consciencism will destroy self-doubt and bring to the African collective personality elements which are imbued with the African tradition of creativity, a sense of cognitive courage, confidence and initiative within the general process of Africa's social transformations which are relationally connected to social conscience, African personality and social practice in the collective decision space.

5.1.1 Social Conscience, African Personality, Social Practice and Philosophical Consciencism

The enveloping of categorial conversion of social actual-potential polarities is a connected continuous map of contours of success-failure outcomes of events of individual and collective decisions-choice activities implemented as either positive or negative actions that people undertake in the process of pursuing their survival, comfort and happiness. The decision-choice activities and implemented actions rest on collective visions and the social personality of the masses from which leadership emerges. The collective decision-choice activities constitute active cognitive processes over a potential that must be actualized by collective positive actions on principles of optimism, hope and hard work and guided by Philosophical Consciencism. The map of the decision-choice contours reflects the personality characteristics of the social collectivity which has taken hold as the operational paradigm in the society, for information collection, processing, and use in the acts of deliberation leading to collective and individual decisions on either positive or negative action. Categorial conversions of social actual-potential polarities are outcomes of Philosophical Consciencism from which the collective personality emerges. The collective personality is the determinant of categorial conversion on the basis of ingrained social consciousness that defines its contents and quality. The social consciousness is derived from and affected by Philosophical Consciencism. The path of the ensuing categorial conversion is shaped by the collective personality of the society. This collective personality operates through the decision-information-interactive processes which assert the resultant direction of conflicting preferences of social forces of decisions and choices in the social and resource spaces. It is here that the structure of people's participation becomes important. It is also here that a democratic decision-choice system or dictatorial decision-choice system finds content and meaning, and social conscience acquires the power of dialectics of action and categorial conversion in transformations of social actual-potential polarities.

The social consciousness that emerges out of Philosophical Consciencism and exerts the preponderating effects on the direction of social decision-choice activities, as well as spins the space of social actions in the positive-negative duality, is the philosophical and ideological system of views of the society, its internal and external relations, logical justification of social actions and the code of conduct. For Africa's social progress to proceed in accord with African preferences, this social consciousness after decolonization must be derived from the totality of African experiential information and culture that define the continuum and unity of Africa's tradition before and after colonialism. Under colonialism the mental make-up has largely been shaped by a system of colonial education intended for subservience and control in support of the resource interest of the imperial system of Western Europe. The masses emerge with this colonial mindset which was intended to maintain the vestiges of colonialism after decolonization. The intention of Philosophical Consciencism, as a system of thought with an African content, is to destroy the incompatible components of imperialist anti-African philosophy and ideology as well as decolonize the system of education, in order to bring about devolution of the power of the colonial mindset that creates negative actions opposed to Africa's positive categorial conversion to an ineffective size. Let us keep in mind that in categorial conversion of actual-potential polarities, no power can be reduced to zero and the devolution process is simply a limiting process to an ineffective size that requires actions of containment.

In the African social decision-choice space, Philosophical Consciencism with an African content must replace the colonial system of views and ideology that have taken hold and de-Africanized the African masses and the collective personality required to act in the best interest of Africa during the period of colonialism. It must also solve the cancerous disease of the neocolonial mindset that is consuming the traditional wisdom, art of thinking and creativity of African leadership at all fronts of African endeavor. It must re-Africanize the mindset and the collective African personality. Philosophical Consciencism must provide a framework of the modern African thought in unity and continuum of African tradition, as well as serve to simplify behavior that affects contemporary African progress in such a way as to provide a degree of societal optimism and hope to the extent that it must win cognitive authority over the African masses and their leaders, and move them to see their decision-choice actions in terms of African collective interests and social vision and not simply in terms of individual interests that violate the conditions, where both collective and individual actions are inter-supportive in continuum and unity within the individual-community duality under the African organizational principle of each for all and all for each in the continual progress of the community in a manner that reduces social tension. Let us keep in mind that the individual-community duality is an instrument to bring about categorial conversion of social actual-potential polarity. It is within this context of Philosophical Consciencism that great African minds will emerge, and great leadership will arise to create great institutions and great policies needed to cultivate conditions of positive categorial conversion of social actual-potential polarities. Here, sustainability of the production of great minds is a necessary requirement for continual creation of great institutions through which progressive positive actions can be implemented against retrogressive negative actions to create categorial moments and categorial transfer functions over all social actual-potential polarities as time proceeds into infinity. The relational structure from great minds to categorial conversion is presented as an epistemic geometry in Figure 5.1.1. It must be kept in mind that over all points of categorial enveloping of actual-potential social polarities, that great minds constitute the primary category from which great institutions emerge as derived categories of being to further produce great minds as secondary derived categories in the process of continual and sustainable social transformations.

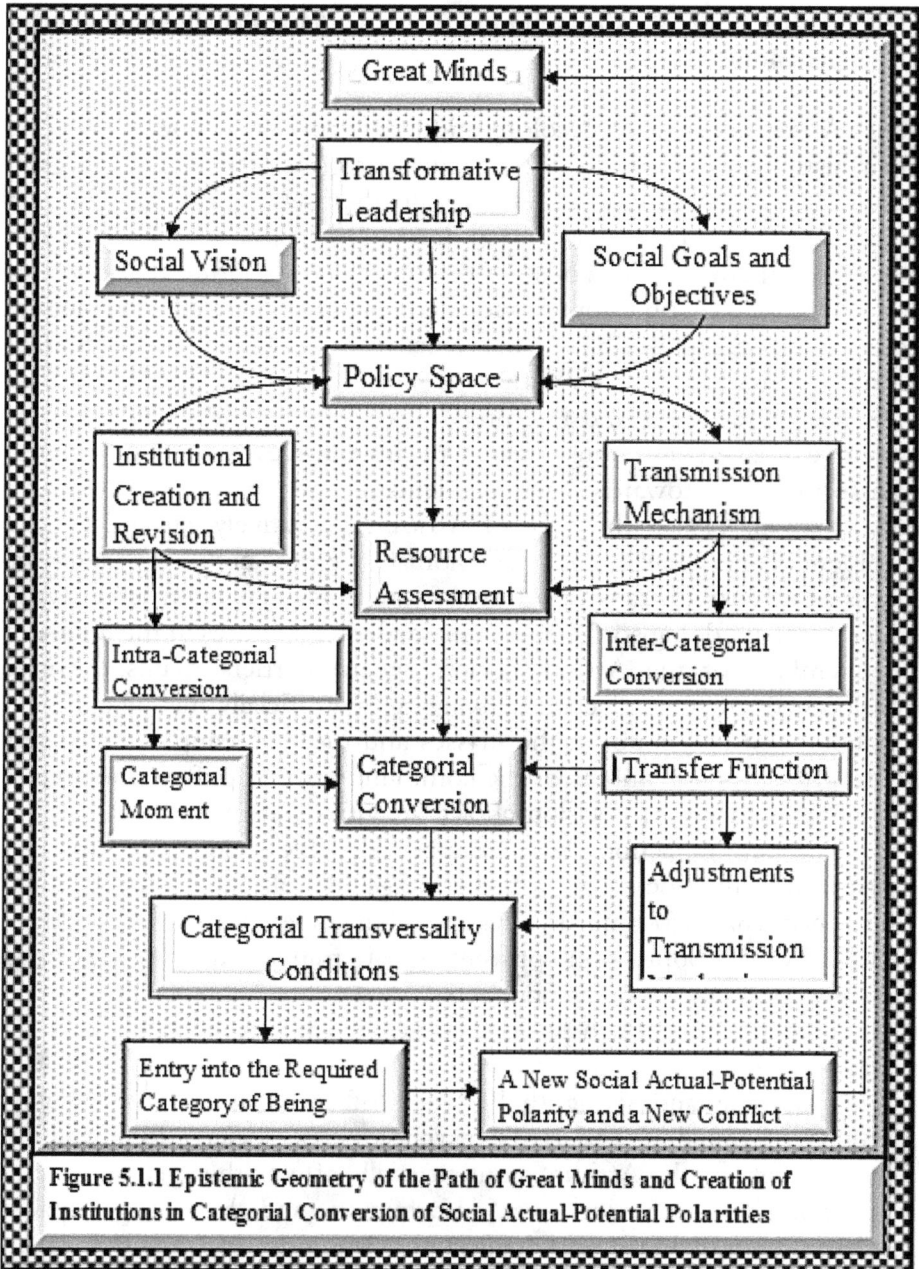

Figure 5.1.1 Epistemic Geometry of the Path of Great Minds and Creation of Institutions in Categorial Conversion of Social Actual-Potential Polarities

5.1.2 Philosophical Consciencism and Africa's Needs Under Decolonization

What Africa needs after decolonization is simply great minds to create great institutions which are the products of great leadership with great minds. Great institutions do not create themselves and they cannot, neither do they create social vision and social goals and objectives nor implement them. Great minds are the products of culture that contains the socio-political philosophy and ideology of action. The ideological component of Philosophical Consciencism, by defining the content of African individual and collective personality, presets the social and individual perceptions and interpretations of facts, truths and reality as cognitive images, and thus creates positive forces for positive social transformation towards independence and complete Africa's emancipation. This is the need of Africa. Alternatively, an anti-African ideology of imperialist and neocolonialist types negates progressive forces as it seeks to transform true African personality from *what it is* to *what it is not.* By doing so the anti-African ideology creates in the minds of the members of the African society special barriers against positive development of the collective personality by artificially manufacturing a psychology of self-doubt in the masses and the leadership. The barriers act to prevent the members of the community from uplifting themselves to the higher level of decision-choice activities , even when the higher levels of social action has become possible and necessary for transforming the society to a higher stage of collective African good. This is not what Africa needs. The neocolonial ideology and anti-African philosophy of oppression prevent great minds and leaders from emerging and by so doing indirectly prevent the emergence of great institutions through which positive social actions can be transmitted for positive categorial conversions.

By defining and fixing the content of the collective personality, methods of thinking, and social psychology of society, Philosophical Consciencism with African content will affect the general social perception and behavior over information and knowledge spaces relevant to individual and collective decisions in the individual-community duality in support of the collective good. At the level of perception, the ideological component of Philosophical Consciencism presents itself as the establishment of the African interest where such interest is taken as a source-free knowledge for social truth such as

African nationalism and hence non-corruptible to ensure Africa's self-confidence as a basic need for her survival and progress. At the level of explanation of social events and outcomes, the philosophical component of Philosophical Consciencism presents itself as a simplified, consistent and self-contained general social theory to the masses and African leadership for analysis and understanding, thus ensuring Africa's need to understand herself and her relationship to the activities of neocolonialism and imperialism on the basis of Africa's experiential information structure, to prevent the emergence of a new system of slavery. At the level of social and individual decisions, choices and practices, Philosophical Consciencism will serve as a guide to social and individual action or inaction thus satisfying Africa's action-needs. At the level of social organization and institutional creation, Philosophical Consciencism presents itself as a unifying force of the members of the African society for positive social actions required to actualize the potential as Africa's need for social transformations. Here, Philosophical Consciencism presents a distinction between the structure of a socio-political organization of social forces on one hand and the structural system of techniques methods of struggle and sanctions on the other hand. Violence and nonviolence are techniques and methods of struggle in the games of categorial conversions of actual-potential socio-political polarities. Both methods and techniques are available to the contestants, who are the ones seeking to preserve the actual and the ones seeking to dissolve the actual in order to bring in a potential within the actual pole. Generally, the ultimate sanction preferred to preserve the socio-political actual under categorial conversion is violence. The nature and existence of power asymmetry forces those seeking for change to select a non-violence sanction as the technique for playing the categorial-conversion game except when such a technique is made impossible.

At the level of Africa's history, Philosophical Consciencism presents itself as a creator of a new African who sees and interprets African experiential information in terms of Africa's socio-economic needs. At the level of categorial conversions of social polarities, Philosophical Consciencism acts as the optimal controller of the development of great minds and creation of great leaders that Africa needs over the trajectory path of Africa's social history. In general, therefore, Philosophical Consciencism seeks to create harmony and preserve stability of Africa's politico-social dynamics in a complex process of continual categorial conversion of social actual-potential polarities under the principles of relational continuum and unity for the Africa's need to uplift the masses of African society. Analytically, if Philosophical Consciencism derives its weapons from the environment of the African experiential information structure and is African-centered, then Africa's need of great minds and great institutions will by logical necessity be met for the needed categorial conversion of actual-potential social polarities and bizarre behavior in decision-choice spaces at the global stage will be avoided.

By changing the nature of the collective thinking and the mode and content of education, one can alter the course of national history through a restructuring of the behavioral foundation of information processing, decisions and choices that affect the social manufacturing of categorial moments and categorial transfer functions in the continual battles between the actual and the potential in the social actual-potential polarities leading to the establishment of new African needs and increasing ascension onto higher planes of independence and social progress not in terms of neocolonial interests but in terms of African interests and welfare. There is an unquestionable need for the African collective personality to be transformed by fundamentally altering the colonial and neocolonial mindset that has crystalized subservience as an acceptable behavior. It is this need of re-establishing the African genius, the true African personality and the African essence that brings into focus the essential need of Philosophical Consciencism the objectives of which are to reshape Africa's collective rationality, mode of reasoning, redefinition of appropriate social behaviors, the peoples' perceptions and preferences, and the manner in which information is accepted and processed in the African social decision-choice spaces of all social activities and events. This need of Philosophical Consciencism is a sufficient condition for positive categorial conversion and progressive social transformation of African societies.

The Philosophical Consciencism providing Africa's social thinking system which consists of customs, rules, laws, and religious ideas with African roots and developed into laws of thought will define the boundaries of acceptable choices and decisions as well as establish broad outlines of socially approved collective and individual behavior and action in the social setup. It will also guide social practices in all endeavors of life and resolutions of conflicts in individual-collective duality in favor of Africa's collective interest and social vision. A successful positive categorial conversion in the course of African social history requires first and foremost the creation and establishment of a new ideology contained in Philosophical Consciencism that militates against the oppressive ideology of colonialism. It must acquire a destructive force against the existing neocolonial one that strips off Africa's dignity in our contemporary times. This new ideology with its embodied new African personality for Africa's social transformation must completely replace the colonial and neocolonial ideology and the corresponding neocolonial collective personality. The development of the inherent basic structure of Philosophical Consciencism must establish new paradigms of African social thought, define new rules of social practice and establish cognitive boundaries in which social decisions are made and social truths are verified, validated and translated into social action. In the context of changing the course of contemporary Africa and its modern history, this new ideology embodied in Philosophical Consciencism is African nationalism with its defined new African personality which projects decolonization and a self-reliance rather than decolonization with a dependency syndrome. The rules of reasoning for the Eurocentric anti-African ideology of racism and inhumanity with its philosophical support of Africa's disunity that have evolved from the period of colonialism to the present period of neocolonialism must be logically discredited and completely destroyed.

The ideology of African nationalism embodied in Philosophical Consciencism will establish a new conceptual framework from which a new African must be born. This new African will be equipped with a new African personality that is to be molded and guided to perfection by the principles of Philosophical Consciencism in accord with the vision of *African nationalists*. In other words, a new African is created and stamped *Made in Africa* by the conceptual system of Philosophical Consciencism. The new African, equipped with a personality molded by Philosophical Consciencism, will see Africa in the light of Africa and her children; will see Africa in the light of collective Africa's progress, welfare and interest within a greater Africa, and Greater Africa as the association of ethnic groups and states welded together by a common African vision that is relationally linked to Africa's collective mission of complete emancipation in unity, justice, peace, liberty, compassion and love among Africans irrespective of their place of residence.

At the core of the ideological component of Philosophical Consciencism is a set of principles to be practiced. They are African traditions, African nationalism, African unity, and African self-reliance, self-motion and self-transformation at the levels of quantity and quality.

251

These instruments are essential to create positive action to bring about independence in African states and to restrain neocolonialism and imperial aspirations of the Western imperial club and any other imperialist advances. The African self-reliance, self-transformation and self-motion find epistemic justification in the theory of categorial conversion, while the African traditions, African nationalism and African unity find their intellectual support from the theory of Philosophical Consciencism which projects a philosophy and ideology of Africa's complete freedom. The adherence to the practice of these principles is an important and difficult requirement for the new African under the conditions of decolonization with an umbilical cord still attached to the imperial system of oppression and terror, but it is an imperative one and the only way to freedom and justice for all Africans. It must be noted that both freedom and justice are defined within the cost-benefit duality where freedom and justice are benefits the cost-supports of which are derived from the sweat of the people. The benefits of freedom and justice cannot be separated from the costs of freedom and justice in such a way that one can abstract the benefits and do away with the costs. They are just like the *Asantrofi-anoma* problems which have been discussed in greater detail in [R1.92][R15.33]. These problems emerge from the universal principles of opposites from Africa's classical antiquity. The analytical definitions and solutions of such problems, implied in Philosophical Consciencism, lie in the application of the fuzzy logical paradigm composing of its logic and mathematical form [R4] [R4.17][R4.18] [R4.29][R4.37][R4.38][R4.41][R4.48][R5.3].

The task before us, as Africans, is not only the drive to create a united Africa, but also to build and market her as great and powerful in a manner worthy of her glorious past, and to create and bring awareness of her mission of freedom, justice and duty of protection for all her children both at home and abroad. In the light of this task, all African states must embrace the highest principle of African nationalism regarding Africa's self-independence and self-governance. Self-independence and self-governance imply self-reliance and self-confidence, without which self-independence and self-governance have no meaning. Lack of self-independence and self-governance implies some form of slavery the degree of intensity and suffering of which will depend on the ruling social relations within production-consumption duality as well as cost-benefit duality. The bringing into being of self-independence and self-governance which are not mirages in the desert requires African collective consciousness from within the African personality. The collective African personality must be such that the African people in each state must resist any form of government imposed on them from without. This includes direct and indirect imposition of irrelevant governments such as colonialism, neocolonialism and crony imperial occupational governments.

The resistance to such oppressive forms of indirect slavery requires the emergence of an enlightened leadership from within African decolonized territories. This enlightened leadership must work with the principles of *Asantrofi-anoma* that reflect conditions within the cost-benefit duality where such conditions present the notion that every decision-choice point is reflected by a cost-benefit composite value. The enlightened leadership in any decolonized territory must rethink the nature of foreign assistance or any domestic promises of fast relief from their immediate suffering by critically analyzing the impact of such relief on Africa's organic goal of complete emancipation and independence. In addition, the members of such leadership should not allow foreign military and intelligence-gathering bases on the African soil. The acceptance of such bases is the work of the colonial, neocolonial and dependency personality produced by venomous anti-African ideology and racism encapsulated in pleasant tasting pills which are deceptively destructive. Any policy of allowing foreign bases on African soil would be anti-African and contrary to the social logic for Africa's independence and sustainable development.

The African states that have servilely acquiesced to foreign bases and are consumed in the beggar's syndrome of the devastating loan-aid paradigm under neocolonial deceptive strategies must understand that these behaviors do not in any form or shape support Africa's emancipation, neither do they support their own proud survival, security and sovereignty. These types of behavior are admonished by African-centered thinking. They stem from the lack of African consciousness and personality which are the products of Philosophical Consciencism. What Africa needs is to generate positive actions from within herself and at all fronts, and to realize that development is a victory that presents a social benefit. This social benefit has its cost support in the sense that the benefit cannot be realized without the cost and the cost cannot be borne by any other people other than the Africans. The cost must come from the sweat of the African people's brow. Independence is a social benefit and the social cost of such benefit is hard work. The joy of such hard work is liberty which has no existence in a beggar's syndrome. This is affirmed by the *Asantrofi-anoma* principle where every decision-choice element contains both cost and benefit in an inseparable relational continuum and unity and one cannot select the benefit and leave the cost [R4.17][R4.18].

5.2 Philosophical Consciencism, Freedom, Justice and Necessity

Africa is at the crossroads of freedom, justice and necessity in the space of wicked pulling thorns of neocolonialism and foreign imperial aspirations of the Western imperial club that seek to dictate Africa's social decisions and her course of history. She must take the bull by the horn and choose the path that serves her interests. This path of Africa's interest involves two pyramidal structures imposed on each other with a relational central-action system. The first pyramidal structure is decision, choice, and optimality. The second pyramidal structure is freedom, necessity and justice derived in the protected womb of Africa's antiquity. The central-action of the pyramidal logical system may be operated either under the principle of Philosophical Consciencism or under the principles of Eurocentric neocolonialism that have competing claims after decolonization. The epistemic geometry of the relational structure of the two pyramidal interactive systems from which a path of African nation building after decolonization must be abstracted under Africa's will and collective preferences is shown in Figure 5.2.1

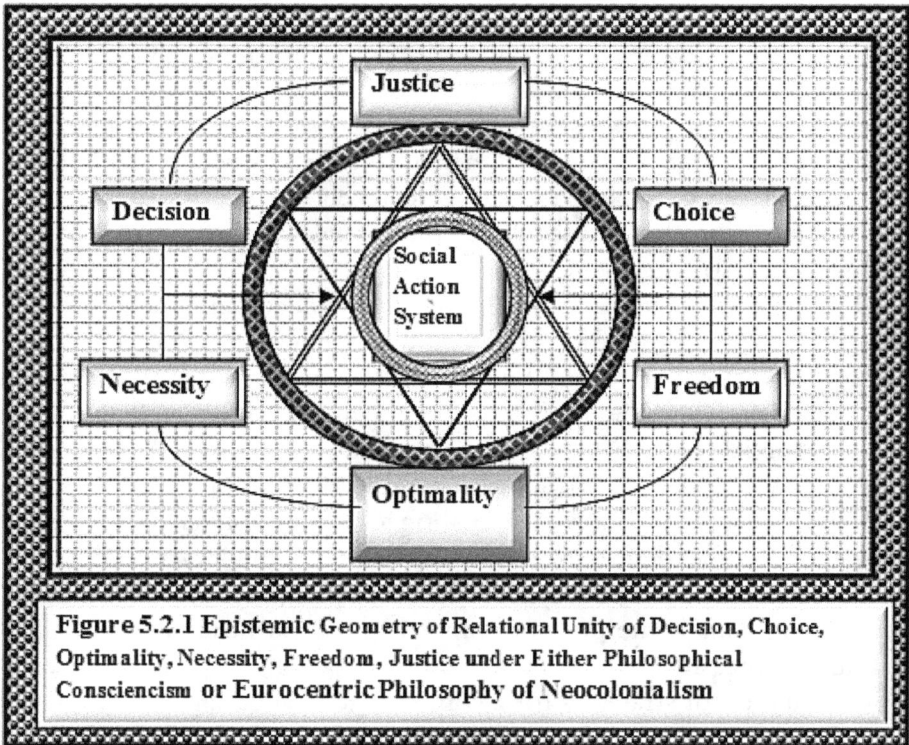

Figure 5.2.1 Epistemic Geometry of Relational Unity of Decision, Choice, Optimality, Necessity, Freedom, Justice under Either Philosophical Consciencism or Eurocentric Philosophy of Neocolonialism

The principles implied by Philosophical Consciencism assert the *necessity* that requires Africa to break away from the old ties to the imperial system of African subjugation where African human and non-human resources were abstracted by the Europeans to service the growth and development of other lands in Europe, South America, North America and Caribbean Islands, to create an African demise and enhance the suffering of her children over the global system of a production-consumption relation and distribution-poverty structure. This is a necessity if Africa wants true freedom and justice. This is the position implied in Philosophical Consciencism as a liberation philosophy and ideology required for the implementation of strategies for the continual success of categorial conversion over the preferred enveloping of nation building in accord with Africa's social vision and national interest. The sufficient conditions for Africa's progress and internal success are the practice of the principles embodied in Philosophical Consciencism under the conditions of appropriate goal-objective formation and internal hard work. The principles embodied in Philosophical Consciencism must be

255

fully believed by those Africans who seek African emancipation and progress, and cannot be sacrificed in any form or shape. Their development and application must be drawn from the African conceptual system and must always keep the vision of Africa's complete emancipation at the forefront of Africa's history. International begging and the carrying of a beggar's basket to receive a gift of water are not part of the principles derived from the intellectual order projected by Philosophical Consciencism.

The principles implied by Eurocentric neocolonialism assert the necessity that requires Africa to be tied to the imperial system of African subjugation where African human and non-human resources will continue to be abstracted by Europeans for the continual service of their socioeconomic growth and development. The implied principles constitute a *necessity* for Africa's lack of true freedom and justice where such qualitative dispositions are enshrined in the golden house of the mirage of independence, freedom and justice. The practice of these principles is also a necessity for the triumph of neocolonialism and imperialism. This is the position implied by Eurocentric neocolonialism as an oppressive philosophy and ideology required for the implementation of strategies for a continual resistance against success of categorial conversion of actual-potential social polarities over the preferred enveloping of Africa's nation building in accord with Africa's social vision and national interest. Under these principles of Eurocentric neocolonialism Africa's socio-economic development is arrested as an appendage to European imperial order and domination supported by a complex regime of violence. Examples of the use of such sanctions of violence are plentiful in Africa and other parts of our globe. The sufficient condition for the growth of neocolonialism is the practice of the implied principles of Eurocentric philosophy and ideology of neocolonialism and imperialism under the ideology of democracy, civilization, Christianization, violence, force and military superiority.

Decisions find expressions in necessity, and choices find expressions in freedom while freedom is expressed within the philosophy and ideology that define the assessment structure of cost-benefit configuration of decision-choice outcomes. Necessity and freedom reside in the categorial-conversion process while freedom and justice reside in decision-choice activities within respective dualities, all of which define the categorial-conversion process toward setting the potential against an unwanted actual in the enveloping outcomes of the social cost-benefit space. The sufficient conditions of the actualization of a potential are also the convertibility conditions that specify the categorial transversality conditions of Nkrumah's Delta. The decolonized African territories have before them two sets of philosophical and ideological principles to use within the social decision-choice space under cost-benefit principles. They are Philosophical Consciencism and Eurocentric neocolonialism

Some of the strategies and techniques used by neocolonialists for creating the sufficient conditions to bring about a neocolonial state are listed by Nkrumah as:

a)   To produce a small educated African "elite" as prospective rulers, whether or not they have the support of the masses.

b)   To educate this "elite" so that they would automatically accept, as part of the natural order of things, the colonial relationship, and defend it in the name of "justice", "political liberty" and "democracy".

c)   To prevent by organisational and ideological means, any concentration of power, without which change is impossible.

d)   While paying lip service to democracy, to exclude by organisational and ideological methods, the representative of the mass of the people for any real control over the State.

e)   To exclude, by all possible means, any teaching which might lead to the advancement and practice of revolutionary ideas

These are the political techniques which neo-colonialism is employing in order to tighten its economic control of the territory through a puppet "elite"

The economics of neo-colonialism is obvious. It gives fake aid to the newly independent country which makes that country virtually dependent economically on the colonial power. Thus it becomes a client state of the colonial power, serving as the producer of raw material, the price of which is determined by the colonial power [R1.214, Vol. 3, pp. 12-13].

5.2.1 Philosophical Consciencism, African Leadership and African Challenges in Social Systemicity

Given these neocolonial strategies to convert any decolonized territory (the actual) into a neocolonial state (a potential) in the categorial-conversion process, Nkrumah asked the question: *How do we proceed then*[R1.214, Vol. 3, p.13] This was the great question of his time, and it still remains the great question that the leadership of every African decolonized and neo-colonized state must reflect on and ask the further question: What is the meaning of independence of a nation and how does this independence relate to freedom and justice? Furthermore, another question still needs to be asked: What is the relational structure established among colonialism, neocolonialism, economic aid, neo-slavery and subservience? The answers to these questions are to be found at the conceptual level in the philosophy and ideology that are embodied in Philosophical Consciencism. At the level of practice, the Africans, especially the leadership must implement the principles embodied in the ideology of African nationalism and its philosophical construct contained in Philosophical Consciencism. This is the continual challenge to Africa and her leadership to reclaim Africa, her freedom and justice from the oppressive political economy of European imperialism.

Ironically, and in a mockery of independence with complete neglect of Africa's modern history of European atrocities, human terror and societal destabilization during the period of the colonial wars and the struggle to reclaim Africa's independence, freedom and justice, some African states, like those that have allowed foreign bases to be established on their territory, manage to bring foreign foxes as the doorkeepers of the chicken houses into Africa while the clock of her destiny slowly ticks with an amazing familiarity and regularity to neo-slavery, destruction and increasing suffering under the deceptive banner of democracy with dubious economic aid, deceptive technical assistance and human right claims complicated by humanitarian military violence to kill the people, give democracy to the dead and bring justice and freedom to the non-living . National security and national freedom are two sides of the same coin. One finds national security in national freedom, and national freedom in national security as affirmed by the central principles of Philosophical Consciencism.

By allowing foreign bases of any kind on an African territory is equivalent to trading off national freedom for illusions of security, and hence the leadership of that country has forfeited both the national freedom and security of the people of that country as well as that of Africa. The structure of the current African ties with imperial Europe reflects a situation of cognitive corruption, moral bankruptcy, visionary darkness, intellectual sterility, slave mentality and political subservience among the majority of the members of the African political and intellectual "elite" that places them in the zone of neocolonial cognitive imbecility strongly objected by the basic core of Philosophical Consciencism as the revolutionary African philosophy and ideology for decolonization and development. These situations and decisions that come with them reflect a schizophrenic personality that is not African which is what Philosophical Consciencism stands for. It is simply a product of mis-education and propagation of anti-African ideology that has molded the members of the African "elite" and leadership into a different strain of Africans trademarked *made in the West* and enslaved by a neocolonial mindset, as well as boxed in the dark wall of the familiar colonial and neocolonial intellectual order.

In these dark walls of the familiar, the African *elites* have taken refuge in the slave quarters of epistemic subservience in which common sense has been rendered a zone of least imagination and have learned to regard the West as the best and Africa as primitive. They actively promote this idea to the masses in their churches, schools, universities, institutions of governance and the likes. They dismantle true African institutions (for example the Kwame Nkrumah Ideological Institute) and replace them with institutions of African demise as prescribed by their neocolonial masters.

They fail to acknowledge a basic fact in the international political economy, which is, that the world is a jungle with its own laws of wickedness for which the events in Africa regarding, slavery, cultural thievery and colonialism, to name a few, are historical testimonies. These historical testimonies cannot be forgotten by any African with an epistemic sense of Africa in global history, especially the Africa's relationship with Europe and the West. The nature of the laws of the jungle and survivability enshrined in resource seeking for power and dominance imposes the principle of recognition of conflicts under the principles of opposites where people with the same national identity are drawn together for understanding, and struggle to overcome adversities of all forms, especially when the laws of the jungle are formulated by imperialists who simultaneously act as the police, witnesses, prosecutors, judges and executioners on the basis of neocolonialism through a complex system of international institutions set up by the imperialists to enhance imperial aspirations to exploit the masses who bear the cost for the benefit of the few. That is how imperialist institutions of neocolonialism work with punishment and suffering in prisons of inferiority with hard labor for the neocolonial masses in the jungle of the international political economy.

The structure of human behavior in the jungle of international affairs, politics, law and economics, demands that Africans, especially the leadership and supporting intelligentsia, firmly believe in the sacredness of the principle of African nationalism, where the authorities of all African states and the growth of national consciousness are derived from Philosophical Consciencism as the cognitive structure, to which all events and outcomes of African social life and international relations must be referenced, and within which the collective personality of every African state is molded and stamped with a trademark made in Africa where the content is truly of an African personality under the full epistemic support of the revolutionary philosophy and ideology for Africa's redemption.

The path to African redemption, given the scientific core of the theory of categorial conversion, faces two interdependent challenges. In the first place, Philosophical Consciencism faces a challenge of epistemic restructuring of the mind of African leadership and the corresponding African personality required for the development of strategic decision-choice positive actions for nation building in addition to the development of strategies and tactics against the development and growth of neocolonialism. In the second place, the African leadership faces the challenge of the practice of positive actions by mind and personality restructuring on the basis of the conceptual framework of Philosophical Consciencism. The challenge facing the members of the African leadership, broadly defined, is to overcome the cognitive difficulties of separating their minds from the prison walls of a colonial and neocolonial mindset, which by logical extension forces them to operate with anti-African negative decision-choice actions within the production-consumption duality into the oppressive zone of neocolonialism in the global resource space. This challenge is amplified by the failure to understand how the nation-building process, composed of socioeconomic development and the creation of social institutional structure on the basis of knowledge, freedom and justice, is explainable by the theory of categorial conversion, where every socioeconomic point presents itself as actual-potential polarity and where each pole has a corresponding duality in relational continuum and unity.

From the view point of African progress towards complete emancipation, the tactical and strategic decisions must be constructed with the guidance of the philosophical and ideological framework of Africentric Philosophical Consciencism. Any other conceptual framework will lead to decision-choice actions that will bring about Africa's further demise and neo-enslavement. The intent of the core principles of the epistemic structure of Philosophical Consciencism is to make an African an African with an African personality and decision-choice practices based on the ideology of African nationalism. The challenge of any African leadership is first to separate from the neocolonial mindset of nation-building and development that are supported by neocolonial philosophy and ideology, and then to understand the content and relevance of the central principles of Philosophical Consciencism with African content, and use them to organize the African human and non-human resources to create *categorial moments* and *categorial transfer functions* to set the potential against the actual over the time path of categorial conversions of actual-potential African socio-political polarities. The challenge is also to use the basic principles of Philosophical Consciencism to avoid moving the African masses into the zone of neocolonial exploitation by setting complete emancipation as the potential against neocolonialism as the actual, freedom as the potential against oppression as the actual, justice as the potential against injustice as the actual, complete African emancipation as the potential against neo-slavery as the actual, and complete African unity as the potential against African disunity as the actual as they are seen in our contemporary international political economy and Africa's difficulties.

The development of a new African personality with the defined African characteristics and in favor of complete African emancipation composed of independence and socioeconomic development is the creation of a decision-choice vehicle. It is also a creation of the knowledge-content of positive action, all of which finds expressions within the revolutionary philosophy and ideology of Philosophical Consciencism to alter the course of Africa's modern history by bringing about African nation building towards complete emancipation under African preferences. Philosophical Consciencism affirms the optimal direction to Africa's complete emancipation by reversing the cognitive forces behind the pattern of decision-choice activities undertaken by both the African leadership and the masses which have always gone against Africa and her children. The core principles of Philosophical Consciencism redefine cognitive foundations of acceptable norms of social practices of African leadership and the masses by restructuring the framework for judging what constitutes justice, freedom, fairness, individual-community rights, appropriate relations and correct democratic practices in international relations and order without being told as to what constitutes a valid decision-choice action by the neocolonialist oppressive forces and voraciously exploiting machine. The core principles of Philosophical Consciencism require us to examine definitions, contents and implications of words and statements, such as democracy, humanitarianism, partnership and many deceptive terms used to confuse and create a deceptive information structure to the advantage of the neocolonialists and imperialists.

It is useful to state here that the core of the theory of categorial conversion of social actual-potential polarity composes of *categorial moments* (intra-categorial dynamics), *categorial transfer functions* (inter-categorial dynamics) and *categorial transversality conditions* that define the point of conversion or transient point from one category to the other. The convertibility conditions require that the categorial moments and the categorial transfer functions must be created and sufficient enough to satisfy the categorial transversality conditions which simply show the point of entry from one category to the other. In categorial conversion of social polarities, the creation of these elements are the work of human information-decision-choice interactive systems on the principles of thought contained in philosophy and ideology that are relevant and derive their weapons and meaning from the societal experiential information structure, which in the case of African conditions is Philosophical Consciencism. The conceptual system of Philosophical Consciencism is to guide African leadership and its masses to create the categorial moment and the categorial transfer function to satisfy the categorial transversality conditions over the enveloping path of the categorial-conversion dynamics of social actual-potential polarities.

The essential analytics is that African nationalism, African personality, and Africentricity that are consistent with the complete emancipation on all fronts of Africa's social life consisting of politics, economics, laws, and technology which are inseparable in concepts and practices are locked in the philosophical and ideological framework of Philosophical Consciencism. The existence of one implies the existence of the others. This is the synthesis of the African experiential information structure which is the true test of Africa's will. The lack of practice of this synthesis has led to the harsh and stubborn realities of current African social life, history and adulterated culture where Africans struggle every day, every month and every year to be what they not , and simultaneously struggle not to be what they are, which is Africans.

5.2.2 Philosophical Consciencism, Independence and Cost-Benefit Analysis under Asantrofi-Anoma Rationality in Decision-choice Systems

The power of Philosophical Consciencism with African content is the acceptance of the general working mechanism of categorial conversion as a process of socioeconomic transformation that finds logic and meaning in Africa's conceptual foundations in relation to the dynamics of quantitative and qualitative dispositions of elemental categories. These conceptual foundations are derived from the principles of opposites composed of polarity, duality, negative and positive dispositions in relational combination to create forces and mechanisms of positive and negative actions for social change where the desired social potential is set against the unwanted social actual as desired. The principles of opposites contain actual-potential polarity, substitution-transformation duality, creation-destruction duality, and construction-reduction duality, all of which are abstractly translated into the cost-benefit space in relational continuum and unity for decision-choice evaluations. Any decision-choice problem in the cost-benefit space is an *Asantrofi-anoma* problem where the cost and benefit dispositions are in relational continuum and unity in such a way that one cannot select the benefits and leave out the costs. The intelligence embodied in the solution to the decision-choice problem as defined is the Asantrofi-anoma rationality. It is this characteristic of decision-choice actions that allows the economist's definition of cost where cost and benefit are always in a relational continuum and unity. The Philosophical Consciencism in general produces a conceptual framework in support of human decision-choice actions on the basis of generalized cost-benefit rationality which in aggregate defines a reasonable social action to be a case where social benefit outweighs the social cost under the Asantrofi-anoma principle of decision-choice action. This principle states that every benefit has a cost and hence there is no free benefit in the socio-natural process. The value of the social cost-benefit calculation will depend on the philosophical and ideological content of Philosophical Consciencism. When the content of Philosophical Consciencism is African-centered, the designed social policies and decision-choice actions of African leadership and masses will be made under the Asantrofi-anoma rationality where on the aggregate the social benefit to Africa outweighs the social cost to advance Africa's interest.

From the view point of Philosophical Consciencism with Africentric content, independence is a benefit; freedom is a benefit and justice is also a benefit. The question now arises as to what are the costs in the sense that independence, freedom and justice which are not really free, and if they were, will they not violate the *Asantrofi-anoma principle* in all the socio-natural decision-choice actions? The cost-benefit calculations meet substantial distortions if the members of African leadership along with the masses accept to operate in a conceptual framework of anti-African Philosophical Consciencism, such as that of Eurocentric neocolonialism as the basis for social policy and decision-choice actions which are then carried through neocolonial domestic and international institutional configurations. Given the *Asantrofi-anoma principle*, the framework of Philosophical Consciencism instructs the members of African leadership and the masses to cultivate an *asymmetric thinking system* through mental cross-referencing of costs and benefits at the both the levels of individual and the society with the individual-community duality [R4.17][R4.18]. The implication here is that the African institutional set up must develop the cost-benefit assessment capacities, creation of cost-benefit information storage capacity and cost-benefit information processing capacity for knowledge output into the Africentric decision-choice systems in the individual-community duality with a relational continuum and unity in the sense that an individual decision-choice action must be supportive of the community and the community decision-choice action must also be supportive of the individual, meeting the conditions of African traditions under the principle of each for all and all for each to chart the path of social progress. It is this Africentric tradition of social formation brought to technological modernity that Nkrumah called scientific socialism. He could also have named it scientific communitarianism.

An analytical point is made where great minds are the works of relevant philosophy and ideology with the weapons of their development taken from the cultural conditions and experiential information structure of the social set-up. In the case of the African society, the required philosophy and ideology to create great minds are embodied in Philosophical Consciencism. This should be a theorem with the lemma that great institutions are the work of great minds. This lemma is followed by a second lemma that progressive national interests with great social vision are the works of great minds. Similarly, great social goals, objectives and policies are the works of great minds and are transmitted through great institutions for great and effective positive decision-choice actions to shape the direction of society and define a progressive path of national history. When Philosophical Consciencism acquires its Africentric content it becomes transformative in the epistemological space and ideologically destructive of intellectual and conceptual systems that supports colonialism, neocolonialism, imperialism and conditions of subservience. It does these by instilling in the African leadership, the masses and social institutions the sense of the African-self, where the leaders and the masses by the internal sweat of African human and non-human resources generate categorial moments and categorial transfer functions over the categorial enveloping of the complex dynamics of actual-potential polarities involving qualitative and quantitative dispositions. The outcome of the complex dynamics are the works of an information-decision-choice-interactive system that is combined with great minds to create the information processing capacity to derive knowledge input into the decision-choice system to create the needed categorial moments from intra-categorial conversions and the needed categorial transfer functions for inter-categorial conversions that move the socioeconomic system over developmental categories. The relational structure of matter, energy, information and Philosophical Consciencism in support of the understanding of the categorial-conversion process is presented as an epistemic geometry in Figure 5.2.2.

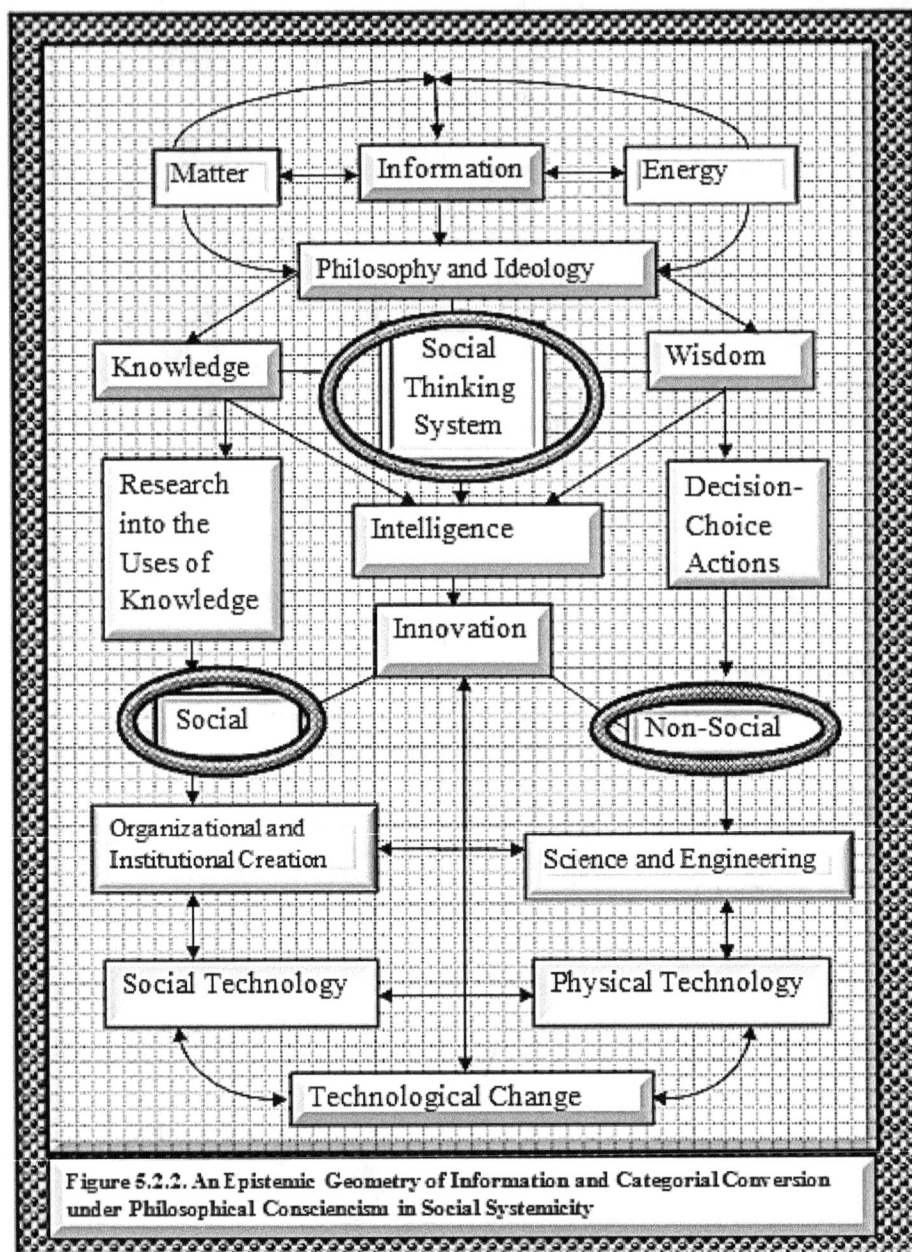

Figure 5.2.2. An Epistemic Geometry of Information and Categorial Conversion under Philosophical Consciencism in Social Systemicity

All experiential information structures from nations point to the fact that every social organization composed of positive and negative forces defines the existence of duality in relational continuum and unity.

266

Similarly, all social transformations and progress are based on implied condition of the existence of actual-potential socio-political polarities that can be operated on by decision-choice instruments of categorial conversions. For any given social organization, there is an internal natural tendency to organize and harmonize all forces of which it is composed away from destructive conflicts and into a unified force of constructive conflict which is directed to create the required stability for social progress, if the organization is to survive. Philosophical Consciencism accepts this position and extends itself to the idea that by creating a new personality true to all African masses and leadership, the instruments for excavating the dark tunnel to Africa's complete emancipation and golden tomorrow become internally available to empower the cultural dimensions of African positive actions. Philosophical Consciencism contains the traditional characteristics of African humanism that help to solve the *Anoma-kokone-kone* problem which represents the general conflict problem of individual-community duality by creatively and ideologically reconciling the conflicts in favor of those positive actions that will favor collective progress. By resolving this *Anoma-kokone-kone* problem, an unshakable ground of African categorial conversions of social actual-potential polarities on the basis of collectivity of brotherhood, sisterhood, motherhood and fatherhood can be nurtured to propel Africans to accept and believe in the laws of unity, internal self-motion, duty and responsibility which come with national independence and sovereignty that can resist hostile forces of neocolonial predation and a repeat of the greatest human atrocities of slavery, forced labor, racism and inhumanities committed by European imperialism against Africa.

The power of Philosophical Consciencism and its usefulness in policy constructs can easily be seen when the structures of the epistemic geometry of Figures 5.2.1 and 5.2.2 are combined to relate great minds to institutional creation, information processing and decision-choice activities to produce positive social actions. The current difficulties in the process of moving African countries in the progressive direction of the internal self-motion of Africa's improvements on the general principles of categorial conversion are merely expressions of poor judgment and decision-choice activities that are reflected in the absence of a compatible Philosophical Consciencism by the leadership, the general masses and institutions through which human development occurs. The absence of compatible Philosophical Consciencism in the African social space is partly attributed to the entrenched bankruptcy of the structure of colonial and neocolonial education that promotes an intellectual and philosophical system which supports neocolonialism, racism and imperialism, and maintains the African neocolonial mindset in the individual and social decision-choice space. Consequently, there is the lack of the required vision, appropriate African thinking and intellectual creativity among the members of current governing bodies and leaderships in institutions of religion, education, culture and governance. The effect of this lack and deficiency in social vision manifests itself as deflated confidence as well as schizophrenic behavior where Africa countries simultaneously claim independence and servitude. The secret to success in the direction of Africa's self-motion is the creation of compatible collective and individual African personalities, where these personalities are consistent with the basic principles of Philosophical Consciencism with African content that will bring an intellectual guidance to create the constancy and unified internal forces towards the realization of Africa's complete emancipation which includes the solution to the race question and the preservation and use of African resources for the collective benefit of all Africans.

All African masses and leadership must understand the reasons for the development of Philosophical Consciencism and the reasons for its application in the African context of the conditions of decolonization as the monograph has tried to show. All Africans, both at home and abroad, must also understand that the Eurocentric colonial education and neocolonial mindset present important intellectual difficulties and barriers in the African collective decision-choice space. They restrict the creative cognitive power of the African genius and hence prevent progressive categorial conversions of social actual-potential polarities through their indirect controlling effects of African decision-choice actions or internal self-transformations from the actual to the potential which will bring about sustainable socioeconomic development and complete emancipation of Africa. Nkrumah understood these intellectual difficulties and barriers and thus offered the alternative of Philosophical Consciencism with African content that must be used to decolonize the African mind and sharpen Africa's collective creativity at all fronts of human endeavor. The African-centered conceptual and logical foundations of Philosophical Consciencism are provided in [R1.92].

The Eurocentric neocolonial mindset and system of education in Africa are largely imitational and a mimicry of the Western way of life for enslaving the African mind which is discussed in [R1.1]. This mindset is good in parroting the philosophy and ideology of Eurocentric neocolonialism. The members of African leadership who develop from their neocolonial education and under the ideological and philosophical guidance of Eurocentric intellectual neocolonialism collectively acquire a mind that never goes through the experience of *critical thinking*, a mind that never understands the deep implications of slavery and colonialism on Africa's modern difficulties, a mind that has no clear concept of the present status of Africa nor a reasonable foresight to *plan* for categorial conversion of social actual-potential polarities, and a mindset that cannot relate and understand the essential meaning of the *time trinity* efficiently and philosophically captured by the *Sankofa-anoma* for the relational continuum of the past-present-future connectivity in relational continuum and unity. Members of the African leadership with such a mind become useful intellectual idiots for the success of the neocolonial deception, and the exploitation of African human and non-human resources. When they have learned to acquire such a mind, they obtain praises from the propaganda machine of the European imperial system of exploitation. By working with this Eurocentric neocolonial mindset, it is easy to justify economic aid, technical assistance, military assistance and the carrying of a beggar's basket across the globe to be filled with water just to arrive at their homes with pure emptiness. Those members of the African leadership who resist in following this imperial intellectual script are punished by torture and death from within and from without. However, the greatest threat to Africa's progress is from within. This threat must be overcome by applications of the philosophy and ideology of Philosophical Consciencism to the colonial and neocolonial mindset of the masses and members of Africa's leadership in all decolonized African states as maintained by the principles in Nkrumaism. The manner and conditions in which this threat manifests itself in African decolonized territories are discussed in [R1.91][R1.92]. I have pointed out that the preferred sanction to engage in the game of categorial conversion of actual-potential socio-political polarities by the imperialist and neocolonialist is violence with hard power. In this respect it must be clearly understood that the choice of non-violence positive direction is always undertaken with constraint of imperialist and neocolonialist instruments of violence of which they claim monopoly of use.

5.2.3 Philosophical Consciencism, the African Mind and the Asantrofi-Anoma Rationality

The central pillars of Philosophical Consciencism for dealing with the problems of socio-political transformations may be seen as: 1) the principle of internal self-motion on the basis of categorial conversion, 2) categorial conversions are the work of conflicts between negative and positive actions, 3) in societies, these positive and negative actions are the work of decision-choice activities on the basis of philosophy and ideology that have taken hold of the society, 4) the development of the philosophy and ideology is drawn from the experiential information structure of the society, 5) since the experiential information structure vary from society to society, the philosophy and ideology developed in one society may be irrelevant in some societies except if they have similarities in social categories and 6) this is the case of colonial and decolonized African territories under a neocolonial mindset. To make progress under categorial conversion of social actual-potential polarities, there is a pure conviction implied in Nkrumaism that no true emancipation composed of justice and freedom of Africa through categorial conversion can come about without a mass intellectual revolt on the part of Africans, at home and abroad, against the neocolonial and imperial ideas and concepts that hold Africans in bondage and cognitive darkness, and serve to maintain a schizophrenic collective personality which destroys Africa's confidence, zaps her courage, and subverts the dignity of African leadership and its masses.

That same schizophrenic collective personality dangerously serves to restrain our perseverance and faith that stand behind Africa's self-motion towards true democracy with freedom and justice under the principles of categorial conversion. This intellectual revolt should be a new thinking ascribed by Philosophical Consciencism that, when it takes hold, must be translated into positive collective action, since unity of thought and collective action are the life blood of every successful revolution towards setting the potential against the actual in the actual-potential social polarity. The act of actualizing the needed intellectual revolt of African masses and the leadership is anchored firmly in the precious and delicate hands of those who show themselves ready to fight for and to conquer an epistemic frontier in order to create a new African intellectual system that would govern Africa's social decision-choice activities, behavior, collective actions and collective self-reliance. This effort requires the rise of a new African intelligentsia with an African collective personality that is trademarked, *made in Africa*. This new African intelligentsia must extend the development of the philosophy and ideology of Philosophical Consciencism to deal with the intellectual deficit of the masses and the leadership. This is how one must see the works in [R1.1][R1.20][R1.35][R 1.38][R1.83b][R1.86] [R1.84][R1.92][R1.97][R1.98][R1.294].

Africa will rise above its current conditions only when we Africans have confidence, courage and faith in ourselves, in that the way to Africa's bright and prosperous future lies in our traditions brought to modernity with full collective self-reliance as is contained in Philosophical Consciencism. The first requirement to achieving this self-faith is not to look to foreign lands, the lands of our predators and destroyers, but to end our servile subjugation and appendage to the imperial system of dependencies and exploitation, and to break the prison walls of familiarity associated with intellectual colonialism. This requires the overthrowing of the Eurocentric neocolonial mindset with a prudent detachment from the Eurocentric personality, its corresponding anti-African philosophy and ideology and all other personalities opposed to the African way of life.

This point defines a situation where Africans become fully aware of their African cognitive traditions. These traditions teach us that the secret of power and success in bringing about an effective change in the current African state and conditions that maintain it are dignity and faith in ourselves and in Africa itself; that the true virtue is the sacrifice of self to the greater and collective good of Africa and the show of compassion for other Africans; that the ability to make good decisions requires knowledge; that the implementation of programs of action demands conviction and confidence; and that the success of such implementation necessitates courage, commitment, perseverance and hard work. The unity of all these is engulfed by a logical and conceptual system referred to here as *African-centered Philosophical Consciencism,* which is a philosophical system derived from African conceptual tradition which retains the African conscience. Africans have not reached this point of development of the African personality required for continual categorial conversions of Africa's social actual-potential polarities if the conceptual system of Philosophical Consciencism developed on the basis of African-centeredness with its supporting philosophical and ideological foundations has not been acquired, mastered and practiced with developmental African self-confidence, where African leadership must

271

drink at the fountain of strength, hope, optimism, courage, survival and progress from the ancient traditions of Africa [R11.85][R1.38][R1.42][R1.79] [R1.180] [R.181] [R1.243] [R1.247] with the uncompromising understanding that:

> A People losing sights of origins are dead. A people deaf to purposes are lost. Under fertile rain, in scorching sunshine there is no difference: their bodies are mere corpses, awaiting final burial [R1.16, p. xiv].

The purpose of Philosophical Consciencism with African content is to help in defining an African social vision around which African interests are established, and the supporting goal-objective set is formed under appropriate social actions. Members of the African leadership on all fronts of human endeavor, working under the guidance of Philosophical Consciencism ask questions regarding what is the African vision to establish an actual-potential polarity, how to create it , use it, and manufacture the material conditions in terms of goals and objectives to support the actualization of the social vision which requires political and economic securities under African cultural unity in defense of African sovereignty and true democracy. One thing that is painfully clear is that the Western imperial club that preaches democracy, territorial integrity and respect for national sovereignty, is one political group that has been accepted by all the members when they are applied to them. Philosophical Consciencism is to provide philosophy and ideology to restructure the African mind to resist the propaganda and deceptive information structure produced and disseminated by the Western imperial club.

5.2.4 The African Intelligentsia, the Clergy, African Personality, and Philosophical Conciencism in Social Systemicity

In these respects of the conditions of African experiential information structure and the need for philosophical orientation away from the Western imperial order, African institutions of learning, culture, worship and the members of all African academies have major interdependent roles to play in fostering the correct African personality formation required in the sculpturing field of African human dignity, point to African origins, open the ears and eyes to the African purpose of emancipation, freedom and human dignity. In previous sections, the cognitive mechanism of the philosophy and ideology of the European neocolonial system in producing an African mindset of an anti-African personality was discussed. In response to African needs, Philosophical Consciencism with African content presents an alternative, creating a battle field and a theater of war of ideas that affect behavior in the social decision-choice space. Without Philosophical Consciencism, confusion is brought to the masses through the content of neocolonial education. Additionally, ignorance is placed on the leadership and the elite in order to dismantle the developmental process of the true African personality from origins required for true sustainable categorial conversions of socio-political polarities. It must be noted that the application of Philosophical Consciencism does not neutralize all internal opposing forces through cognitive controls. However, the creation of social confusion and ignorance in African decolonized territories, which operates through the predators' manipulation of information, distortion of knowledge and creation of propaganda, can be continually countered by the epistemic forces of the philosophy and ideology of Philosophical Consciencism. This may be done by elevating the masses above the zone of cognitive confusion and pointing the leaders to operate above the epistemic zone of ignorance as defined by the neocolonial mindset. By pointing the leadership, clergy and elite to the origins and equipping them with alternative tools of reasoning, information collection and processing, there is the hope that they will understand and practice the African resolution to conflict in the individual-community duality in favor of the community. This resolution brought to the modern times of continual technological improvement is what Nkrumah called *scientific socialism.*

The intelligentsia and the clergy together constitute an important force in the societal dynamics for general social transformation. By their impact on culture and the evolving collective thinking of the people the members of the intelligentsia and clergy play critical roles in defining the direction of social change and the lines of resistance and struggle. The intelligentsia defines the societal thinking system and its progress on the basis of which perceptions about social truth and realities are formed by the individual and the collective and translated into decision-choice actions in the social decision-choice space. The clergy, on the other hand, defines the framework on the basis of which perceptions are formed by the individual and the collective about the spiritual reality of life and the deity systems of the community. The intelligentsia and clergy thus have preponderating effects on the structure on which the national belief system (information and non-information supported) rests. By affecting the national belief systems, the intelligentsia and clergy affect the cultural character and the revolutionary spirit of the nation through the philosophy and ideological principles that they hold. In this way, they can play a role either as facilitators or as impediments to the decision-choice activities that are required to bring about categorial conversions of actual-potential social polarities.

The important role that the intelligentsia and the clergy play whether in Africa or elsewhere must be seen in terms of the relationship between perception and reality or truth. Perception is a human model of reality or truth on the basis of held ideology and the supporting philosophy. It is defined by human sense formation about events through the cognitive process. Truth and reality are independent of human cognition and will exist whether we know them or not. Truth, therefore, is its own defense while perception has its defense in logic. Thus, perception may or may not be accepted by the individual and/or the collective as representing the truth. If the logic is weak, the perception may be revealed to either the individual or the collective or both to diverge from the reality even if it coincides with reality, and hence may be rejected as an input into decision making. In the world of human operations, it is the formed perceptions about truth but not the actual truth or reality that motivates and instigates human action. Thus, by affecting the perception formation of the individuals and the collective, the intelligentsia and the clergy also affect the individual and the collective decisions and actions of the society as well as the nation's destiny and history. This relationship among perception, truth and the activities of the national intelligentsia and clergy must be clearly understood if the evolving social system is to retain its authenticity and rationality. It is through the relational structure of perception and reality that Philosophical Consciencism creates its connecting vehicle to induce categorial conversions in socio-political polarities.

The imperial predators know the powerful role that the intelligentsia and the clergy can play, either against or in favor of their predatory activities. They are also aware that an awakened and uncorrupted intelligentsia and clergy of a subjected or targeted territory can become problematic for their smoothed predatory operations even when they have successfully established a neocolonial state and politically imperial cronies. The smoothed and uninterrupted predatory operations of the imperial destroyers and predators require support from the domestic intelligentsia and the clergy acting as neocolonial cronies in the subjected territory. To obtain such support, the predators corrupt the local intelligentsia and the clergy, by massaging their brains with nauseous conceptions of life and their pockets with leftovers and meaningless perks, in order for them to accept and promote the philosophy and ideology of neocolonialism. In other words, the predators turn the members of the intelligentsia and clergy into supporting imperial cronies whose personal interests are placed before Africa's national interests. This is done directly through a system of scholarships and technical assistance dubious aid or indirectly through the ruling imperial cronies and is reinforced through the established international system of rules and regulations in addition to latent ones for scholarly publications in imperial academies. In this way, the predators work to turn the African intelligentsia and clergy into intellectual useful idiots whose pre-assigned role is to shape the perceptions of the African masses, on the basis of neocolonial philosophical and ideological mindset, into accepting illusions, the destructive activities of the imperial predators and false conceptions of the spiritual and material world as the reality and truth that will comfort the masses from their pain and suffering under the principle of do not blame your problems on history, such as colonialism, slavery, racism and imperial social destructions .

CHAPTER SIX

PHILOSOPHICAL CONSCIENCISM AND SOCIO-POLITICAL DECISION STRATEGIES FOR CATEGORIAL CONVERSION IN SOCIAL SYSTEMICITY

In this chapter a presentation is made to examine how Nkrumah combined categorial conversion and Philosophical Consciencism to present a model of socio-political and socio-economic transformations for a complete emancipation of Africa. In the process, critiques and analysis appraisals of the model will be offered. This is a theory on application of theories. The Nkrumah's model of socio-economic development is abstracted from Nkrumah's domestic policy in Ghana and his African policy. It is argued that this socio-economic model presents the best path for Ghana and Africa's sustainable development as well as a complete emancipation given African unity. The critique and analysis are presented through an establishment of economic-theoretic foundations of a viable social formation where the social system is composed of three basic elements of economic structure, political structure and legal structure. An economic and political case is then made in support of African cultural unity, sovereignty preservation and territorial integrity. The logic of the theoretical structure may be called the theory of social progress and nation building under Philosophical Consciencism.

The theoretic analysis places distinctions among three concepts of economic development, social development and nation building. The distinction allows us to present the essential differences and similarities among these three concepts of transformations in socio-economic progress through the guidance provided by the intellectual map of African-centered Philosophical Consciencism. Central to the theoretic analysis of the policies of social progress and nation building under Philosophical Consciencism is human-capital development composed of education and health factors. The education and health factors are developed and mapped unto the categorial-conversion space to create categorial moments and categorial transfer functions to be applied to the economic structure, political structure and legal structure for sustainable internal nation building. The chapter is concluded by demonstrating how the theory of social progress forms the essential logic of Nkrumah's domestic and African policies for sustainable social progress towards complete African emancipation under Philosophical Consciencism. The conclusion is extended to the reasons why Nkrumah insisted on African unity as an important vehicle for Africa's socioeconomic transformations process in terms of preservation of democracy territorial integrity and the respect of African sovereignty.

The concept of nation building is central to Nkrumah's policies under Philosophical Consciencism. The central idea is that no sustainable economic development can take place without a nation with political stability and domestic control of her sovereignty. The reason for the nation building emphasis here is that all the new African states were put together by the colonialists under violence in the imperial *terror-dome* to suit their colonial resource interests and imperial expansionism. After decolonization, the African states were not nations. The first prerequisite of sustainable socioeconomic development is the reconstitution of the decolonized territories into nations with the ability to protect her citizens. The policy of nation building must be supported by a policy of socioeconomic development. Nation building, socio-economic development and social welfare impartments constitute social progress where national sovereignty protection, adherence to international laws, human justice and respect of territorial integrity are seen through Nkrumah's African unity policy, while the problems of the African Diaspora are dealt with by non-African policy.

6.1. Social Bases for the Analysis of Development Policies under Philosophical Consciencism

To establish the theoretical basis of socio-economic development policies for any African territory, one must enlist the essential characteristics and needs of the country and cultural endowments at the dawn of the country's decolonization. These characteristics and needs must then be related to Africa's conditions within the global power distribution as well as the global politico-economic structure including the conditions of neocolonialists under Western imperial aspirations and aggressions. Furthermore, one must reconstruct the understanding of the concept of social progress, its measurement and characteristics that define the path to general orientation to nation building. The enlisted characteristics, endowments and needs will constitute the initial conditions for the controllability of the dynamics of both the qualitative and qualitative disposition of the social system. With reference to Ghana situation and by logical extension Africa, Nkrumah understood this initial condition with the statement:

> We shall measure our progress by the improvement in the number of children in school, and by the quality of their education; by the availability of water and electricity in our towns and villages, and by the happiness which our people take in being able to manage their own affairs. The welfare of our people is our chief pride, and it is by this that my Government will ask to be judged [R1.207, p. 51].

From this statement Nkrumah has an implied social vision, national interests and social goal-objective set as the problem foundation for African social action. This statement must be related to the concept of internal transformation on the basis of independence, self-reliance, nation building and its cognitive foundations as they reflect categorial conversions of African society from stage to stage in a socio-economic plane where each stage must be transformed in progressive motion under the domestic controls of African sovereignties over the social decision-choice space. These are consistent with the politico-cultural policy dynamics of Nkrumah, some of which have been explained in the previous chapters of this monograph and [R1.91]

6.2 On the Conception of the Theory of Social Progress in Social Systemicity

From the above Nkrumah's statement and numerous ones from his speeches and philosophical works, Nkrumah saw socioeconomic progress in terms of nation building and social welfare improvement of the masses who constitute the backbone of history and the material creation of the society. He thus changed the terms of reference as well as the indicators of socio-economic progress from that of economic tradition, the terms of reference of which are the rates of output growth and intertemporal profile of per-capita income to that of socio-cultural tradition the terms of reference of which are the general welfare and inter-temporal profile of the happiness and dignity of the people are the foundation of economic progress and nation building. In this way, socio-economic production is to benefit the people but not for profiteers. This is particularly relevant to the African situation. The terms of reference of social progress, if viewed in terms of nation building and the success of this nation building are measured by a *composite index* that includes the levels and rates of increases of indexes of education, health, social infrastructure, social happiness and the degree of social confidence of African people to manage their affairs constitute an important departure from accepted paradigms of theories of economic development [R15.14]. The implicit conceptual system is referred to in this analysis as *the theory of social progress* with the components of socio-economic progress, politico-economic progress as well as progress in national identity. The theory of social progress is about social transformation of the qualitative disposition with the support of quantitative dynamics. This theory of socio-economic progress through nation building requires the creation of a decision-making process that allows the establishment of:

a) Social goals and objectives;
b) Internal policy space;
c) Institutional casting that is appropriate for policy transmission;
d) Mode of policy implementation;
e) Resource-economic options that will be available in support of the plans and the agenda of nation building and
f) Social information structure

The structurally logical foundation in support of the development of the theory of social progress through nation building may be presented as an epistemic geometry in Figure 6.2.1. The socio-economic decision-choice rationality consistent with Nkrumah's application of Philosophical Consciencism to categorial conversion emerges out of his understanding of the detrimental conditions of colonialism, the nation-reconstruction job ahead after independence and the conception of nation building into a *unitary state*. The theory of categorial conversion requires the creation of a logical process that involves the establishment of social categories and conditions of categorial conversion [R3.7] [R313]. Generally, the theory of nation building is composed of sub-theories of political theory, legal theory and economic theory under which other sub-theories may be constructed with the logic of their interconnectedness [R13.8] [R13.9]. The sequential structure of the subsystems of the theoretical guidance of the path of Nkrumah's policy behavior under Philosophical Consciencism may be presented as in Figure 6.2.2. These conceptual system and subsystems are presented here in explaining Nkrumah's policy rationality and why this policy rationality offers African states the best strategic combination for the maintenance of independence, sovereignty, an optimal path towards Africa's emancipation and the best direction in building *Great Africa*. It also exposes the flaws of external policy recommendation through technical assistance, advices and material aid.

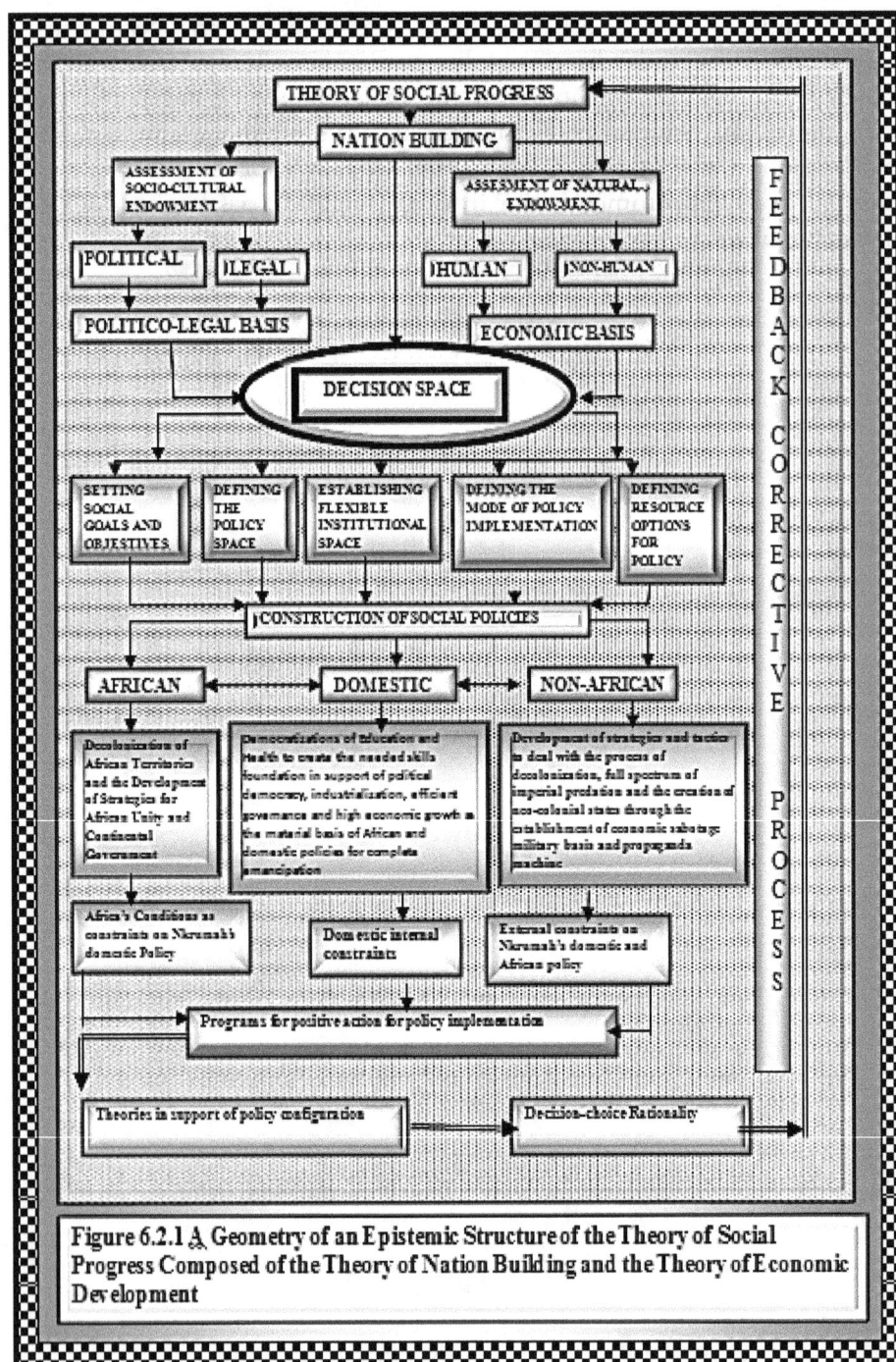

Figure 6.2.1 A Geometry of an Epistemic Structure of the Theory of Social Progress Composed of the Theory of Nation Building and the Theory of Economic Development

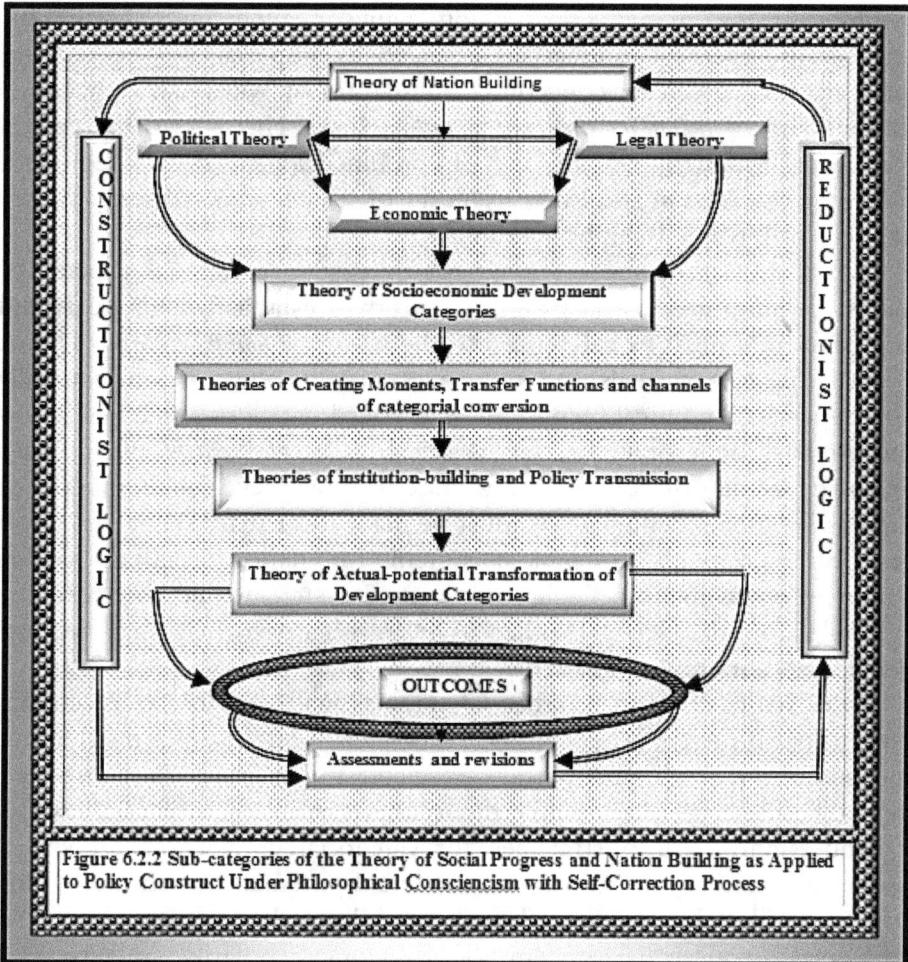

Figure 6.2.2 Sub-categories of the Theory of Social Progress and Nation Building as Applied to Policy Construct Under Philosophical Consciencism with Self-Correction Process

The initial conceptions of the theory and sub-theories and their applications require of them to take into account factors of resource availability, composition and physical endowment, technical preparedness of the available labor force, the damage of national spirit from colonialism and occupation, executive capacity, the confident level of the population in independent creativity, inventiveness and responsibility in dealing with large scale construction and restructuring programs. Additionally, they must take into account the philosophical and ideological appropriateness to generate and utilize appropriate knowledge for nation building in addition to examining the suitability of available institutions for policy transmission while dealing with constraints imposed by external interests as revealed by imperial and neocolonial maneuvers. The central conceptual foundation of Nkrumah's policies is the building of *a unitary state and a united Africa*. This is also the central conceptual foundation of the use of combined logics of the theories of categorial conversion and Philosophical Consciencism to present a theory of social progress as seen by Nkrumah.

On the basis of these factors for initializing a rational process of nation building and social progress Nkrumah established his goals and objectives that must be referenced to the *vision of a unitary state and African Unity*. The set of goals and objectives may be divided into mutually interdependent political, economic and social categories that will allow the establishment of a) *political freedom*, b) *economic freedom* and c) *social freedom* consisting of cultural and welfare improvement for the people of the newly independent state. It must be made clear that Ghana was a test case of Nkrumah's conceptual system for nation building toward complete emancipation. Such a conceptual system would then become available for practice by other African territories as they become decolonized or are decolonized.

One main concern in this Chapter is with regard to economic rationality on the basis of which African social policies are pinned under the applications of Philosophical Consciencism. In other words, we seek to outline the type of economic theory that explains as well as allows us to understand African strategies for categorial conversions of socio-economic and politico-economic actual-potential polarities. This approach, like the conceptual path of Nkrumah, operates from the Africentric conceptual system of categories, polarity, duality and relationality with categorial substitution-transformation processes of social states under the principles of opposites where negative and positive actions are matter-energy-information connected through the activities of the people who are the backbone of decision-choice action for social change. The logical conditions of the theory of categorial conversion are such that we consider every socio-political stage as belonging to a category of degree of social progress. Each category is at a crossroad of actual-potential social polarity under categorial conversion. The actual is defined by a characteristic set composed of negative and positive characteristic sub-sets that establishes its residing duality in relational continuum and unity. The numbers of characteristics contained in the set and sub-set define the quantitative disposition. The actual is always unique with an identity established by the relative negative-positive characteristic set. This uniqueness defines the qualitative disposition of the social set-up. The potential element to replace the actual element, however, is not unique but appears in many qualitative forms which are in competitive struggle against each other to negate the actual. Each potential element is, however, uniquely defined by its qualitative disposition and the residing duality in relational continuum and unity. The game of categorial conversion is about either the defense of, or a change of the existing qualitative disposition of the social setup. In other words, a theory of social progress or nation building is about a theory of the dynamics of qualitative dispositions of the character of the social set-up. This theory of dynamics of qualitative disposition is always supported by a theory of quantitative dynamics. In every actual pole and every potential pole there is a residing duality composed of negative and positive characteristic subsets in relational continuum and unity under tactical and strategic actions seeking to acquire dominance through negation to bring about categorial conversion where the social actual is destroyed, and in its place there is a manifestation of a new actual from the potential involving a new qualitative disposition.

From this conceptual system for the understanding of the development of appropriate socio-economic and politico-cultural policies for social transformation, Philosophical Consciencism sees political policies, economic policies and legal policies as constituting a set of interdependent relations in continuum and unity on the basis of which nations acquire their progress. This is how Nkrumah saw the logic and practice of policies on the basis of the theories of categorial conversion and Philosophical Consciencism for revolutionary changes in Africa. From the theory of social change and actions on nation building emphasis is placed on the primary role of decision-choice activities in the political structure for inducing social transformation dynamics through the behavior of residing dualities for categorial conversion of the actual-potential social polarities. Obtaining political freedom by acquiring the control of the political structure and national sovereignty provides the power to make independent domestic social decision-choice activities in order to change both the legal and economic structures which will ultimately affect the path of social progress to bring domestic economic and social freedoms under democracy and justice.

The theory of social progress on the basis of self-reliance demands that socioeconomic decisions must be controlled from within Africa itself by controlling the political structure that installs the power to make social decisions for shaping the progress path of the nation building. This conceptual understanding of the primary category of political structure, and hence political freedom was also the rationale behind Nkrumah's drive for the creation of a unitary state and African political union, and hence his slogan "*seek ye first the political kingdom*". The first step towards a people's freedom is the control of their national sovereignty, institutions that encapsulate it to guarantee the domestic control of the power to make social decisions and choices which must be guided by a revolutionary mindset provided by Philosophical Consciencism that is African-centered. The control, decision-choice refinements and further development reside in the political structure which contains the national sovereignty and power. Sustainability of such control rests on the protection of the political structure. It may be noted that the emphasis is on obtaining political freedom that will ensure the security of African political structures through strong political union of the states such as in Ghana, by means of which the Lego-economic structures will be shaped by appropriate political decisions and policies to engineer social progress. The engineering of the social progress and nation building depend on developing organizations of the masses to maximize social effort and peoples' social action, and on casting relevant institutions to ensure effective policy transmission mechanisms. It is within this epistemic structure that social history may be viewed as an enveloping of success-failure outcomes of decision-choice activities of the labor process, which when guided by African-centered Philosophical Consciencism induces the needed necessary and sufficient conditions for categorial conversion of actual-potential social polarities in favor of African progress towards complete emancipation composed of political, economic and legal freedoms. Philosophical Consciencism has two important tasks to perform. The first task is to provide a cognitive framework for instrumentation of control elements that establish the necessary conditions of categorial-conversion moments and transfer functions within the transformation process. The second task is to provide another cognitive framework for the management of the command and control of the conversion elements that establish the necessary conditions in order to create the needed sufficient conditions of the transformation process so as to satisfy the categorial transversality conditions for the categorial conversion.

To ensure an effective implementation of constructed policies requires a change from colonial subservient behavior and vestiges left over by imperially colonial *terrordom* to a new social attitude towards actions of nation building rather than actions in support of oppressive colonial and neocolonial machinery. To this end, ideological education is required to alter patterns of accepted behavior produced by colonialism and imperial oppression with violence. This explains the ideological education policy of Nkrumah and the establishment of the Kwame Nkrumah Ideological Institute to which extensive account is given in [R3.7] [R1.91][R1.203]. The connecting point here is that, it is not simply education the level of the decolonized territory, like Ghana, that is important, but the content of the education and the philosophy that supports it and its value creation. The powerful role played by ideology in helping to establish social vision, national interest and the formation of the supporting social goal-objective set also cannot be underestimated. Generally, ideology defines the boundaries of acceptable individual and collective social behavior within the struggles to either set any social potential element against the actual or to defend the actual against any social potential element in the categorial conversion of socio-economic and politico-legal actual-potential polarities of any social formation. Nkrumah provided discussions on this role which is also further extended in [R3.7] [R3.13].

6.3 The Rational Path of Socio-Economic Policies under African Centered Philosophical Conciencism

Having decolonized and secured the political structure of an African territory like Ghana by categorial conversion of socio-political polarity, Nkrumah recognized the need to develop the economic basis to support both the political and legal structures as well as deliver social progress. The theory for development on the basis of international social actions will constitute the thought on the basis of which the social system is moved from one qualitative state to another. For sustainable stage-to-stage relational continuum and unity the constructed socio-economic development theory as a basis of action must follow the theory of categorial conversion where every state in the transformation enveloping is seen as a category defined by an actual-potential socio-economic polarity. The movement of one category to the other is governed by a qualitative motion created under tension from the residing dualities. The qualitative motions that govern the controls of intra-categorial conversions and inter-categorial conversions of the categorial stages to bring about nation building and social progress are the products of socio-economic policies under *decision-choice rationality*. This decision-choice rationality must be constructed from the thought system of Africentric Philosophical Consciencism which is designed to keep Africa's vision and African interests at the forefront of African progress in the defense of the African people. In other words, the crafting of the needed socio-economic policies for Africa's social progress and nation building must be rational and supported by a scientific reasoning that is consistent with African conditions as seen through African-centered revolutionary philosophy and ideology contained in Africentric Philosophical Consciencism as seen by Nkrumah and is being advanced here.

6.3.1 People, Thought, Rationality and Scientific Reasoning in Social-Decision Space

The theory from which rationality and scientific reasoning are couched begins with the fundamental idea that people are the backbone of nation building and national progress which is directed towards the general improvement of the collective welfare and happiness. It is by the sweat of the people that great institutions and great nations are constructed. The people are also the beneficiary of the success and failure of their effort and inaction. The people, therefore, constitute the foundation of freedom and justice by creating the contents of ideology that supports fairness as a social value in the cost-benefit distribution of social activities within the individual-collective duality. It is on the basis of this Africentric fundamental idea of African tradition in the social construct that the whole process of socio-economic development and nation building is refered to as the *labor process*. The rational path towards social development and nation building is shown in Figure 6.3.1.

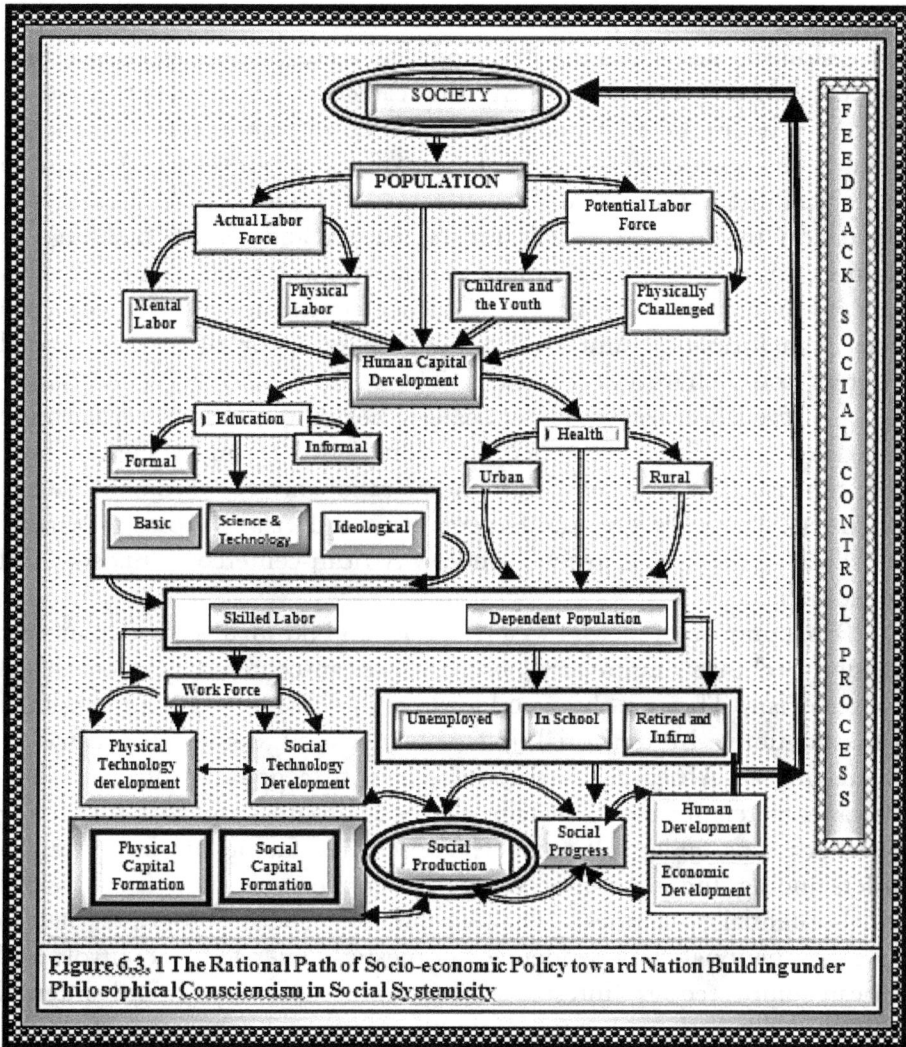

Figure 6.3. 1 The Rational Path of Socio-economic Policy toward Nation Building under Philosophical Consciencism in Social Systemicity

The rational path of policy constructs takes the population as the foundation of the society where the people are the driving force as well as the beneficiaries of social progress through the labor process. The categorial conversions of natural actual-potential polarities are the natural processes defined in the ontological space. For example, the transformation of life to death is an ontological process. They may be called ontological categorial-conversion processes. The categorial conversions of social actual-potential polarities are labor processes defined in the epistemological space. They may be called epistemological categorial-conversion processes. The labor process creates the conversion moments and the corresponding categorial transfer functions for transforming societies from one category of social progress to another. The more educated and healthier the population is on the average, the greater the quality index of the labor force. The more ideological the educated and healthy labor force is, the greater the conversion moment of social transformation and nation building. In addition, the more the education and ideological contents are African centered under Philosophical Consciencism, the more successful are social transformation and nation building in favor of African social progress and welfare improvement. The central building blocks of the theory of the labor process are education, health and ideological alertness regarding the collective essence and conscience of any society. These were fundamental to the decision-choice rationality of Nkrumah's policy constructs. These are also fundamental to the decision-choice rationality of social policy constructs under African-centered Philosophical Consciencism.

Three important factors weigh heavily on Nkrumah's socioeconomic decision- choice rationality under African-centered Philosophical Consciencism. They are the grand goals of a) decolonization and national domestic unity, b) territorial independence and domestic control of sovereignty of Africa and African Unity with a grand control of Pan-African sovereignty and c) sustainable social progress and nation building for all Africans under self-reliance. These three factors are important in understanding the nature of categorial conversions of African social actual-potential polarities. The sustainable social progress, nation building, national independence, domestic sovereignty control and African unity require internal self-motion, self-reliance and internal creativity with national confidence without external impression. In order to bring about sustainability of African social progress with national independence, African control of her sovereignty and African unity requires the manufacturing of intra-categorial conversion moments and intra-categorial transfer functions that are internally generated to provide the qualitative law of motion that will govern the transient process of categories of social progress through the fulfillment of the required categorial transversality conditions [R1.92] [R1.203].

The appropriate law of motion that will govern socioeconomic transformations depends on capital, labor, technology and relevant social infrastructure. The development of capital, social and physical technologies depends on labor, its level of quality and the nature of socio-political organization of societies with implicit or explicit defined social vision and national interest under a particular philosophy and ideology that provide a mindset of decision-choice activities and social action. The assessment and valuation of the initial national endowment in decolonized Africa territories revealed that Capital, social and physical technologies were both limitative and limitational due to the absence of appropriate work and labor forces (see [R13.10a] [R13.10b]). The problems of capital and technological limitativeness and limitationality can obtain sustainable decision-choice solutions only through the improvement in the quality of the domestic labor force in terms of *education* and *health* to create the needed human capital. In other words, the problems of capital and technological limitativeness and limitationality reside in the limitativeness and limitationality of African human capital. Furthermore, under the theories of category formation and categorial conversion, labor constitutes the primary category while capital constitutes a derived category with the two linked by technology as the information-knowledge structure in the categorial-conversion process under the analytical structure of construction-reduction duality. Similarly, true democracy, freedom, justice; rational democratic institution-building and participation by the population in the national life in every nation are impossible without an educated and enlightened population with a social conscience.

6.3.2 Philosophical Consciencism, Contents, Curriculum of African Studies and the African Collective Decision-Choice Space.

Here, the contents of education are always at the mercy of the nature of Philosophical Consciencism that the nation holds. When Philosophical Consciencism is African-centered, then the contents and nature of the education in the creation of African human capital will be supportive of African progress through nation building. It was through this understanding that Nkrumah actively promoted African studies. The curriculums of current African studies in various universities around the world are irrelevant to the needs of categorial conversion of African social actual-potential polarities. In fact, these curriculums at various African institutions of higher learning serve the interests of the neocolonialists and imperialists but do not serve the interests of Africa. They are developed with the controls of neocolonial Philosophical

291

Consciencism without the guidance of African-centered Philosophical Consciencism. Under the guidance of African-centered Philosophical Consciencism, the knowledge obtained from African studies must permeate all areas of knowledge studies and enrich them with the need to identify Africa's socio-economic and politico-legal problems and find solutions to them. In this way, even the studies of mathematics, physics and other disciplines acquire their African content from African education for the improvement of African nation building and enhancement of social progress as they were in the African traditions and not to improve the skills of African cronies for further management of neocolonial oppression of the African people.

6.3.3 Philosophical Consciencism and Human-capital Development in Social Systemicity

To Nkrumah, therefore, the immediate important policy was the concentration on human capital development through an effective *educational policy* to raise the skill levels at all fronts of social transformation dynamics supported by an effective *health policy*. From the policy of nation building and sustainable social progress education and health policies are inseparable and inter-supportive. This is particularly and extremely important relative to the nature of the African experiential information structure which contains elements of slavery, European atrocities, racism and the European crude promotion of an ideology of European superiority, African inferiority under the action principle of christianization and civilization. Nkrumah's education policy was divided into *general education* and *ideological education*. The general education was to be supported by an *ideological education* designed to imbue into Ghanaians and other Africans, a true African personality, a sense of African awareness regarding the ancient African wisdom of the responsibility of each to the welfare of all or alternatively *each for all and all for each* which emerges from the African conceptual system of a solution to the positional problem in the individual-community duality relative to social actual-potential polarity [R1.92][R1.203] which the imperial terror sought to destroy through the destruction of traditional institutions. This value principle for organization and governance of the African social collectivity guided by scientific decision and management is what Nkrumah called scientific socialism. This approach is a human-centered rather than a profit-centered approach to nation building and sustainable social progress under true democracy and not democracy in a mirage.

The economic-theoretic underpinning of this human-centered approach to social progress and the decision-choice rationality have come to be known in our contemporary times as the human-capital theory of socio-economic development. In this way, the policy construct of social progress is seen as driven not by the *capital process* but by the *labor process* as it should be, since capital itself is a product of the labor process. It is on this basis and under the guidance of Philosophical Consciencism that Nkrumah asserted that: *The people are the reality of national greatness* [R1.203, p.103]. This statement of Nkrumah has been extended with the addition of *the vision of a people is their history yet unfolded. It is a potential to be actualized. This vision is anchored in the intellectual life and collective mind of the people* [R1.91, p. 199]. The long-run solution to Africa's sustainable social progress, composed of human and economic developments with independence, freedom and justice is first to democratize *education* and *health*. This approach finds its rationale in the *theory of the labor process*. It is also the cognitive rationality of Nkrumah's policy approach to the definition of African social vision, national interest, social progress and nation building through the substitution-transformation process on the basis of internal self-reliance.

The labor process is seen as manufacturing the categorial moments and categorial transfer functions that will act to set the potential against the actual in the dynamics of social actual-potential polarity by satisfying the categorial transversality conditions where the desired potential is actualized and the unwanted actual is potentialized through the human decision-choice process and its implementation. Additionally, there is recognition under Philosophical Consciencism that knowledge without collectively ideological consciousness is dangerous and collective social ideology without knowledge is vicious. Thus social progress requires both technical and ideological education. Here, the content of education, the organization of learning and the integration of knowledge into the general decision-choice space are very essential to achieve the required categorial conversion with a minimum risk of failure. Given this cognitive foundation, let us examine some of the aspects of economic rationality from Nkrumah's general domestic policy as an African case study. We must keep in mind that there is an element of inseparability and relationality between Nkrumah's domestic and African policies relative to the activities of the imperial incumbency. The domestic policy, the African policy and the non-African policy are designed around Nkrumah's goals and objectives to which we now turn our attention.

6.4 Nkrumah's Domestic Goals and Objectives under Philosophical Consciencism

Nkrumah's sociopolitical agenda and the required economic support system are extensive, covering the whole of the African Diaspora and the major elements of the accumulated Pan-African agenda in relation to the *race question*. His domestic goals and objectives are intimately linked to the Pan-African agenda as well as inseparable from his African unity ambitions. It is thus not easy and straightforward to ascertain his domestic goals and objectives from his goals and objectives of decolonization, sovereignty control, African unity and reconstruction of a new Africa. However, Nkrumah's goals and objectives may be classified into political, social and economic dimensions. The economic goals and objectives are viewed as the vehicles through which certain social goals, nation building, African social progress, complete emancipation and true sovereignty can be attained and maintained.

From the epistemic foundations of Philosophical Consciencism, economics is thus viewed as the material basis of life and social progress in relational continuum and unity under a defined information structure. It is like the economics of natural transformation whose material basis is matter that is linked to energy through the ontological information structure. The relevant material basis of life finds its development in social construction and nation building. As a material basis of life it merely defines future social possibilities and the resources to reach and maintain them over the categorial-conversion enveloping. From the material basis arises economic freedom that will support political and social freedoms. Economic freedom is thus viewed as a means but not an end. Without economic freedom, sociopolitical freedom is mere emptiness. In other words, independence without economic freedom is vacuous residing in a dessert mirage. The grand goal in the new African social order to be established in accord with the Nkrumaist scheme of things is complete freedoms for all Africans at home and in the Diaspora. Complete African emancipation has three goals of political, economic and social freedoms. From the position of strategic decision-choice actions, these economic, political and social goals and objectives are further divided into strategic and tactical ones. The strategic decision is made up of interconnected consequences of tactical choice actions to bring about categorial conversion of African social actual-potential polarities under the epistemic guidance of Philosophical Consciencism.

To be able to implement actions toward the grand goal, it is necessary that the goal of political freedom must be secured at first. This is the first *positive action imperative* demanded by collective social action for Ghana's emancipation as well as the emancipation of other colonized African places. This required the dismantling of the colonial machinery of exploitation and as well as the destruction of the imperial *terror-dome* of physical, mental and cultural violence. In other words, the decision-choice actions are to bring about the categorial conversion of socio-political polarity in order to set decolonization as a potential against colonialism as the actual and establish domestic sovereignty. On the viewpoint of reconstruction of Africa and nation building on the basis of the African conception of humanity, social progress and colonialism are incompatible, just like the viewpoint of sustainable peace, freedom and injustice are incompatible. The relational rationalities of the political, economic and legal structures that will lead to social freedom and progress is presented in Figure 6.4.1.

The imperative of securing political freedom and the logical grounds of action on which this imperative rests are anchored in the guarantee of the decision-making power that the political freedom offers. This imperative acknowledges the idea that he who controls the power to make socio-economic decisions for a people also defines the content and the path of their history. Here lies the epistemic explanation of the actions of imperial predators to engineer the overthrow of democratically elected governments with weak military power and resource-rich territories that are opposed to imperial predation and neocolonialism on their nations and people.

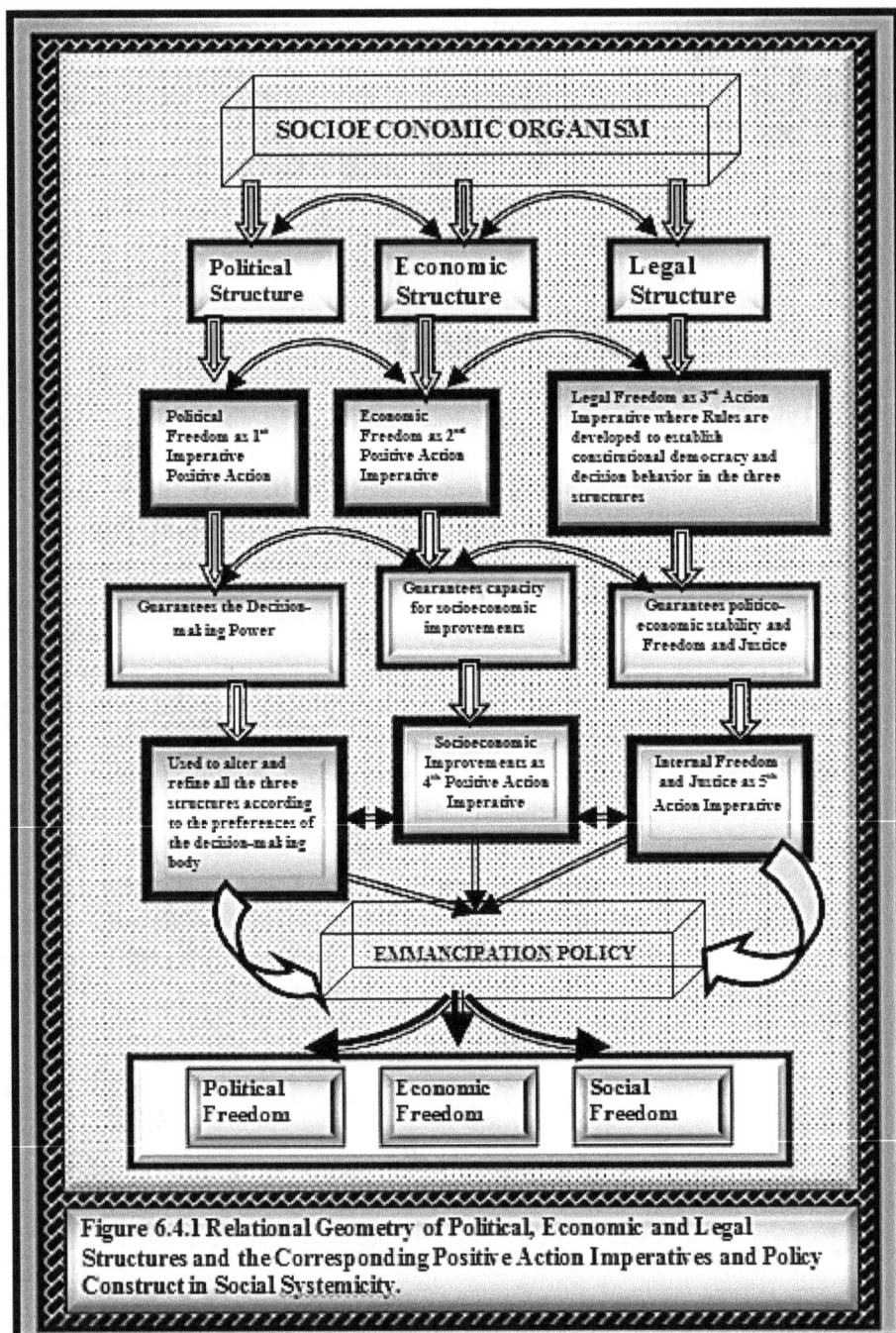

SOCIOECONOMIC ORGANISM

| Political Structure | Economic Structure | Legal Structure |

Political Freedom as 1ˢᵗ Imperative Positive Action

Economic Freedom as 2ⁿᵈ Positive Action Imperative

Legal Freedom as 3ʳᵈ Action Imperative where Rules are developed to establish constitutional democracy and decision behavior in the three structures

Guarantees the Decision-making Power

Guarantees capacity for socioeconomic improvements

Guarantees politico-economic stability and Freedom and Justice

Used to alter and refine all the three structures according to the preferences of the decision-making body

Socioeconomic Improvements as 4ᵗʰ Positive Action Imperative

Internal Freedom and Justice as 5ᵗʰ Action Imperative

EMMANCIPATION POLICY

| Political Freedom | Economic Freedom | Social Freedom |

Figure 6.4.1 Relational Geometry of Political, Economic and Legal Structures and the Corresponding Positive Action Imperatives and Policy Construct in Social Systemicity.

296

It is this historical materialist imperative that led Nkrumah to assert *seek ye first the political kingdom* [80, p. 50], which became the principal slogan of the policy strategy of Nkrumah. He recognized that *political power is the inescapable prerequisite to economic and social power* [R1.207, p. 78]. In other words, Ghanaians and other Africans will not have the power to decide their destiny and achieve the grand goal of Africa's and Ghana's complete emancipation, sovereignty and freedom with fairness and justice without political independence and domestic control of sovereignty. Additionally, without political independence of African territories the questions of defense of sovereignty and African unity are mute and reside in the dustbins of imperial and neocolonial oppression. For the set of positive actions of economic objectives to have a revolutionary impact and lead to the required socio-political transformation of African societies they must constitute a program of thought and actions where the decision-information-interactive processes are examined with cognitive intensity at all level of Ghanaian social practice, and be intimately linked to Africa's complete emancipation through Ghana's *African policy*. This must be so for all decolonized African territories. Both Ghana's domestic policy and African policy must be crafted in relation to *non-African policy* in such a way as to support the grand goal of Africa's complete emancipation. The domestic policy, the African policy and the non-African policy constitute a policy trinity. All decolonized African countries must have this structure of policy trinity if Africa is to retain her sovereignty and embark on meaningful and sustainable nation building from within. For all the individual policy trinities of decolonized African states to be effective in preventing neocolonial insurgence and imperial harassment of African human and non-human resources, the Africans must have one epistemic understanding and be coordinated where Africa speaks with one voice, *The Voice of Africa*. This unified epistemic understanding is provided by Africentric Philosophical Consciencism. The nature of the policy trinity as abstracted from the social policy space is presented in Figure 6.4.2. This was the decision-choice foundation of the theory of diplomatic decisions and this must be the guiding principle of other African diplomatic decisions where Africa's interest is always first.

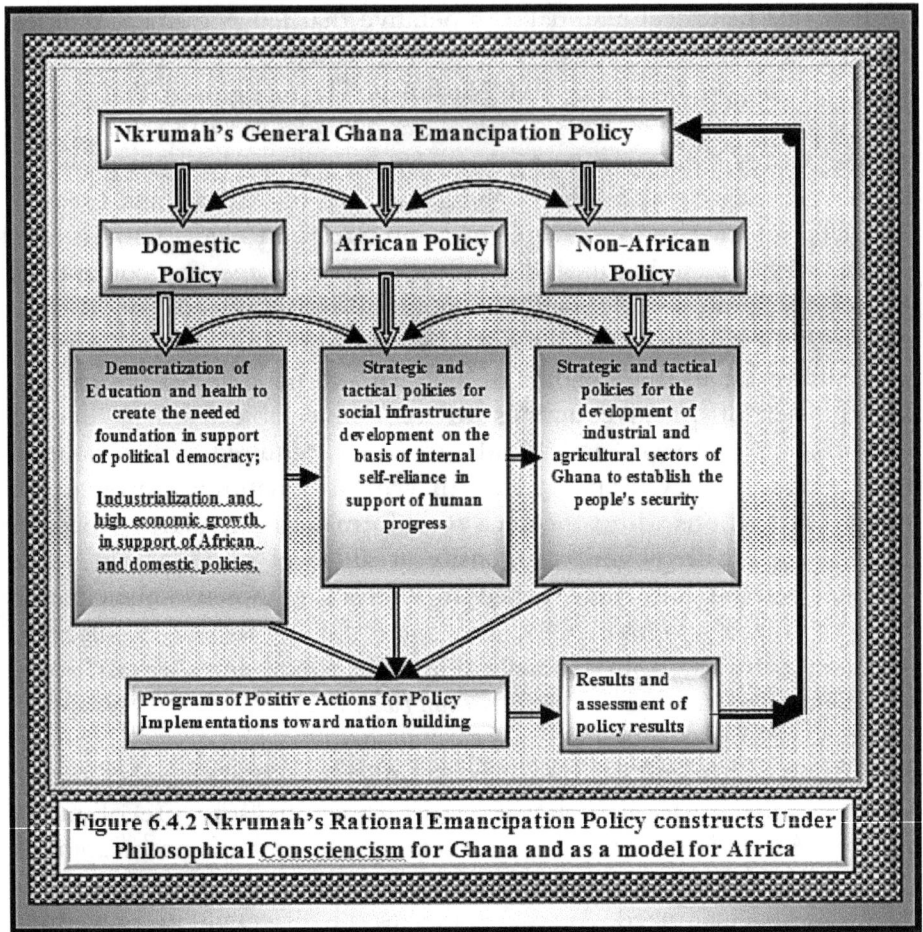

Figure 6.4.2 Nkrumah's Rational Emancipation Policy constructs Under Philosophical Consciencism for Ghana and as a model for Africa

From Nkrumah's general emancipation policy rationality under African-centered Philosophical Consciencism, a generalized structure may be presented for use by other African decolonized territories where Africa is guided by a common African intellectual mindset relevant to complete African emancipation and not a mindset that enhances neocolonial conditions in Africa and strengthen the imperialist machinery of oppression as a continuity of colonialism, slavery and subservience. A generalized structure of comparable relational emancipation policies of African decolonized territories under the principles of relational unity may be presented as an epistemic geometry in Figure 6.4.3.

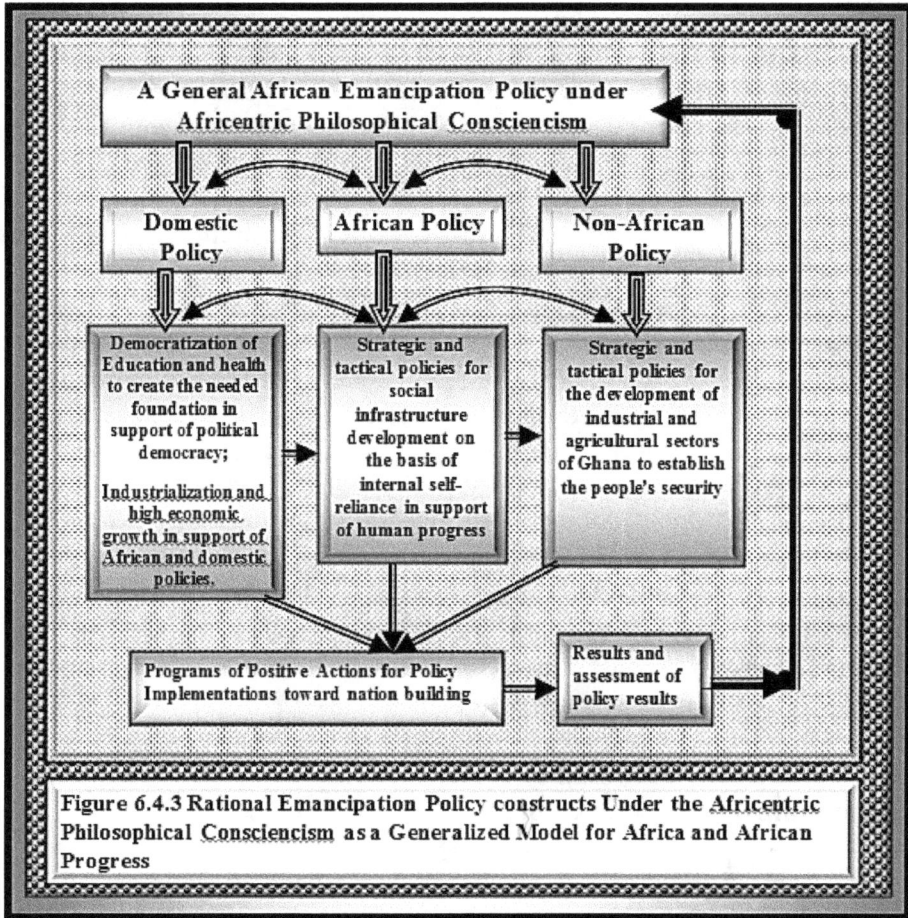

Figure 6.4.3 Rational Emancipation Policy constructs Under the Africentric Philosophical Consciencism as a Generalized Model for Africa and African Progress

6.5 Decision-Choice Rationality and Nkruma's Policies under Philosophical Consciencism in Social Systemicity

Nkrumah's model of decision-choice rationality under the mindset of African-centered Philosophical Consciencism for the defense of territorial integrity, African sovereignty, social progress, nation building and administrative management of national affairs is made up of political, economic and legal rationalities that are the result of the three organizational sectors of the social system. These three rationalities correspond to political, economic and legal decision structures that constitute the decision-choice foundation of the society in relational continuum and unity. The political rationality involves the intelligence of

299

decisions and choices in the political structure of the society in which social decision-making power is vested in terms of organizational rules, modes of governance and codes of power distribution with individual and collective power participations. The economic rationality involves the intelligence of decisions and choices in the economic structure that constitutes the foundation of life in terms of rules and institutions of resource allocation, use and cost-benefit distribution. The legal rationality, on the other hand, involves the intelligence of decisions and choices in the legal structure that constitutes the foundation of control and harmonization of behavior of members of the society in terms of appropriate rules of the national organizational system of its governance of the legal, political and economic structures. The type and the character of any of the three rationalities associated with a national leader and the leadership of governance depend on the social conscience, ideological orientation and their philosophical support that affect the conceived vision of nation building, social progress, the path of national history and the collective personality of the citizens. Let us examine these rationalities under African-centered Philosophical Consciencism that form the logical justification of Nkrumah's policies and their extensions to decolonized African territories in some detail.

6.5.1 Political Rationality and Domestic Policies under Philosophical Consciencism

At the level of political rationality and in accordance with Nkrumah's goals and objectives, Nkrumah's decision intelligence reveals that without political control Africa's social vision, national interest and aspirations cannot be defined. The lack of this definition simply implies that the goal-objective set in support of the relevant vision, interest and aspirations under African will and preferences cannot be formed. In this way, categorial conversions of socio economic polarities for increasing qualitative transformation of African territories will not be African-preference driven. Furthermore, without political control of the national institutions and assets sustainable nation building from the internal social forces in accord with Ghana's vision as well as the linkage of this vision to Africa's complete emancipation is impossible. Such impossibility in the social progress is also true when the political structure with the decision-making power is externally controlled after decolonization. This is the case of neocolonialism.

This type of political rationality points to the direction to completely dismantle colonialism, imperialism and neocolonialism, as well as the system of internal collaborators under the imperialist *terror dome,* in order to bring into place true democracy by creating favorable internal conditions on the basis of the efforts of African people. It also points to the direction where true sovereignty in Ghana and any other African decolonized territory can only be enhanced and secured with African unity under some form of continental government. In this policy process, African Unity acquires a strategic decision importance in ensuring the maintenance of territory-specific independence and sovereignty as well as true peace and stability required for individual nations' social progress. African unity becomes part of the instruments in the manufacturing of categorial moments and transfer functions to ensure the satisfaction of categorial transversality conditions to qualitatively move the decolonized African territories over increasing higher socio-economic categories without external impressions. The political rationality further points to the direction where true political democracy can only be attained after decolonization by democratizing education, information and health. These are landmarks of the governance of Nkrumah's domestic policy in Ghana where technical education was made free from primary school to the institutions of higher learning, and where the national health services were made free and available to all residents. Here we find a different look of the concept of democracy where democracy is not synonymous with voting but many essential elements of life must be democratized in accordance with natural democracy regarding the four elements of air, water, fire and land that form the foundation of African philosophy under the principles of opposites in terms of polarity, duality and negative-positive characteristic sets with relational continuum and unity [R1.92] [R15.15].

It may simply be pointed out that the power to shape national history through decision-information-interactive processes is vested in the political structure of the society and not in either the economic or legal structure. Without the internal control of the political structure and sovereignty, the nation is at the mercy of external forces and the interest of the nation is appended to the interests of those external forces that directly or indirectly control the political structure, sovereignty and the corresponding social decision making power of the state. In fact, the people of such a nation cannot claim the history of the nation because they have no ownership and any talk of sovereignty is simply an apparition. In this way, the freedom to internally shape the national history and the peoples' destiny through the development of economic forces to serve the people's interest according to their legal structure is externally arrested and taken away from them. The processes of categorial conversion of socio-economic categories of this class of nations, such as neocolonial states, are indirectly arrested while those of the class of colonial and imperially occupied states are directly arrested. The right of a people to collectively decide their own destiny, to create freedom and justice, to develop their productive forces according to their needs, and to distribute them under conditions of fairness is guaranteed only through their control of the political structure that holds the sovereignty and the social decision-making power. Thus, the political rationality for emancipation at both the individual states and the union of Africa is simply *Seek ye first the political kingdom*. Unquestionably, this situation was clearly understood by Nkrumah under the framework of African-centered Philosophical Consciencism. In turn, this political rationality must be understood by all the leaders of decolonized African territories who have the vision of nation building and social progress for Africa.

The basis of the problem of human poverty, hopelessness, misery and external dependency of a nation is economic. The basis of the solution to this problem is political. It, thus, logically follows that the people's control of the political structure is unquestionably the first and indispensable step towards emancipation and redemption of African territories, such as Ghana, where such emancipation and redemption will guarantee a path to freedom and justice that can be linked to other African states. One cannot speak of a policy towards African Unity and the possibility of forming a continental African government when the territories are under colonial occupation and imperial *terror-dome*. This internal political control is a necessary condition and not sufficient condition to bring the solutions of social progress and nation building into being. The sufficient condition is found in the *collective conscience, collective ideology* and *collective personality* of the society on the basis of which perceptions are formed, information is processed, knowledge is constructed, decisions are made and choices are implemented to alter the unwanted situations and to bring about categorial conversion of socio-economic actual-potential polarities. The *collective conscience, collective ideology* and *collective personality* required to create the sufficient conditions and to manufacture categorial moments and transfer functions to satisfy the categorial transversality conditions of categorial conversion in the African case is Africentric Philosophical Consciencism.

The collective ideology must be African and the collective conscience and collective personality must in the essence be African based on the general epistemic structure of Nkrumah's policy construct for African decolonized territories. The conception of the vision of society and the structure of nation building is only possible when the political power is domestically controlled to generate the nature and form of the political organization which then dictates the appropriate ideological regime, while the ideological conditions shape the morphology of the socio political organization that may take hold of the society as well as define the social power distribution in the social decision-choice space and the regime of political rationality that may be defined over the political structure. The regime of the political rationality imposes the form of the legal rationality to which we now turn our attention.

6.5.2 Legal Rationality and Domestic Policies under Philosophical Consciencism

From the understanding of the urgency and the imperative of the control of the political structure with the defined political rationality, questions arise as to the nature of the rationality of the rules and regulations that will enhance or restrict the people's decision-choice activities in the politico-economic production space. This is the legal rationality that is defined in the legal structure. Here, the vision of the society as conceived in the political structure with its decision-choice rationality shapes the legal configuration that includes social property rights within the individual-community duality as well as define the legal rationality which is the decision-choice intelligence in creating laws and rulemaking to set the boundaries of acceptable behavior, and how such rules and laws are administered to ensure social stability and national harmony. Every social system has a legal structure composed of explicit and implicit laws, rules and regulations and legal decision-choice rationality with a supporting philosophy and ideology. At the level of legal rationality, Nkrumah's decision-choice intelligence reveals that legal precepts and emerging rules of individual and collective behavior in the three structures of politics, law and economics must have their roots in the culture of the people if the evolving and constructed legal precepts are to allow flexible circumference of full participation in the national life of the people. For consistency of African social formation, the construct of the legal rationality, just as that of political rationality, must follow the epistemic framework of African-centered Philosophical Consciencism. The flexible circumference of democratic participation in the national life and nation building is established by the diameter of the political rationality. The more such political rationality is rooted in the people's culture, the more enhanced the evolving diameter and the greater the flexible circumference that is spun. In this way, both the political and legal rationalities provide solutions to the conflicts of preferences in the individual-community duality, where such solutions are shifted to the community preference structure to generate an increasing space of democratic decision-choice actions with greater participation by the members of the community as it has always being the Africentric tradition in the African social formation under the principle of *each for all and all for each*.

The legal rules must also be related to the national interest and aspirations where freedom and justice operate, without which the concept and practice of democracy is pure emptiness. There is no democracy in abstract. Democracy is a set of algorithms for computing and resolving conflicts in the structure of community decision-choice activities under diverse individual preferences without violence as seen in the Africentric conceptual system for the various relational points in the individual-community duality. This is also the principle on the basis of which Nkrumah claimed that the structure and contents of Africentric scientific democratic socialism. True democratic social organization is an evolving process from within the culture of a particular society. It cannot be imposed if it is to be sustained. It must draw its nutrients from the cultural confines of the society if it is meant for full participation in the national life and collective vision. The development of the supporting philosophical and ideological frameworks must be inspired by the culture and experiential information structures of the society which in this case are the African historical conditions and traditions. Without legal rules that are culturally based under African-centered Philosophical Consciencism, harmony and unity of African societies cannot be created and maintained. Nkrumah's implied decision intelligence in the legal rationality points to the direction of *each for all and all for each* where the community is the primary element of legal rationality and the individual is a derived category, to provide guaranteed freedom, justice and fairness in the collective and the social organism. It also points to the road of democratization of responsibility, benefits, freedom and justice in accordance with the collective interest of the society where such democratization is embedded in the legal rationality. Within this epistemic structure the work of the individual must draw its strength and value from the community in terms of the behavior of individual-community duality [R1.92] [R1.197] [R1.203]. Similarly, the community behavior must find support from the members within the framework of legal rationality.

The African-centered thought system that has guided Africentric social formation and governance defines legal rationality where individual interest rests on community interest. Here individual freedom, while forming the legal foundation of individual action, does not determine collective freedom in the legal space in accordance with the African traditions which must be incorporated into the construct of African-centered Philosophical Consciencism. This is one of the puzzles in the foundations of African philosophy regarding social formation. It is on the basis of this type of legal rationality that the coat of arms of Ghana is inscribed with the motto *Freedom and Justice*. The type of regime of legal rationality adopted imposes a form of economic rationality that must be consistent with the political rationality by logical extension. Let us now turn our attention to economic rationality.

6.5.3 Economic Rationality and Domestic Policies under Philosophical Consciencism

Given the politico-legal rationalities and the internal controls of the political and economic structures, freedom and justice must be maintained by the development of the material basis conditional on *economic rationality*. At the level of economic rationality, decision-choice intelligence is structured in three mega-dimensions under Nkrumah's application of African-centered Philosophical Consciencism. The first dimension is the rational recognition that the people are the backbone of the national creative process at all levels of human endeavor, nation building and social progress. It is by the people's effort that social vision is defined, national interest is constructed and the social goal-objective set is formed. The people are the actual and potential wealth of a nation. It is also by the people's sweat that the conditions of an actual are destroyed to leave room for new conditions and the conditions of the potential are created for the actualization of the potential within the categorial conversions of actual-potential social polarities. Additionally, the youth are the national asset for the defense of today and the construction and actualization of tomorrow's potentials. The second dimension involves the rational recognition that the people's creative force involves *mental* and *physical qualities* given their quantity. The mental quality is seen in Nkrumah's decision intelligence under African-centered Philosophical Consciencism as *mental labor* whose improvement can come about by *education and training*. The content of such education and training and the methods of delivery are extremely important if education and training are to achieve the intended social goals.

The *physical labor* is seen as raw labor whose improvement can be brought into being by good *health services* and establishment of institutions through which health-services policies are delivered. The third dimension is the rational organization and utilization of mental and physical labor towards the production and reproduction of life in the creative process of nation building under the conditions of Africa's vision of the world. The third dimension can come about by a rational construct and application of a system of production forces through the development of industrial, technological and agricultural policies that are consistent with the categorial conversions of actual-potential polarities of social stages and national character of the society. In other words, to organize human and non-human resources to execute the policy program to actualize the potential in the vision to replace the existing socio- economic actual, such as higher socioeconomic development for lower socio-economic development, or fairness for unfairness in the distribution of cost-benefit configuration.

The relational structure of the labor process as projected by Nkrumah's policy structure under African-centered Philosophical Consciencism is presented in Figure 6.5.3.1. This policy construct is consistent with sustainable development composed of nation building and societal-welfare improvements through the labor process given Africa's social vision of complete emancipation and the supporting national interest. This economic rationality of categorial conversions of socioeconomic actual-potential polarities is applicable to all the decolonized African territories under the continual threat of neocolonialism through the insurgence of the remaining forces of imperial aspirations in the decolonized African territories. This economic rationality requires the participation of all in the cost and benefit of categorial conversion of the socioeconomic actual-potential polarity under the Africentic principle of each for all and all for each in the individual-community duality in relational continuum and unity.

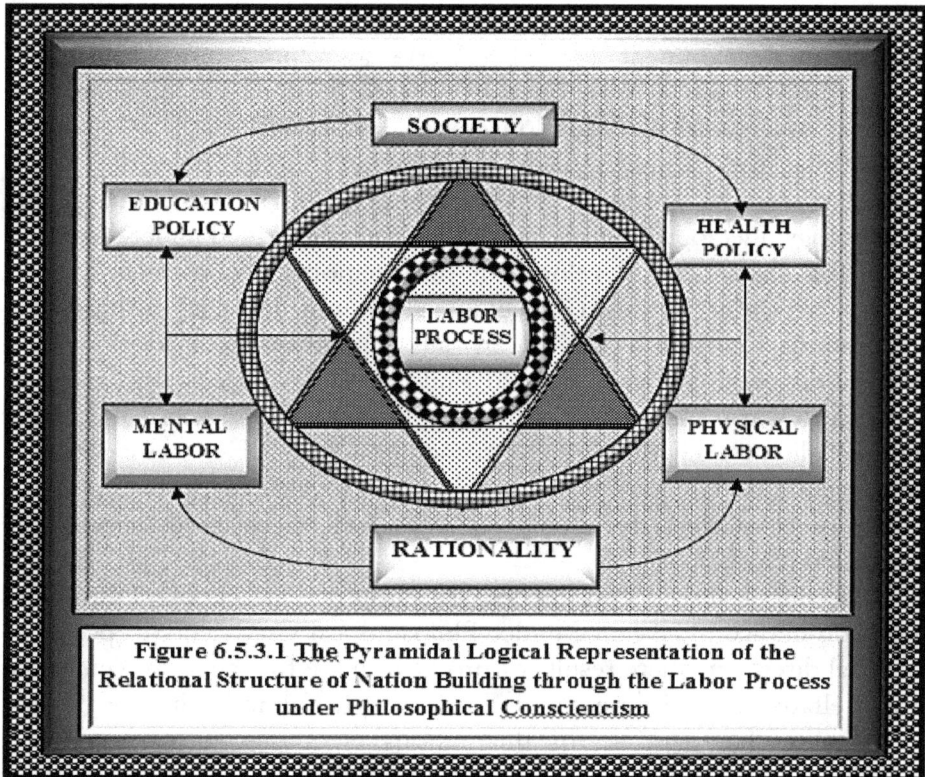

Figure 6.5.3.1 The Pyramidal Logical Representation of the Relational Structure of Nation Building through the Labor Process under Philosophical Consciencism

The socio-economic rationality points to the direction where, by democratizing education and health, the system leads to democratize responsibility, freedom, justice and fairness for all in the effort of nation building composed of social deconstruction and reconstruction under the time trinity of the *Sankofa-anoma* principle of information- decision-interactive processes toward Africa's redemption. At the level of socio-economic practices, relative to social goals, objectives and economic rationality, and guided by the framework of Philosophical Consciencism, the basis of nation building from within requires that education and health facilities be provided by the state in all provinces, rural areas, towns and cities. The underlying decision-choice intelligence translates into the health and education (human-development) policies of Nkrumah's vision of Ghana and Africa where other Africans were given opportunities to benefit from this grand social design of Africa's tomorrow of nation building and social progress under categorial conversion. In other words, the theories of Categorial Conversion and Philosophical Consciencism provide a complete explanation to socio-economic development as well as serve as a prescriptive intellectual framework to set African on the right path of nation building with Africa's sovereignty after decolonization.

At the level of policy practice, the socio-economic rationality points to an integrated approach where the policies of education, health, technology, ideology, industry and agriculture are unified under a national plan for sustainable development and national reconstruction from within in order to resist neocolonialism and imperial insurgencies. Education and health backed by an African-centered ideology and supported by an African-centered Philosophical Consciencism must be directed towards the welfare of the people under the principle of each for all and all for each through the recognition that Africa's emancipation and redemption require African solutions to Africa's problems in the fields of economics, technology, culture and science with appropriate social ideology.

At the level of production-consumption practices, Nkrumah's decision-choice intelligence under Philosophical Consciencism points to a social direction where results of production are for the improvement of the welfare of all the people and not simply for individual profit and enjoyment. The wisdom of this rational synthesis is that economic development of a nation is brought about through the peoples' effort and creativity in all dimensions of human existence. The efforts must be organized on the basis of the African tradition of the communal principle where each is for all and all is for each in relational continuum and unity. The communal principle finds social expression and human inspiration from African humanism. This is the audacity of African tradition and the criminality of neocolonialism and the supporting imperial oppressive philosophy that contains racism, war and violence. Besides democratization of education and health and the creation of corresponding facilities, macroeconomic rationality compelled Nkrumah to organize the workers and the youth so that economic development should proceed from within the nation where internal factors would be the basis for categorial conversion, and external factors would be the conditions that point to the direction of the categorial conversion (see

[R1.122]) on the basis of scientific applications in organization of human and non-human resources.

This approach is supported by an economic rationale that development and nation building are labor processes that rest on the population of a nation (see [R15.14]). The effective utilization of the labor process must be scientific and draws its strength from African cultural foundation. In the African context, therefore, the policy of the labor process must be abstract from the understanding of the behavior of the dynamics of individual-community duality in relational continuum (communalism) in order to create conversion moments and transfer functions to effect the required categorial conversions of socioeconomic actual-potential polarities in the substitution- transformations of categorial dynamics under African domestic control of sovereignty, where political and legal powers are in the hands of domestic social decision makers.

African ideology contained in African culture and supported by Philosophical Consciencism is the center of categorial conversions and science is the center of its application to bring about conversions from one category of progress to another category in the nation-building process towards Africa's complete emancipation. It is within this context of African communalism under the principle of African humanism and supported by the application of modern science on the basis of decision-choice rationality abstracted from the behavior of individual-community duality that Nkrumah spoke about *scientific socialism* which he might have called scientific communalism which is the African tradition of the solution to the problem of individual-community relative freedom brought to scientific modernity [R1.203]. Extensive philosophical and sociological expositions to the foundations of scientific communalism and its relationship to Nkrumah's terminology of African scientific socialism have been given in [R1.90a] [R1.91] [R1.92]. It must be made clear that *categorial conversion* presents a framework of general dynamics in quality-quantity dispositions that require the creation of *categorial moments*, *categorial transfer functions* and *categorial transversality conditions*. Philosophical Consciencism presents an integrated decision-choice framework to manufacture the categorial moments and transfer functions to satisfy the categorial transversality conditions in the categorial-conversion process of social actual-potential polarities.

It is precisely this substitution-transformation logic in combining African Ideology contained in Philosophical Consciencism and modern science why Nkrumah's education policy emphasized ideological, education and general education on the basis of science and technological education. To implement the policy, Kwame Nkrumah Ideological institute at Winneba, Kwame Nkrumah University of Science and Technology at Kumasi and University of Science Education at Cape Coast were established. These educational facilities were then supported by the development of a general university education with medical sciences at Legon-Accra. The policy of higher education was aggressively supported by an awesome development of secondary institutions of general, science and technical education. To build the foundation for true participatory democracy with African content, education was made free. Essentially, the development of democratic social formation that is to take root in the culture and social traditions of the people requires equipping the citizens with the ability to access information and cognitive capacity to process the information into knowledge for collective decision-choice activities with conflicting preferences through the democratization of education.

At the policy level of health services, hospitals were built to create internal facilities for medical services by Nkrumah based on the practical guidance of Philosophical Consciencism in that *thought without action is empty and action without thought is blind* [R1.203]. In other words, a theory without practice is empty and practice without supporting theory is irrational. This included fixed and mobile facilities whose goals were prevention, cure and treatment. The mobile hospitals were to service rural areas where hospitals were not accessible. The services were made free. These policies were supported by an aggressive policy of information dissemination through mobile cinemas and Africanization through the mobilization of domestic cultural elements all made free to support a work-and-happiness policy.

The policy of organizing the workers and youth on the basis of the framework of Philosophical Consciencism must be composed of activities of skill training, work-ethics, self-assuredness in action, creativity in thought and practice, freedom and justice in the management of African social formation under relational continuum and unity for development, complemented with ideological orientation of good citizenship, nation building and love for Africa. The ideological orientation is particularly important since the entrenched colonial collective and individual mentality of the nation after decolonization must be replaced with an African-centered ideology that is culturally consistent with the African way of life in order to point to a correct direction of the use of education and skills in the service of the nation and in support of domestic and African policies of all decolonized African territories such as Ghana, rather than the use of education and skills to support the colonial and neocolonial machinery of exploitation and oppression as happened in the period of colonization and colonialism under the imperial terror-dome involving complete abstraction of Africa's fruits through instruments of enslavement of Africans.

The national demands of skilled workers with good work ethic, citizens with African consciousness and the external conditions of imperialists led Nkrumah to follow a rational policy of development of the organization of a portion of the Ghana labor force into the Workers Brigade, where workers were trained in discipline at all levels, responsibility to self and the community and good citizenship with an African personality and African consciousness [R1.122] [R1.209] [R1.211]. On this policy objective Nkrumah stated: *They* [the workers] *are being given the elements of skill which will enable them to find employment in agriculture and industry* [R1.203, p.126]. This kind of organization of workers was rationally appropriate for a country where about ninety percent of the population was locked in under-self-employment on agriculture without the experience and discipline of having to work for an organized system with rules for all. Both the Workers Brigade and Ghana Young Pioneers may be viewed as policies of an apprenticeship program of national importance. They were meant to enhance the categorial-conversion moments and categorial transfer functions required to effect social transformation and nation building through the behavioral dynamics of social actual-potential polarity.

The need for sustainable socio-economic development that will translate into sustained social progress composed of human and socio economic development requires the creation of conditions that will allow the youth to be ideologically connected to the working masses, social vision, national interests and nation building in a way that allows the youth to develop a sense of self and national pride in Africa and Ghana. From the view point of Nkrumaist thought and under Philosophical Consciencism, we are Africans first before we define ourselves in terms of state-specific or ethnic-specific conditions. Thus, as it has been argued somewhere, African nationalism supersedes state-specific nationalism which overrides ethnic-specific affinity and tribal-specific attachment in accordance with the framework of the contents of Philosophical Consciencism if the decolonized African territories are to be serious in finding solutions to the problems of categorial conversions of social actual-potential polarities [R1.90a][R1.91]. It is easy to change the ideological habits of society by starting with the youth by creating a new mindset with African content. To this end Nkrumah established the Young Pioneers as a youth movement,

> "...which is designed to give them training in citizenship within a society which will be tooted in co-operation and not acquisitive competition. For this end Africa needs a new type of citizen, a dedicated, modest, honest and informed man. A man who submerges self in services to the nation and mankind. A man who abhors greed and detests vanity. A new type of man whose humility is his strength and whose integrity is his greatness." [R1.202, p. 130].

Both Nkrumah's youth policy and workers policy within the general labor policy where education and ideology are combined to enhance the needed categorial conversion moment and transfer function to satisfy the categorial transversality conditions for categorial conversions of social of socioeconomic actual-potential polarities are not understood by Nkrumah's critics as well as African leaders of yesteryears and today. Nations are built by the people who constitute the essential matter-energy structure in a unified continuum and relational unity under an information structure. At the level of higher education at the universities, the practice of combining ideological education and scientific education takes an unprecedented form under the ambit of Philosophical Consciencism. The policy of African studies in support of domestic and African policies was introduced through the establishment of African Studies in the University of Ghana. This was the first of its kind. To make it effective it was mandated that every student in any Ghanaian university must pass the examination in African Studies in addition to his or her area of study in order to receive certification. The simple idea behind the ideological policy is to make an African an African through the understanding of the African-self. The model of African studies at the University of Ghana became duplicated throughout the world and completely lost the ideological significance in the categorial conversion process of stages of social progress through the enveloping of categorial conversions of social actual-potential polarities.

The logic of the socio-ideological deconstruction of the colonial and neocolonial intellectual system and the reconstruction for social progress after decolonization, is that members of the Workers Brigade and Young Pioneer Movement must be imbued with African-centered ideology and personality to allow them to think in terms of African nationalism in the use of their education, knowledge and skills in the reconstruction of Africa as Africans emerge from the oppressive terror-dome of imperialism and colonialism where there was no principle of democracy. In other words, to change the frame of reference that allows the use of education, skills and knowledge that Africans possess. Instead of using knowledge and skills to service colonial machinery and the interest of the imperial predators, they must now be put the skills to the service of Africa and nation building through social reconstruction of specific states and Africa in intra and inter support towards complete African emancipation. The rational path for organizing the workers and youth in Nkrumah's model of Ghana's progress and Africa's emancipation and redemption was motivated by the principles of *Sankofa* [R1.219] as applied to the internal and external conditions of Ghana's history in particular, and African history in general as the past, present and the future are connected.

As Nkrumah observed: *Our cruel colonial past and the present-day intrigues of neocolonialism have hammered home the conviction that Africa can no longer trust in anybody but herself and her resource* [R1.202, p. 127]. This Nkrumah's statement is fundamental in examining Africa's socio-economic reconstruction, technological development and the casting of relevant institutional configuration in support of Africa's policy process for categorial conversions of socio-economic actual-potential polarities under African sovereignty. The power of this observation has not been appreciated by other African political leaders who are running around the globe with beggar's baskets in humiliation to beg for water, and merely discover after arriving in their respective states that the baskets are empty of their substance, and hence the cycle of begging must continue without honor and in full disgrace of their citizens. What a shame and what an anti-African ideology this begging process projects onto the African decision-choice space of social transformation under the neocolonial ideology.

The integrated policy model of Nkrumah within the conceptual framework of Philosophical Consciencism is for a complete emancipation based on Africa's self-reliance, internal transformation and continental unity. His application to Ghana is such that Ghana's domestic policies must closely follow her African policy for integration and unity that Africans must take advantage of the beauty of Africa's diversity, and Africa's strength must come from the unity of her people and not from colonial divisionism. The solutions to Ghana's internal and external problems just as those of other African states can be found on the African continent and in African unity under the African-centered principle of each for all and all for each within the Africentric conceptual system of the principles of the opposites. No other choice is available to us looking at the strategies of the imperial predators and their insatiable appetite for more and more resources, global dominance, and military power with violence under the dubious concept of Western humanitarian killings, democracy and violence without consideration of the interests of other countries. This was the case of the emergence of the African holocaust that took place under the European imperial terror-dome and the unprecedented abuse of humanity through slavery, violation of territorial integrity and sovereignties. The path of the human-capital theory of national progress and nation building may be ascertained from Figure 6.2.1 in this Chapter.

6.6 Some Reflections

We have presented the socio-economic and politico-legal theoretic foundations of the rationale to explain Nkrumah's organic policy and sub-policies for sustainable African social progress under Philosophical Consciencism with African contents. The structure of the policy group is composed of a sub-policy of nation building and a sub-policy of socio-economic development under the moral principle of African communalism brought to scientific modernity. It constitutes a model for Africa's decolonized states. In understanding the explanatory and prescriptive rationale of the policy group one must construct the initial decolonization and post-colonial questions. The question faced by Nkrumah is how Ghanaian society and African social formations should be organized after decolonization to bring about sustainable social progress with freedom and justice under African control of sovereignty to ensure territorial integrity. The second question was how to protect the decolonized territories from neocolonial violence and the activities from the imperial terror-dome that will seek to negate African control of sovereignties and territorial integrity. These questions are also questions for all African leaders with an African conscience. The questions as posed here are discussed in many dimensions in [R1.197][R1.198][R1.202] [R1.204]. They initialized the vision of Nkrumah and must initialize the visions of other African leaders now if Africa is to move forward in national reconstruction and social progress for the masses with traditional African pride. These questions are stated in two parts in the essay and summarized here for clarity.

1. How does African build a unitary state of each decolonized African territory after decolonization to create a nation and to mobilize the nation's human and non-human resources for independence and complete emancipation?
2. How should African decolonized societies, and hence African social formation, be organized to internally bring about sustainable nationhood and social progress with freedom and justice under Africa's own social vision under Africa's democratic ideal?

To answer these two questions a policy trinity composed of political, legal and economic structures under relational continuum and unity was constructed by Nkrumah within the conceptual framework of Philosophical Consciencism. This policy trinity acknowledges certain fundamental realities about colonial, de-colonial and post-colonial realities about African societies as they appear in imperial terror-dome and the intense work of neocolonialist to disrupt Africa's progress. These factual realities as part of African experiential information are summarized here to give a logical understanding of the questions and the theoretical foundation of Nkrumah's policy construct as a model for African decolonized territories.

1 African colonial territories were put together by the colonizers through terror and governed by violence through the method of divide and rule by elevating one ethnic group above others and most of the time arming it as imperial collaborators.
2. Africa was partitioned into colonial states along the lines of imperial resource interests for stability of a system of competitive imperialism without due regard of Kingdoms, ethnic and tribal relations and established territories under occupation.

3. After decolonization these African decolonized states emerged not as harmonious nations but as ethnic and tribal political nightmares. The states were ethnically divided without national identity.
4. Colonialism and imperialism did not and do not develop under democratic principles and their governances are based on principles of oppression, terror, injustice and slavery, where the overriding principle of governance is simply the imperial resource interest through subordination to the detriment of socio-economic development of colonial territories. The imperial resource interests have no respect for free international trade.

They always override any decent moral behavior and international laws based on territorial integrity, sovereignty rights and the rights of others except those of the imperialists.

These are some of the colonial legacies and factual elements of neocolonialism. The Nkrumah's policy trinity is to design a decision-choice system to deal with these stubborn political, legal and economic problems of the colonial legacies and neocolonial insurgence on the continent of Africa. The policy trinity is recognition of the fact that political stability needed for socio-economic development requires that the nations of the African states must be reconstructed from the ashes of African traditions that were put in a hold and /or were sometimes damaged by colonialism, and imperialism and slavery. The policy trinity must be constructed immediately after decolonization by creating conditions that will allow the new nations to have collective identity and personality designed from Africentric Philosophical Consciencism that will create the pathways to decolonize the African mind for nation building.

The creation of a unitary state demands a carefully designed policy of nation building where ethnic and tribal groups haphazardly put together by the colonialists working in the imperial terror-dome are integrated into nationhood, and ethnic and tribal affinities are welded into state-specific nationalism to create a unitary nation within the frame work of African-centered Philosophical Consciencism [R1.91]. In simple language, any decolonized African territory is initially not a nation. The oneness of each nation must be built without which the state will fall into political instabilities through tribal and ethnic conflicts as well as other internal wars with neocolonialists and imperialist pulling diabolical strings for the demise of the African progress. The principle of territorial unity was Nkrumah's recognition which was also projected to the continent at large in terms of African unity in order to avoid intra-state and inter-state wars within Africa. This basic fact of a complete emancipation policy of African existence was not understood by other African leaders who do not appreciate it. This policy recognition when carried out will put a powerful dagger into the heart of the imperial system and its operations in Africa. The potential outcome under the application of the conceptual framework of Philosophical Consciencism was exactly the reason for imperialist terror against Nkrumah in particular, Nkrumah's Ghana in view, Nkrumah's system of thinking in general and any African leader that adheres to this Africentric framework. If this analysis is theoretically faulty or non-convincing, please cast your empirical spectacle over the continent of AFRICA today and paint a picture of what you see.

The policy of nation building is first domestic after decolonization and must be backed by the policy of socioeconomic progress as seen from Nkrumah's policy trinity in a relational continuum and unity under the guidance of Philosophical Consciencism. But this model of construction of a unitary state must not stop there. There is a fundamental recognition that nation building, techno-scientific advances and sustainable socio-economic progress are under the labor process. The effectiveness of the labor process in nation building requires social investment policy in social infrastructure, education and health. Thus, the politico-legal and socioeconomic rationalities of Nkrumah's policy towards sustainable social progress initialize the nation to infrastructure policies, education policies and health-service policies in terms of social investments. One can view this Nkrumah's policy in the states as a house-building strategy and in Africa as a castle-building strategy where domestic social investments concretize the foundations on the basis of which both the nation and Africa are to be built.

Under Philosophical Consciencism the domestic policy of nation building of every African decolonized territory must be supported by Africa policy and non-African policy strategies. The African policy is divided into decolonization policy and African unity policy. The decolonization policy is a strategic one that is geared in support of decolonization of individual African territories under different colonial bondage and imperial terror in order to secure sovereignty with freedom and justice. The African unity policy is also a strategic one geared towards the integration of the decolonized territories in order to strengthen and protect the freedom, independence, territorial integrity and sovereignties of the African states though unity. The non-African policy has its own strategic value in working to prevent neocolonialism in Africa and the transformation of the old imperial terror-dome into a new imperial terror-dome, as well as effectively connecting with the African Diaspora towards the general African emancipation.

CHAPTER SEVEN

CONTENTS AND CURRICULUM OF AFRICAN STUDIES FOR THE CONSTRUCT OF AFRICAN-CENTERED PHILOSOPHICAL CONSCIENCISM IN SOCIAL SYSTEMICITY

The subject matter of Philosophical Consciencism has been discussed in terms of its theoretical foundations and the role it plays in categorial conversion. Philosophical Consciencism relates to knowledge production and information interpretation of reality as it is perceived. The knowledge production, information interpretation and perception of reality then become inputs into decision-choice activities involving instrumentation, controllability and management of the categorial-conversion process through commands and controls. The outcomes of the decision-choice activities become the new reality whose perception leads to restructuring and creating an enveloping of the history of reality on one hand and the perception of such reality on the other hand. In this respect, few concepts of reality must be kept in mind as one examines actions on reality. The conceptual types of reality are ontic reality in the ontological space, epistemic reality and logical reality defined in the epistemic space. The type of relationships formed among these conceptual types of reality will define the experiential information structure for any social set-up. It may be noted that the drive to change any social reality is not what the social reality is, but the perceptions that decision-choice agents attach to the reality and the conditions that maintain it. Such perceptions and the corresponding decision-choice actions are, in the last analysis, the work of Philosophical Consciencism which is the philosophy and ideology that have taken hold of the ruling members and is supported by a substantial portion of the masses of the society.

In general, Philosophical Consciencism must be constructed with the use of epistemic tools of that society as well as the experiential information structure of the society. The epistemic tools and the experiential information structure may be internally constructed or externally induced. In the case of Africa's current conditions, it is because of the relevant needs and true knowledge of the African experiential information structure along with the correction of anti-

African distortions why rigorous African studies in theory and practice tend not only to arise but become critically essential for creating the path of African progress to form a continuity and unity in the spirit of the Africentric time trinity of *Sankofa-anoma* (the interconnectedness of the past, the present and the future). The unity of the time trinity (past-present-future connectedness) encompasses three other important unities of African cultural unity [R1.84], African socio-political unity [R1.91][R1.202[and African epistemic unity [R1.86][R1.192]. The African cultural unity defines African experiential information structure in the action space. The African socio-political unity defines Africa's tomorrow in terms of survival, security and progress in the action space. The African epistemic unity defines African collective thought system in the epistemological space. As a thought system epistemic unity must be created to guide the activities in the action space and also be intimately connected to African-centered Philosophical Consciencism. Here, an important question tends to arise with respect to the development of Philosophical Consciencism that is African-centered and draws its weapons from the African experiential information structure. What is the relationship between Philosophical Consciencism and African studies in terms of African epistemic unity? This question and the answer that may be constructed will occupy our attention in this chapter.

Social transformations are the products of the arts and sciences of decision-choice systems in creating and implementing relevant policies to bring about the needed negations in actual-potential polarities constructs. Policy implementations take place through relevant social institutions. The structure of decision-choice system, the policy constructs and the casting of relevant institutions are the works of the social thinking system that has taken root in the socio-cultural set-up. The social thinking system is a derivative of philosophy and ideology embodied in social-specific Philosophical Consciencism which constitutes the primary category of thinking, learning and knowing. It is this sequential process that provides an epistemic understanding with respect to the idea that any great society is a network of great institutions, and great institutions are the creative works of great minds who are also the creators of the relevant *revolution in thought* that establishes the dynamic content of the social-specific Philosophical Consciencism. For Philosophical Consciencism to be an epistemic revolution, it must constitute a program in thought and action where such a program is embedded in the curriculum of research, learning, knowing, teaching and dissemination in the epistemic process. The African-centered Philosophical Consciencism is derived from an African organic principle composed of the frameworks of polarity and duality with relational continuum and unity. The continual development of Africentric Philosophical Consciencism must be confined to African thinkers who are true believes with confidence, and also emancipated from the colonial and neocolonial mindset that opposes the African socio-cultural principle of opposites for the resolutions of conflicts of preferences in the individual-community duality within the continual process of the Africentric social formation and transformation. In this way, the general African society, under the guidance of Africentric Philosophical Consciencism, will create social institutions that will be an integrated network of African society from within Africa where there is an unshakable understanding that the problem-solution duality is internal to Africa and its command-control dynamics must be dealt with from within Africa if the benefit configuration of political independence is to be maximized, and the cost configuration of political independence is to be minimized to produce a greater and greater social welfare as the basic vision of social formation in accordance with the Africentric principle of *each for all and all for each*.

7.1 The Change of Traditional African Conscience

There is an understanding on the part of Nkrumah from his literary works that the African decolonized territories have new societies that are different from the old African societies before Africa was colonized. There have been cultural and moral interruptions with discontinuities and perhaps shifts here and there produced by Islamic and Euro-Christian activities, epistemic parameters and influences due to invasion, colonialism, slavery and religious violence. The Islamic and Euro-Christian activities were set up, guided and are still guided by a refined African conceptual system in relation to their needs and intentionalities. The traditional African social formation with an egalitarian view of persons and social management under the principles of philosophical humanism has been corrupted and defiled by both Euro-Christian and Islamic traditions where the emphasis is shifted from social collectivism with the central focus on collective interest and freedom to individualism with the central focus on individual interest and freedom. Here, the African concept of God and the pathway to It, which are accorded democratic universality from within the individual-community duality as relational continuum and unity, are replaced by a concept of God and the pathway to It that are controlled by human made rules and regulations for power, dominance and exploitation. In this respect, the individual and collective conscience in relational unity and continuum has been dichotomized into dualism with a relational excluded middle and separation. The individual working with the principle of individualism sees his or her fellow person as an instrument to exploit for his or her advantage. The concept of collective existence is no longer useful in this philosophical individualism. Philosophically, this is diametrically opposed to the African way, and produced and is still producing schizophrenic collective personality in the space of African progress where the social decision-choice activities to promote African social vision, independence and interest count the most within the space of decolonized territories.

Through colonialism and strategies of occupation and colonial control of sovereignty, a new mindset was developed to produce African anti-African mindset and self-hatred in a substantial part of the general African population to support the validity of Eurocentric anti-African ideology with its supporting philosophy. This Eurocentric colonial mindset develops within the relevant core of the African individual and collective social-decisions a colonial conscience which is created to support European imperialism and African human and non-human resource exploitation for the benefit of Europe as well as the related slavery production systems of the Americas and Caribbean Islands. Here, the exploitation of African society whose labor transforms nature into articles of use and bears the cost of the transformation is not the society which receives the benefits of the transformation. This colonial mindset produced by colonial non-African-centered Philosophical Consciencism after decolonization has gone through metamorphosis to reproduce itself as the neocolonial mindset through neocolonial Philosophical Consciencism which is more insidious than the original colonial mindset. Both of them are produced by the same education system and perform the same task of racism, human degradation, exploitation and oppression African. Also both kinds of colonial and neo-colonial Philosophical Consciencism are deadly intellectual viruses that attack the minds of the people under colonialism and neocolonialism and place them in a program with a mindset that induces its victims to operate in a thinking box, where they argue and make decisions in support of their own oppression and suffering as well as pay for them. The most damaging production of conceptual guidance from both colonial and neocolonial philosophical Consciencism in the African decision-choice space is the separation of the individual from its collective in terms of distribution of preferences and interests. Here, the individual or group of individuals having been conceptually separated from its African conceptual roots acts in the same manner as a colonialist or neocolonialist, where the community is seen as an instrument of domestic subjugation for individual welfare in terms of wealth accumulation and not the community welfare in terms of general community's socio-cultural capital accumulation in accordance with African conceptual tradition. The Africentric principle of *each for and all for each* is replaced by dog-eat-dog principle of *each for him/herself and poverty and suffering to the rest*. The principle of individualism taken to an extreme without any regard to the community.

The cost-benefit distribution is completely at variance with respect to democracy, justice, freedom and fairness in human action under colonialism, neocolonialism and imperialism. The colonial mindset creates within an African and by extension the African collective a dislike for colonial government, and this engrained dislike is extended to all governments after decolonization. This colonial mindset favors categorial conversion of African societies with external control of domestic sovereignty in favor of the welfare of the colonialists, neocolonialists, imperialist and their allies. Working with this colonial and neocolonial mindset, under the supreme principle of individual interest and freedom, the members of African leadership, working in colonial and neocolonial tradition personalize the social decision-choice actions and the corresponding resources for their individual interests and welfare at the expense of the people and the collective. The principle of individualism, composed of individual liberty, pursuit of individual welfare and happiness with little, if any, concern for the collective, is given a divine blessing as the primary moral category for which everybody must emulate. The notion that social formation is a collective and the progress of social formation requires collective action. The supreme guide of individual interests and benefits, as copied from colonial and neocolonial Philosophical Consciencism, obscures the meaning of national independence from the individual moral action to the collective progress. National independence loses its collective meaning and then becomes equated to personal independence to pursue the practice of enhancing individual self-interests and greed in the domain of individual arrogance.

The colonial and neocolonial mindset under the principles of individualism has nothing to contribute to African domestic progress in terms of positive categorial conversions within the African domestic control of its sovereignties under decolonized African states. In fact, this mindset acts as an important negative checks on African progress with tendencies of negation of domestic control of African sovereignty into indirect foreign control through the establishment of neocolonial states. In this respect, the set of rules of the domestic economy is justified under neocolonial ideology and the supporting oppressive imperialist philosophy where the driving force of the individual and collective to work within the guidance of these rules as well as the people to enforce them operate under crises of conscience at complete disadvantage to the African people. The African leadership operating under the colonial and neocolonial mindset fails to understand the simple collective meaning of national independence where,

> Independence is of the people; it is won by the people for the people. That independence is of the people is admitted by every enlightened theory of sovereignty. That it is won by the people is to be seen in the successes of mass movements everywhere. That it is won for the people follows from their ownership of sovereignty. The people have not mastered their independence until it has been given a national and social content and purpose that will generate their well-being and uplift [R1.202, pp.105-106].

In fact, under colonial and neocolonial mindset African independence and national sovereignty become lost and acquire the character of a mirage in the desert. It is here, that conflicts and contradictions emerge within the individual-collective duality and why or how the resolutions of these conflicts and contradictions affect the behavioral transformations of African actual-potential polarities after decolonization. The nature of resolutions will depend on the dominant Philosophical Consciencism that is held by the individual and the collective in the social decision-choice and action spaces where such individual and collective actions will shape the direction of categorial-conversion of African actual-potential polarities towards emancipation composed of economic and political freedoms. It is useful to keep in mind that African transformation and progress will not come by itself as time passes by. It will only come by setting social vision, defining national interest, creating social goal-objective sets, making social programs and implementing them. All these require hard work, human toiling, provision for excellence, overcoming adversities, subduing and expanding new frontiers of social and natural environments and creating and recreating new institutions. All these human activities are defined and established in the decision-choice space and transformed into action through the individual and collective mindset. For the success of African transformations, the individual and collective mindset must be guided by African-centered Philosophical Consciencism which draws its weapons from African traditions and African experiential information structure. Let us keep in mind that the African-centered Philosophical Consciencism is composed of African philosophy and ideology for decolonization, development and complete emancipation.

The problem, here, is simply the colonial mindset carried from colonial education, and neocolonial mindset that has been developed after decolonization, both of which allow the African people, especially the educated class, to accept the neocolonial ideology under imperialism as a way of African intellectual and social life. The neocolonial mindset has become the ruling intellectual king controlling African decision-choice behavior in the African socio-cultural space by indirectly controlling thinking, decision-choice activities and social action, where the African experiential information structure has been transformed into an imperialist and neocolonial propaganda bullhorn though education and mass-media machine. This neocolonial mindset must be changed through a process of the development of philosophy and ideology to destroy the neocolonial African mindset and replace it with a true African mindset that is liberated from the sphere of the governance of neocolonial ideology and the supporting imperialist oppressing philosophy.

In terms of the language of the theory of categorial conversion, the neocolonial mindset is the actual in the African socio-political space and must be negated by the creation of necessary and sufficient convertibility conditions in the actual-potential polarity to negate the neocolonial mindset and send it into the potential space by actualizing the African liberating mindset through the categorial-conversion process under the guidance of African-centered Philosophical Consciencism. Here, the neocolonial mindset is the actual pole and the new liberating African mindset is the potential pole which together create a place of actual-potential polarity of conflicts and wars of ideas between imperialist forces and liberation forces, the resolutions of which will define the epistemic and progress paths of Africa's future. The problem here is that each African is completely defined by either independent or neocolonial rationality in the cognitive space where the so-called educated African has become victim of his/her education that has placed him/her in the familiar box of the neocolonial mindset. Instead of his/her education freeing him/her from his/her chains of colonialism, it has dragged him/her into an irreducible chain of intellectual dependency and decision-choice slavery. This irreducible chain of intellectual dependency and decision-choice slavery has produced conditions of transformation blindness, international muteness and intellectual dumbness placing Africans in the zone of intra-African fighting and warring with imperialists and neocolonialists enjoying the game of pulling vicious international and domestic policy wires for Africa to destroy herself from within in order to nullify Africa's independence and sovereignties. The objective of the imperialist and neocolonialist game of pulling vicious international and domestic policy wires is simply to maintain racism, neocolonialism and imperial dominance for acute resource exploitation. Having defined the relevant poles in the epistemological space, it is necessary to associate these poles with negative and positive poles and then identify the nature of the negative and positive dualities and the structure of the negative and positive duals for social actions in the African decision-choice space.

With respect to the conditions of the decolonized African territories, the neocolonial mindset is the negative pole with corresponding negative duality. The liberating mindset, a mindset with African focus and conscience, is the positive pole with residing positive duality. It is useful to keep in mind that the categorial conversion is always on the actual which must be potentialized by the development of decision-choice strategies and tactics to increase the positive characteristic sub-set and reduce the negative characteristic sub-set in the negative duality to transform it into positive duality and thus actualize the potential which is the true African mindset (see [R1.90b]. This actualization is the creation of a new ideology that is backed by a philosophy which is derived from the African thought system and experiential information structure to solve the crisis in the African conscience. The set of propositions within which a philosophy in support of an African liberating ideology retains Africentric humanism and collective freedom is what Nkrumah called *Philosophical Consciencism*. It seeks to solve the crisis in the African conscience by reconciling the conflicts and contradictions in African tradition, Euro-Christian tradition and the Islamic tradition as has been explained. For the purpose of generalization of Philosophical Consciencism in relation to the development of the sufficient conditions for socio-natural transformations, Nkrumah's Philosophical Consciencism is referred to as African-centered Philosophical Consciencism with African content. All other forms of Philosophical Consciencism will be called *non-African-centered Philosophical Consciencism* with non-African contents. The Philosophical Consciencism defines a framework and the boundaries in which socio-political and economic decision-choice activities are undertaken.

The epistemic totality of Philosophical Consciencism is a structure of cognitive program for knowledge production and use, logic of institutional creation, strategic social-policy formulation and creative-destructive positive-action process to continually set the potential against the actual in the dynamics of social systemicity. In this respect, Philosophical Consciencism is the backbone of socio-economic transformation actions in the actual-potential polarity where the transformation actions are social decision-choice dependent. It is this interdependency between Philosophical Consciencism and social decision-choice system with relational continuum and unity that allows one to claim that national social history is an enveloping of success-failure outcomes of decision-information process.

7.2 Socioeconomic Knowledge, Imperialism, War of Ideas and Independence in Social Systemicity

A discussion about the structure and role of abstract ideas in the theory of categorial conversion has been presented in [R1.90b][R15.15]. The construct of the perceptive experiential information structure is influenced by culture, it is argued, while the role that the experiential information construct plays is influenced by socio-cultural intentionality which is then mapped into national interests and social vision which are in turn mapped onto the social goal-objective set. The structural relations of national interest, social vision and goal-objective set are discussed in [R13.8] [R13.9]. It is this socio-cultural intentionality that influences interpretations of derived epistemic results for the perceptive information structure. It is also this intentionality that generate deceptive information structure composed of misinformation, disinformation and projected onto the propaganda space by adversaries. Different cultures

327

present different types of intentionality of the same epistemic element. These types of intentionality from different cultures may violently create opposing interpretations of the same social event and even the same scientific event as they relate to peace, justice, freedom and cost-benefit distribution. The intentionality leads to direct social interpretive actions to distort interpretations of scientific and non-scientific events by creating *deceptive information structure* and *possible worlds* as well as instruments to promote them as truth in order to deceive and manipulate social decision-choice actions for control. The intentionality types are different for neocolonialists and imperialists who seek oppression, injustice, and the control of other people's destinies to enhance their benefits from the members of the colonized and neo-colonized, and shift the imperialist costs to the colonized and neo-colonized while enjoying the benefit from their oppressive decisions.

7.2.1 Oppression, Liberation and Intentionality of Knowledge Production

The intentionality of colonialists and neocolonialists regarding knowledge production and use is about control of other people's destiny strictly for dominance and enhancement of resource seeking. This may be called *oppressive intentionality* where knowledge is sought to guide different phases of the colonial, neocolonial and imperialist tactics and strategies to sustain oppression. The intentionality of the people of colonized and neo-colonized states under imperialism is about the vision of their liberation, freedom and justice for the control of their collective destiny. This may be called *liberation intentionality* where knowledge is sought as an input into decision-choice actions to fight imperialist oppression and design strategic actions to support freedom, justice and destiny control. The oppressive intentionality and liberation intentionality find themselves as constituting duality in the imperialist-decolonization polarity that contains the continual negation-of-negation struggle. Each of these types of intentionality is translated into national interests that are placed in ideological capsules with philosophical protective belts in support of a particular social vision. The different types of intentionality generate ideological battles which are supported by wars of ideas in the oppression-liberation duality and projected into the propaganda space. The output of the wars of ideas is disseminated by mass the media machine of a propaganda bullhorn by the neocolonialists and imperialists for categorial conversion of colonialism-independence polarity. The objective of ideological battles and wars of ideas is to win minds in the game space of the social actual-potential polarity by increasing the negative characteristic sub-set though the decreasing of the positive characteristic sub-set and vice versa . The success of the ideological battles and wars of ideas in favor of retention of the actual pole by maintaining the conditions that support it is to at least maintain the magnitude of the relative negative-positive characteristic set in favor of the actual. Alternatively, the success of the ideological battles and wars of ideas in favor of the intentionality of the potential pole and hence the negation of the actual pole by destroying the conditions that support its existence, is to alter the magnitude of the relative negative-positive characteristic set in favor of the potential.

The liberation intentionality is an intentionality of resistance against imperialist intentionality of unjust exploitation of human and non-human resources with socio-natural environmental destruction leading to oppression of a population. In this battle and wars of ideas, the imperialists are always organized by creating different kinds of domestic and international institutions through which their policies are executed. They call the external institutions of their creation, international institutions which are tightly controlled to support the imperialist domestic institutions and policies of oppressive intentionality. The imperialists understand that the success of their oppressive intentionality internally depends on them through their domestic decision-choice actions and the manner in which these actions are projected in the global political economy as well as the inter-supportive actions of the imperialists' behavior in the global space of production and consumption. These decision-choice actions are guided by their imperialists' ideology supported by an oppressive philosophy under the imperialist intentionality to which all the participating imperialists subscribe to. The imperialists know that they are their own strength and weakness in the game through their power, and have an estimate of the power of their victims and the territories to colonize, neo-colonize or occupy. They have the concurring organization in terms of ideology, intentionality, and capacity for violence supported by imperial discipline, surprise, global deception and integrated complex surveillance systems. They amass, overwhelming power to physically, intellectually and spiritually subdue and concur, and directly or indirectly enslave for dominance and resource abstraction. This imperialist power is built on fear, envy, suspicion and dislike for competition.

7.2.2 The Western Imperial Intentionality, Colonialism and African Enslavement

The Western members of the Western imperial system understand the meaning of power and its capacity to alter social arrangements. The development of imperialist institutions to create the power for war-making, resource-seeking, colonization, neocolonialism, territorial occupation, human oppression and resource abstraction from other nations has its roots in the imperialist ideology and supporting philosophy which is derived on the basis of imperialist information structure and intentionality. For example, Ambrose I. Lane Sr. reflecting on the early USA social formation in relation to enslavement of Africans had this to say about England's Queen Elizabeth I who came to power in 1558:

> The Queen and her slave traders justified slavery because, they said, Africans were living "without God, law religion, or commonwealth." Therefore slave traders were "benefactors" doing the Africans a favor by carrying them off to a Christian land where their bodies might be decently clothed and their souls made fit for heaven [R1.69a, p 4].

This statement reflects pure ideological nonsense and philosophical imbecility locked in royal ignorance about the developmental history of Christianity, the true history of Africa and the history of the Bible in addition to the true foundation of the so claimed European civilization which is now called the Western civilization. This royal ignorance is amplified by ignorance of the cradle intellectual history of Europe and by extension the West and the World in terms of its roots which are traced to Africa [R1.24] [R1.25] [R1.26] [R1.35] [R1.38][R1.56][R1.58] [R1.83b] [R1.86][R1.92][R1.178][R1.179][R1.180][R1.245][R1.247][R1.248][R1.290]

The reflection of the England's Queen Elizabeth I is amplified by similar statements from Europeans. One such statement is by Graves and Psti:

> ...Negroes are doomed to serve men of lighter color was a view gratefully borrowed by the Christians in the Middle Ages; a severe

shortage of cheap manual labor caused by the plague made the reinstitution of slavery attractive [R1.35, p. 604].

The above statement by Graves and Pati, in addition to the reflection on the England's Queen Elizabeth I is further magnified by European Jewish traveler Benjamin of Tudela with:

> There is a people...who work, like animals, eat the herbs that grow on the banks of Nile and in their fields. They go about naked and have not the intelligence of ordinary men. They cohabit with their sisters and anyone they can find.... They are taken as slaves and sold in Egypt and neighboring countries. These sons of Ham are black slaves [R1.35, p. 605].

These anti-African venoms are supported by depersonalization and dehumanization of Africans by Western imperialists to justify African enslavement as expressed by Captain Canot in 1800 about enslavement of West Africans:

> A man, therefore, becomes the standard of price. A slave is a note of hand that may be discounted or pawned, he is a bill of exchange that carries himself to his destination and pays a debt bodily; he is a tax that walks corporately into chieftain's treasure [R1.70b, p.xx].

The British imperial violence against Africa is also reflected by a witness before the imperial parliament who states in 1788 that:

> An immediate Abolition of slavery is incoherent with the Interest and Policy of Great Britain, that from the Silence of the Planters it may be concluded, they place such confidence in the Wisdom of the British Senate as renders and serious Opposition Unnecessary [R1.70b, p. xxi].

These statements are echoed in Portuguese ideology and the practice that this ideology manifests itself in Portuguese colonies as reflected by Nkrumah:

> The 'assimilado' or 'civilizado' system, whereby an African may, by process of law, become in effect a 'white' man, if he comes up to certain European standards, demonstrates yet another aspect of the Portuguese brand of colonialism. Quite apart from the arrogant assumption of racial superiority implied in the idea that every African

would wish to become 'white', is the insidious effect of a policy aimed at deliberately trying to turn Africans into Portuguese. I am reminded of the Africa from Lourenco Marques who said: "The Portuguese think that it was a mistake on the part of God to make the African. Their assimilado policy is an effort to correct this divine error" [R1.202, p. 12]

Statements and reflections like these constitute part of European and by extension the Western experiential information structure on the basis of which Western ideology is constructed with an imperialist philosophy of oppression, intellectual forgery and violence against Africa and her people. The Western imperialist ideology and the supporting philosophy of oppression and violence are to justify partial truth and false claims about Africa and her people in order to justify commitments of human-right abuses through violence and slavery. This philosophy and ideology is referred to in this monograph as *Eurocentric Philosophical Consciencism*. It is philosophy and ideology of militarism, territorial aggression, colonialism, territorial occupation, oppression, human-rights violations and underdevelopment creation through institutional destruction, replacement of social stability with chaos, human suffering and many similar behaviors. One may also examine books like [R1.69b][R1.70b] [R1152][R1.143][R1.169a][R1.239]R1.290]. The fundamental principle of individualism without the due respect to the commons under a capitalist mode of production flows directly from this oppressive philosophy that has a moral bankruptcy in the space of collective good. This Eurocentric Philosophical Consciencism in social formations emphasizes the individual as the center of society and social action around which the community must evolve.

The point of emphasis is that the European colonial-neocolonial Philosophical Consciencism with European contents defines a program continuity of epistemic actions for imperialism, domination, racism and exploitation of Africa and her people. The epicenter of this Eurocentric Philosophical Consciencism is an oppressive intellectual program involving power system where the management of the command-control actions are undertaken with instruments of war, violence and mass destruction. This oppressive intellectual program can only be challenged and negated by an alternative intellectual program which must constitute a revolution in thought and action. This alternative intellectual program is Philosophical Consciencism with African content that stands on firm and solid Africentric grounds to engage in the battle and war games of ideas to decolonize the African mind, liberate the African conscience and set free the African creative forces that have been frozen in the ice-chambers of racism and oppression.

7.2.3 African Intentionality, Decolonization, Freedom and Emancipation

Generally and historically, the Western imperialist intentionality of information collection and knowledge production about Africa is to find justification to support their anti-African ideology in order to demonize, oppress and destroy the African philosophical and ideological basis of collectivism and humanism. The African intentionality of information collection and knowledge production about Africa is to produce knowledge for decolonization, freedom and complete emancipation with African control of her sovereignties. This African intentionality requires the construct of a knowledge system that will discredit the Western oppressive intentionality in creating lies, erroneous claims about Africa and outright forgery that are locked in epistemic imbecility and shielded by ideological and philosophical walls of a deceptive information structure. This is the African liberation intentionality of the search for information and knowledge about Africa to negate the Western imperial intentionality of the information search and knowledge about Africa to discredit Africa's contributions to humanity.

To negate this process of the Western imperial false claims about Africa, it is necessary to set the African general and social history right including the intellectual history of Africa and its contribution to global intellectual history. This task has commanded some of the great minds of African and sympathizers like the works in [R1.34] [R1.38][R1.37] [R1.42][R1.73] [R1.84] [R1.86] [R1.96] [R1.97][R1.100]. It is here that the subject matter, curriculum and the role of African studies acquire an ultimate importance and operational meaning to Africans and African education. The task is to prepare the primary and secondary material for the construct of African-centered Philosophical Consciencism in order to create the required convertibility conditions for categorial conversion of socio-political and socio-economic polarities. The categorial conversion of the socio-political polarity is in relation to the neocolonialism-independence polarity with foreign indirect control of the African sovereignties, while the socio-economic polarities are about underdevelopment-development polarities for social transformations within domestic control of African sovereignties. In reflecting on the content requirements, curriculum design, research activities and philosophy of teaching of African studies in African institutions of learning, it is useful to keep in mind that education is about something; and this something is thinking and knowing For this thinking and knowing to acquire transformative power, they must be related to human conditions and, in fact, to freedom, justice and the fight against either internal or external oppression and subjugation. The power of thinking and knowing is to liberate oneself from the oppressive confines of ignorance and torture from the familiarity of slavery of irrational social thought system.

When the subject area of African studies is considered in relation to African emancipation rather than in relation to Western imperialist oppression, a transformative intellectual polarity is created and ready to be subjected to the laws of categorial conversion in the information-knowledge space. In this respect, the content of general education are at the mercy of the nature of the Philosophical Consciencism that the nation holds. The content is used as a set in this epistemic structure. The content of education is, in general, to create social intellectual capital that is relevant in producing solutions to specific and general problems of a particular social formation. It is here, that an epistemic discipline is required to create and maintain a momentum for change with the effect that the principles concerning social vision and national interest cannot be compromised if social success is to be achieved. The poles of the intellectual polarity have their residing dualities defined by relative negative-positive characteristic sets. The centeredness of the Philosophical Consciencism plays preponderating effects on the intensity of the tension and conflicts in the polar dualities and hence the direction of categorial conversion.

When the Philosophical Consciencism is African-centered under the control of African intelligentsia whose members are liberated from the principles of imperialist anti-African ideology and any other principle that opposes African essence, then the content and nature of African education in the creation of African intellectual capital will be the support of African progress through the internal activities of nation building to achieve the African vision and advance African interests. These conditions of intellectual support were clearly understood by Nkrumah, and it was through this understanding that Nkrumah actively promoted the subject area of African studies and the manner in which the knowledge of the African studies can form inspiration and motivation as well as integrate into all other areas of knowledge to act as the convertibility conditions of the socio-political polarity and the socioeconomic polarity [R1.202][R1.200]. It is through this understanding that one must find the epistemic potency of Nkrumah's statement:

> The history of human achievement illustrates that when an awakened intelligentsia emerges from a subject people it becomes the vanguard of struggle against alien rule [R1.202, p.43].

The system of education in Africa must not only train members of the intelligentsia but an awakened intelligentsia. The members of the awakened intelligentsia are those who are intellectually trained and have liberated consciences derived from the crisis of the African conscience within the African experiential information structure. They are deeply involved in the understanding and expansion of the African experiential information structure. In this respect, the curriculums of current African studies in various universities around the world, and especially those in Africa, are irrelevant to the needs of the development of the African-centered Philosophical Consciencism needed to effect categorial conversions of African socio-political and socio-economic actual-potential polarities at the level of both global political arrangements and domestic socioeconomic transformations. To develop an awakened intelligentsia, the whole education system composed of organization, administration, curriculums, teaching, research and learning must undergo radical restructuring for complete Africanization. In fact, the contemporary African must undergo a revolution in values and culture, where such revolution must be anchored in African tradition as defined and supported by the foundations of Maat [R1.157] in order to overcome the mimicry of oppressive foreign values of colonialism, neo-colonialism and imperialism which have come to destroy the African cultural and spiritual unity.

A properly Africanized curriculum of African studies in African universities and the supporting chain from primary schools will allow some members and the emerging members of the African intelligentsia to generate a knowledge structure that will bring relevant education and allow Africans to construct an African thought system with African intellectual weapons on the basis of the African experiential information structure from antiquity to the present. Let us keep in mind that under the principles of opposites, the world is composed of actual-potential polarities which in turn are composed of problem-solution dualities in relational continua and unity that reside in a never-ending categorial-conversion process. The relational continuum and unity of a problem-solution duality simply expresses itself as every problem has a solution that can be found, and when it is found, it creates a new supporting problem with a new search for its solution. It is this understanding of the principles of opposites in polarities with residing dualities under relational continua and unity, expressed by continual categorial conversions of actual-potential polarities, where the actualization of a potential is always in a temporary equilibrium relative to a contestant in the potential. Here, and in this respect of the behavioral conditions implied in the principles of opposites, the theory of *socio-natural transformation* composed of the sub-theories of *categorial conversion* and *Philosophical Consciencism* presents a principle of never-ending problem-solving and problem-generating activities in nature and society. After the success of each transformation, the new actual becomes a contestant for a new order contained in the potential through the establishment of a new actual-potential polarity with new conflicts, contradictions and new problems. The nature of this process is stated by Nkrumah as:

Revolution has two aspects. Revolution is a revolution against an old order and it is also a contestant for a new order. [R1.202, p.34].

Here, Nkrumah, unlike Marx [R15.34] and Schumpeter [R15.45] [R15.46], emphasized the role of ideology and its supporting philosophy rather than the material conditions of the people in the transformation activities of the actual-potential social polarity. The reason for Nkrumah's emphasis is rooted in the idea that it is the human decision-choice activities that create the sufficient conditions for transformation and these decision-choice activities are under the guidance of Philosophical Consciencism collectively held by the society. The material conditions of the people are simply the outcomes of the works of social decision-choice activities that draw their directions from Philosophical Consciencism. The material conditions are not self-caused; they are derived social category from the primary social category of philosophical Consciencism which gives rise to a particular set of socioeconomic decision-choice activities over the epistemological space. In other words, the material conditions express the need to change while the decision-choice activities under philosophical Consciencism bring about the needed transformation. An alternative way to see the Nkrumah's emphasis is that social progress is a labor process at all times and this labor process is continually under the guidance of Philosophical Consciencism in the collective decision-choice space. The outcomes from the collective decision space go to update a particular experiential information structure which creates an epistemic framework for updating the Philosophical Consciencism which will provide conditions for revisions of the decision-choice activities.

The African past-present-future conditions are such that Africa does not simply need her own internally produced educated population and intellectuals in all fields of knowledge production. What Africa needs is an awakened intelligentsia from her intellectuals and educated class in all fields of knowledge production. The members of this awakened intelligentsia must be equipped with African-centered Philosophical Consciencism to understand that knowledge production must be related to African social intentionality which is African emancipation at all fronts of human endeavor. Such emancipation contains African vision and interests around which all social policies must be developed. The requirements to fulfill this African intentionality must guide curriculum development of African studies in African universities, colleges and other institutions of higher learning and research relative to African-centered Philosophical Consciencism and its continual refinements. The objective here, is to develop African-centered thinking and sow the seeds of African-centeredness within the African in order for the African to understand that the solutions to African problems are within Africa and solvable by Africans, and that when one problem is solved, a new problem, more stubborn in nature emerges within the categorial conversion process. In fact, this is the beauty of life as maintained with the African-centered principle of opposites composed of polarity and duality with relational continuum and unity. It is this never-ending problem-solving process within the universal system that provides the African thought system with its epistemic strength where every state is in temporary qualitative equilibrium where permanency is guaranteed only to change.

The current curriculums of African studies in various institutions of higher learning are geared towards producing knowledge that serves the interest of neocolonialists and imperialists and not the interest of Africa. These curriculums are developed under the controls and epistemic directions of neocolonialist and imperialist Eurocentric Philosophical Consciencism and not with the guidance of African-centered Philosophical Consciencism which holds the center and stability of the African conscience. The curriculums of African studies developed under the guidance of African-centered Philosophical Consciencism are to produce and teach relevant knowledge about Africa, where such knowledge is to be used in the social decision-choice space to create the needed sufficient convertibility conditions of categorial moments and transfer functions to satisfy the categorial transversality conditions of the categorial conversion process of both the socio-political and socio-economic actual-potential polarities. This relevant curriculum development of African studies must inform its content, methods of teaching, directions of research, the utility of learning and its influencing effects in and on other areas of knowledge production and application in the African solution-problem space.

The knowledge obtained from the curriculums of African studies with African-centered intentionality acquires special status in African educational enterprises where the obtained knowledge about Africa must permeate through all areas of knowledge studies, education, training, research and teaching to enrich its recipients with African content, emancipation, intentionality, social vision and African interests. An African-centered system of philosophy of education, knowledge production and methodology is then established where the contents of education and learning are made relevant in dealing with the competing forces in problem-solution dualities within the general African social progress from within, and with the objective of creating true collective freedoms and justice. In this way, all areas of teaching and research on African education acquire a collective purpose, African intentionality and social relevance. With respect to this Africanization of education, even the subject areas such as mathematics, physics, chemistry and engineering which are traditionally considered as non-ideological acquire an African-centeredness and Africentic ideology in their actual and potential applications for the improvement of African nation building and social progress as they were in African historic traditions. It is this Africanization process, under the guidance of African-centered Philosophical Consciencism that will create the required forces to shift the negative-positive relative characteristic set in favor of African positive categorial conversion of all relevant social actual-potential polarities with supporting negative-positive dualities.

The aspects of the relevant contents and curriculum development of African studies which must inform the African thought system in all areas of knowledge and in relation to African-centered Philosophical Consciencism are specified by Nkrumah in a number of places [R1.213] [R1.214]. It must be kept in mind that the knowledge obtained in the African-centered curriculum of African studies has a purpose for the organic African progress over the path of the time trinity of the past-present-future continuum and unity. This knowledge must enrich the mind of Africa, help solve the crisis of the African conscience, become an input into the development of African-centered Philosophical Consciencism, influence the direction of general education, strengthen the African personality, cure the disease of the colonial and neo-colonial mindset, raise African confidence and empower Africans for self-transformation induced by categorial conversion. The African-centered Philosophical Consciencism in turn must shape the development of the methodology of knowledge production as well as the intentionality towards African freedom and justice from oppression and injustices by the imperialists and the supporting domestic imperialist cronies. It must be clearly kept in mind that the goal of the development of African-centered Philosophical Consciencism is to provide an epistemic framework that will shape the African mind and thinking both in theory and practice in all areas of knowledge, their teachings, research and learning under individual and collective decision-choice activities with full connectivity to African traditions.

The objective of Philosophical Consciencism is not simply about providing learning and training under an educational process; it is about developing the African mind, thinking, and methodology of knowledge acquisition and dissemination that will constitute the essential properties of the African personality in the social decision-choice space in order to craft solutions to African problems under never-ending categorial conversions of actual-potential social polarities through the behaviors of their residing dualities. In general learning is good; but what the learner learns is the overriding factor for personality development. The question, therefore, boils down to what the learner learns and how is it useful to his or her existence and progress. Useless knowledge relative to one's environment is more dangerous that no knowledge. Under these conditions, the *grand problem* of Africa that African-centered Philosophical Consciencism has to solve is the set of problems of decolonization of the African mind with entrenched credulity, behavioral mimicry and lack of independent thinking, in order for it to be possible to instill the true African personality without apologies. The solution to this grand problem is the responsibility of all Africans working with African-centered Philosophical Consciencism and remembering the reflection of Tekyi on racial unity.

> Never before were the signs of the times clearer than the present as to the coming together of Africans throughout the world. Instinctively it is being felt that in race solidarity is coming strength of a people who have once again in the cycle of the ages to contribute substantially to the new civilization that is about dawning. When you speak of unity men prick their ears as if a dangerous dogma were being preached. And yet this idea must surely be born of ignorance. It seems to be forgotten that the whole universe makes for unity, and that the forces which attempt to oppose it must themselves end in disaster [R1.171, p.402].

7.3 Curriculum of African Studies and the Art and Science of Thinking in Social Systemicity

For African-centered Philosophical Consciencism to help solve the grand problem of decolonization of the African mind, the curriculum of African studies must acquire African-centeredness. In reflecting on the relevant curriculum of the subject area of African studies, a number of conceptual ideas about the whole area of knowledge production and its dissemination through teaching comes to mind. These ideas, their usefulness and preponderating effects on the management of social systems and their dynamics are discussed in [R3.7] [R3.10] [R3.13]. Knowledge production is a qualitatively and quantitatively dynamic, feedback and self-correcting system within the knowing-ignorance polarity with its residing dualities under the principles of relational continuum and unity. Any curriculum development of African-centered African studies for teaching, learning, research and application must take these properties of knowledge production into account given the goal of designing an epistemic frame to solve the problem of intellectual neocolonialism, slavery and living in the zone of a steel box of cognitive familiarity. The curriculum of the African studies in African institutions of learning must be completely Africanized where its content must influence the methods of Africanization of curriculum development of other areas of knowledge production. The properties of knowledge production, their social utility and cognitively transformative character of

African conscience from colonialism and enslavement to decolonization and complete emancipation demand that African education should not simply transmit what is claimed to be known and demonstrations of its validity within logical structures. African education must become transformative within dualities to effect the categorial conversions of socio-economic and socio-political actual-potential polarities where the African positive actions are set against the negative actions towards complete emancipation. The understanding of the self-correcting and feedback knowledge-production process under Africanized curriculums and African-centered Philosophical Consciencism is to produce an internal organization and thought that knowledge is incrementally attained through hard work in solving African problems of different types in human knowing-ignorance polarity with residing dualities under relational continuum and unity where mistaken claims and correction are inherent properties of the process.

The developed curriculum of African studies must reflect the general principle of Africanization of education. The Africanization simply means that the goals and objectives of education must reflect the identification of African problems and the search for solutions to them. Education must be made relevant to African social formation and the management of the dynamics of its qualitative and quantitative dispositions under internal conscious actions of categorial conversions of social actual-potential polarities. The curriculum development in general must relate to the social vision of education in the needs and dynamic transformation of African societies with continual preservation of African interests. The Africanization of the curriculum of African studies is to service the production of the information requirements for the development and refinement of African-centered Philosophical Consciencism. The Africanization of the curriculums of other areas of knowledge development is to produce information requirements for the production of a unified African conscience in the use of knowledge in the service of Africa, her development and welfare improvement of her people.

The concept and the meaning of Africanization of curriculums as they have been presented here are such that no area of knowledge production, research, teaching and learning can be exempted. Scientific knowledge in terms of empirical and axiomatic information structures is not acquired for its own sake but must be made relevant to Africa's progress, nation building, efficiency of administration, and Africa's defense. However, the acquisition of knowledge about the behavior of natural forces is not group-specific. The uses to which such knowledge is put in the service of society is group-specific. This group-specificity of the application of knowledge is influenced by the content of Philosophical Consciencism. If such influence of the use of knowledge is under African-centered Philosophical Consciencism, then the knowledge will be applied in finding solutions to African problems with the continual categorial conversions of actual-potential social polarities over the categorial enveloping in the time trinity of past-present-future space. The objective of Africanization of the curriculum is to produce a knowledge system with emphasis in thinking rather than simple mimicry of claimed knowledge by foreign knowledge systems and aping of behavior inconsistent with the African way of life. African-centered knowledge production, education and dissemination of what is claimed to be knowledge must form a continuity of traditions to make education relevant to the masses. The whole education process must be transformative in all aspects of African life and African personality with complete emphasis on creativity and independent thinking. In this respect, it is useful to reference the statement of J. Ki-Zerbo:

> The African is highly creative in art and culture, and in these fields as in architecture, sculpture, drama etc. the African University cannot be indifferent. It must learn to create, not copy; it must turn out producers and creators, not just consumers of goods and cultures of other lands. In politics, sociology, law and economics, the African must derive original and more suitable system of organization and ideology....We must not hesitate to introduce courses that will promote the ideal of African unity. Nor must we hesitate to appoint as university teachers African traditional scholars; for who is a scholar but one who knows? We must not hesitate to permit that cross-fertilization of ideas and debate, without which the African university would be like the idealists portrayed by Peguy when he quipped that 'they would have clean hands; but the pity is that they have no hands.' [R1.298, p.26]

This stamen of J.Ki-Zerbo's has been reflected on in [R1.92, pp. 1-32]. This is crucially related to the challenges of the members of the African intelligentsia, the intensification of the degree of African consciousness, and the difficulties of engaging the science and art of the development of African-centered thinking that links the modern to tradition in relational continuum and unity. It must be kept in mind that

342

the progress path is also a solution path which is the child of thinking under conscience and experiential information structure operating in the collective and individual decision-choice space.

7.4 Contents of Educational System, Research, Teaching and Learning in African-Centered African Studies

The contents of education system, teaching and learning in African studies, the problem definitions and specifications for research in African studies and the methods of research, teaching and learning must have an African purpose for the knowledge derived from African-centered Philosophical Consciencism whose framework is to encourage intellectual democracy under the principles of free thinking and individual minds but within the organic principles of African collectivism and humanism. African-centered Philosophical Consciencism defines a framework of education that promotes free research and epistemic inquiry to produce knowledge that will support collective progress within the individual-collective duality under the principles of relational continuum and unity. The content of education about Africa in African-centered African studies must lead to the production of knowledge that is continually relevant in improving the collective African understanding of Africa, her history from antiquity to the present and the welfare of African people in the time trinity of past-present-future continuum and unity. For the collective African understanding of Africa to acquire transformative character it must be enhanced by an epistemic recognition that the knowledge produced must serve as input into collective decision-choice actions to solve economic, cultural, technological, information, scientific and non-scientific African problems over the categorial enveloping of African progress.

The education from African-centered African studies must show the urgency and necessity of a combination of love for Africa and knowledge about Africa in pointing out the right way of African progress through the categorial-conversion process. In this respect a method of education that promotes mimicry and credulity of what is claimed to be knowledge by other lands will lead to African cognitive decay whose cure lies in the encouragement of conditions of free inquiry under the practice of the principles of doubt and art of thinking that must be related to the African experiential information structure, solutions to African problems and conditions of African progress. The knowledge obtained from

African-centered African studies is to enrich and enhance continual development of African-centered Philosophical Consciencism which is to provide guidance for individual and collective actions over the African social space in relation to education, economic production and cost-benefit distribution of all policies related to African vision, interests and welfare of her people.

African-centered education is unique in its characteristics. It is directed toward an intellectual revolution against colonial education which was intended to produce educated Africans who have an affinity and service commitment to the social structure of imperialists and neocolonialists. To create these kinds of Africans, it was necessary and important that the colonial education, and by extension neocolonial education, create mindsets that allowed the educated Africans to be devoiced from their cultural roots and their people of whom their education has instructed them to despise with venom as uncivilized under the Western anti-African ideology. The colonial and neocolonial education has produced and continues to produce an African educated class whose members are corrupted with Eurocentric intellectual personalities that deny that:

> ...all the imperialists, without exception, evolved the means, and their colonial policies, to satisfy the ends, the exploitation of the subject territories for the aggrandizement of the metropolitan countries. They were all rapacious; they all subserved the needs of the subject lands to their own demands; they all circumscribed human rights and liberties; they all repressed and despoiled, degraded and oppressed. They took our lands, our lives, our resources, and our dignity. Without exception, they left us nothing but our resentment, and later, our determination to be free and rise once more to the level of men and women who walk with their heads high [R1.202, p. xiii].

Armed with these Eurocentric personalities the members of the educated Africans love the imperialists and neocolonialists more so than they love Africa and develop a hatred for their African roots. In this way, the Eurocentric intellectual personalities that have taken hold in Africa are completely inimical to Africa's independence, progress and complete emancipation with global freedom and justice. These Africans with the Eurocentric intellectual personalities under complete control of colonial and neocolonial mindsets fight the hardest to join and serve international institutions of their own oppression rather than think to link in the

African hands to create organizations and institutions to separate from the imperial terror-dome in order to enter the golden gate of complete emancipation. The underlying forces of these Eurocentric intellectual personalities are to allow indirect external destruction of Africa's true categorial conversions of socio-political and socioeconomic actual-potential polarities through their effects on the individual and collective decision-choice actions in the residing polar dualities

Eurocentric intellectual personalities in the African social set-up must be negated in their individual and collective existence by the creation of African-centered education with African internally constructed subject matter and implied African curriculums if Africa is to progress from within to promote African sustainable vision and social interests. This African-centered education with the corresponding African-centered curriculums and methodology of teaching, research and learning under the guidance of African-centered Philosophical Consciencism is to create conditions for the establishment of African collective intellectual personality for the manufacturing of wise individuals, and the collective in African communities, where the African is imbued with true liberty and African personality, creativity and state artistry under the Africentric principles of collectivism and humanism which are true of African tradition. The principles of collectivism and humanism as epistemic characteristics of good societies from the African conceptual system are such that it is the community as a collective in which social good must take hold to establish the community freedoms within which an individual freedom is established for the good of the collective on the basis of which African statecraft and management are developed for social efficiency. In this respect, there is the principle of each for all and all for each under African collective progress whose measure must relate to how the social system takes care of the weak. Equipped with African intellectual personalities, the members of educated Africans will have African intellectual mindset that will allow them to behave in Africa's interest and social vision in the decision-choice space with the principle of collective freedom and humanism rather than individual freedom and individual interests where social cost-benefit distribution is unjust in consumption-production duality. Working within the framework of African intellectual mindset as a continuation of African tradition and connected to the African cultural roots, the members of the African educated class will combine love of Africa and knowledge of Africa to pioneer the pathway to African emancipation and redemption in order to set examples to other Africans. In this framework as set in place by implied conditions of Africentric Philosophical Consciencism, the members of the general African leadership will undertake decision-choice activities that emphasize collective good and freedom rather than individual good and individual liberty within the individual-collective duality.

7.4.1 Teaching, Learning, Knowing and the Contents of Educational System

The content, teaching and learning of African studies are reflected in the acceptable and claimed knowledge in the subject area. The content of African-centered African studies, unlike the colonially imperial content, must reflect the knowledge of Africa within the time trinity of pre-colonial Africa, colonial Africa and decolonized Africa. The acquisition of true knowledge of pre-colonial Africa is not simply to establish the African interpretation of her glorious past and its foundations of other global societies and their civilizations, but to redefine the essential characteristics of the African collective personality for Africans in decolonized Africa. The knowledge of colonized Africa is not to establish the Western imperial evil, atrocities, human-rights abuses, terror and other indignities as well as provide a historic account of African experiential information, but to find out what went wrong, learn from it and teach it to the African masses and her children in order to intellectually arm them to fight against its possible reoccurrence. The knowledge of decolonized Africa is to understand the difficulties and problems that were created by the imperialists' aggression, domination, slavery, and occupation in order to provide relevant cognitive materials for continual research, teaching and learning, the results of which will become useful inputs into the decision-choice process for the management of the command-control activities of Africa's social transformations and defense of sovereignty.

Here, the effective duty of any teacher and a learner from both educated and formally non-educated classes and the lasting effect of this teaching and learning are reflected in teaching and learning to think and not simply teaching and learning what is claimed to be knowledge that is expressed in orthodoxy. In this way, the African intellect is stimulated to be active and not left in a passive state to mimic other peoples thinking which may be inimical to Africa's progress, her personality and survivability. By teaching learning and thinking within tradition and non-tradition under the principle of doubt using the existing knowledge structure, African education does not become a victim to the devastating disease of mimicry and training in acquiring certain kinds of skills that may be conditionally irrelevant to African problems within the dynamics of the qualitative and quantitative dispositions. Reflecting on the African time trinity of pre-colonial, colonial and decolonized periods in relation to African personality, mind set and their relevance to categorial conversions of African social actual-potential polarities, Nkrumah had this to say:

> The personality of the African which was stunted in this process can only be retrieved from these ruins if we make a conscious effort to restore Africa's ancient glory. It is only in conditions of total freedom and independence from foreign rule and interferences that the aspirations of our people will see real fulfillment and the African genius finds his best expression [R1.214, Vol. 5, p. 131].

The Nkrumah's reflection defines the audacity of the African personality, the dynamism of the African genius, the hope and success of ancestral linkage, love for Africa and the creativity of knowledge applications in all fronts of human endeavor with a complete focus on Africa's principles of collectivism and humanism in relation to a democratic structure of freedom and justice for all [R1.157][R1.159] [R1.160].

The success of actualization of Nkrumah's reflection will require African-centered African studies to provide important assessment of essential factors of the African experiential information structure where these factors must be reinterpreted in terms of the African needs of the present and projection into tomorrow and beyond. In this sense, the Africa's glorious past must inform Africa's present and guide Africa's vision of tomorrow with un-quenching inspiration and taste for freedom and justice over generations. For this inspiration to take hold and become permanent over generations the subject matter of education must incorporate African-centered analyses and interpretations of the African-centered experiential information structure to do away with the colonial and neocolonial mindset in which the art of mimicry is taken to an extreme level of cognitive imbecility where, the African essence is completely destroyed and replaced with a deformed non-African essence.

Some of the content from the subject of study of African studies for research, knowing, teaching and learning are reflected by Nkrumah. At the level of history, art, culture, science and statecraft, he stated:

> ...Valuable pieces have already been unearthed, including evidence of the origin of man in Africa. We have made our contribution to the fund of human knowledge by extending the frontiers of art, culture and spiritual values.
>
> Democracy, for instance, has always been for us not a matter of technique, but more important than technique – a matter of socialist goals and aims. It was however, not only our socialist aims that were democratically inspired, but also the methods of their pursuit were socialist.
>
> If we have lost touch with what our forefathers discovered and knew, this has been due to the system of education to which we were introduced. This system of education prepared us for a subservient role to Europe and things European. It was directed at estranging us from our own cultures in order that we, more effectively serve a new and alien interest.
>
> In rediscovering and revitalising our cultural and spiritual values, African Studies must help to redirect this endeavour. The educational system which we devise today must equip us with the resources of a personality and a force strong to meet the intensities of the African presence and situation.
>
> Education must enable us to understand correctly the strains and stresses to which Africa is subjected, to appreciate objectively the changes taking place, and enable us to contribute fully African spirit for the benefit of all, and for the peace and progress of the world.
>
> African Studies is not a kind of academic hermitage. It has warm connections with similar studies in other countries of the world. It should change its course from anthropology to sociology, for it is the latter which more than any other aspect creates the firmest basis for social policy [(R1.214, Vol. 2, p. 125].

At the level of language and interpretive thought systems, the authenticity of the African-centered information structure must be in the relational structures of the arts with due respect to details of the representation, sculpture, painting, musical idioms, symbolisms, musical instruments, proverbs, artistry of story-telling and linguistic codes, all of which may be studied to abstract similarities and differences as well as the underlining thought systems that produced and maintain the African cultural unity. In this framework, the works of Francis Bebby, Kwabena Nkatia, Diop, and others are examples in this direction where the philosophy of musical codes and instruments are presented for further refinements [R1.32] [R1.157] [R1.193] [R1.194] [R1.196] [R1.86]. At the level of philosophy, thought and mathematics, the works in [R1.8][R1.12][R1.19][R1.30a][R1.37] [R1.86]][R1.92] [R1.133] [R1.134] [R1.157] present some direction.

Here, the compilation and interpretations of the African experiential information structure must reflect the African creativity of the African genius and intentionality of the African interest and social vision to protect her human and non-human resources through collective African actions with true emphasis on the cultural, political and epistemic unity of Africa as has been established in [R1.8] [R1.84] [R1.92] [R1.202][R1.243][R1.249] . In this respect, the task of true African education must be socially distributive, and its development must be under the guidance of African-centered Philosophical Consciencism to develop the sufficient conditions of unity and categorial conversions of African socio-political and socioeconomic actual-potential polarities. In terms of the research and construct of the epistemic unity of the African conceptual system in philosophy, logic and mathematics, one may examine the Pharonic, Dogon, Akan and other African traditional conceptual systems which are based on principle of opposites composed of polarity, duality, negative-positive forces which are used in this monograph as well as in [R1.92] [1.90b] [R3.10] [R3.13]. Here, Dr. J. Ricord, in his preface to de Lubicz's book entitled *The Egyptian Miracle* had this reflection on philosophy and mathematics:

> The outcome of his study of Pharaonic mathematics confirms and surpasses what we already know through the work of his predecessors, and the spontaneous collaboration of his stepdaughter, Lucie Lamy, enabled him to present this mathematical thought in all its details. This achievement is all the more astonishing in the light of claims that philosophy and science as we know them were invented by the Greeks.
>
> Indeed, it is easy to forget that Moses, and Pythagoras, among , received their entire culture from Egyptian temple, but much more difficult, it seems to me, is categorically to deny this fact [R1.243, p. 3].

The study of science and mathematics of the Dogon people, the Akan people and the people of the Nile-valley civilization, West Africa, East Africa, North Africa and South Africa will enhance the African information structure to allow members of the African-centered educated group and intelligentsia to continue the philosophical and logical unity of the African conceptual system to produce an African mindset and personality under the guidance of African-centered Philosophical Consciencism. It must be emphasized that, personalities are shaped by knowledge and ideology which are influenced and in most cases established by education and its content. Here, the contents of education are always at the mercy of the nature of the Philosophical Consciencism that the nation holds. When the Philosophical Consciencism is African-centered, then the content and nature of the education in the creation of African human capital will be supportive of the African progress through nation building. The work of African Studies is to create love and knowledge about Africa. It was through this understanding that Nkrumah actively promoted African studies. The curriculums of current African studies in various universities around the world are irrelevant to the needs of categorial conversion of African social actual-potential polarities. In fact, these curriculums at the various institutions of higher learning serve the interests of the neocolonialists and imperialists and not the interests of Africa.

These curriculums are developed with external controls of neocolonial Philosophical Consciencism and without the guidance of the African-centered Philosophical Consciencism. They are designed to directly and indirectly control the African collective decision-choice space of African progress through a system of aid-given, useless system of technical supports and African attachments to fake institutions of imperialism and neo-colonialism. Eicher reflecting on this neocolonial dilemma of African education had this to offer.

> African education is intimately linked with the international aid and education industry, and that the donor/client relationship has inhibited the development of African institutions and the capacity of Africans to develop educational policies which are socially relevant and financially feasible, for the last quarter of this century [R1.298, pp.27-28].

These aid-driven and externally connected curriculums developed for African education are not useful to the needs of Africa yesterday, today and tomorrow. Under the guidance of African-centered Philosophical Consciencism, the knowledge obtained from African-centered African studies must permeate all areas of knowledge studies and enrich them with the need to identify Africa's socioeconomic and politico-legal problems, and find solutions to them without external dependency in order to affirm the meaning and content of independence [R1.91]. In this way, even the studies of mathematics, physics, chemistry, biology and others acquire their African content for the improvement of African nation building and social progress as they were in the African traditions and not to improve the skills of African cronies for the administration and management of neocolonial oppression of African people who subscribe to philosophical individualism under the principles of self-interest. The path of African categorial transformation demands adherence to philosophical collectivism and humanism [R1.159] [R1.160] with a moral foundation drawn from Maat [R1.157]. It may be useful to keep in mind that the same international aid-driven and developmental association to a system of dependencies that continue to prevent Africa's complete emancipation and created neocolonial states have been discussed in [R1.91] [R1.204].

7.4.2 The Concepts of Controllability, Convertibility and their Relational Structure in Social Systemicity under Philosophical Consciencism

Categorial-conversion processes in socio-natural elements are defined and maintained by *categorial controllability* and *categorial convertibility* in a defined information-decision-choice system. The analytical framework of the explanation of categorial conversion and its applications in socio-natural processes are built on two interrelated theoretical sub-structures. One theoretical sub-structure relates to *convertibility conditions* while the other sub-structure relates to *controllability conditions*. The convertibility conditions are studied by *the theory of categorial conversion* after the establishment of the *theory of category formation* which provides a logical explanation and justification of the existence of categories that relate to *matter, energy* and *information* under the principles of relational continuum and unity. The convertibility conditions indicate the *necessary conditions* for transformation of socio-natural elements. The controllability conditions are studied by the *theory of Philosophical Consciencism*. The convertibility conditions characterize the internal dynamic conditions of *forces at work* and *production of energy* under a defined information structure that together establish qualitative motion which moves one element from one category to another or transforms one category to a different category in the

351

universal socio-natural elements and categories. The controllability conditions characterize the internal control mechanisms of strategies and tactics, and counter strategies and tactics that produce the dynamic conditions of *forces at work* and *production of energy* within a defined information structure under principles of opposites in a game space with negative and positive characteristic sub-sets that together shape the direction and produce the resultant qualitative motion which then moves one element from one category to another or transforms one category to a different category in the universal socio-natural elements and categories according to the control and counter-control decision-choice structure under the principle of opposites, in which actual-potential polarities are established for the socio-natural elements defined in the space of negative-positive dualities with relational continuum and unity. The controllability conditions indicate the *sufficient conditions* for transformation of socio-natural elements.

The *theory of category formation* is essential to the development of the *theory of categorial conversion* which is essential for the development of the *theory of Philosophical Consciencism*. Without the justified existence of categories, one cannot even think of language and categorial transformations. This monograph deals with the theory of Philosophical Consciencism. The theories of category formation and categorial conversion are dealt with under a separate monograph [R1.90b] The convertibility and controllability conditions are the foundation for understanding the dynamics of change in natural categories and the application of which this understanding can be brought to bear on society and human artificial creations such as all kinds of engineering and production-consumption systems under the principles of opposites, where the focus is always on matter, energy and information structures, all in relational continuum and unity. The principles of opposites in relational continuum and unity on the basis of which self-transformation is explained depends on the existence of *actual-potential polarity* which gives meaning to primary and derived categories where the primary category is the actual and the derived category is an actualized potential in a never-ending process of continual transformation. In this case every actual and every potential are in temporary qualitative equilibria. It is the controllability conditions that create the sufficient conditions for transformation. The controllability conditions are defined in the decision-choice space and under Philosophical Consciencism, composed of ideology and supporting philosophy, define the types of decisions, choices and actions that create the transformation enveloping of categories of social states. Whosoever controls the Philosophical Consciencism controls the path of transformations of the socio-natural elements and this is true of all social transformations where ideological conditions take precedence over material conditions. It is not the material conditions that are important but the perception of the material conditions in motivating actions in the decision-choice space. This perception is the product of the Philosophical Consciencism. It is here that African-centered Philosophical Consciencism addresses itself to the African perceptive reality from the African experiential information structure as seen through the curriculum and content of African studies. There are two types of curriculum of formal nature of the academic at the lower and higher level and informal nature of civic and social. Both types must be guided by African-centered Philosophical Consciencism.

EPILOGUE

SOME CRITICAL REFLECTIONS ON AFRICAN PROBLEMS AND SOCIO-ECONOMIC TRANSFORMATION IN SOCIAL SYSTEMICITY

The theory of Philosophical Consciencism is generally applicable to all socio-natural transformations. However, this epilogue is presented with special attention to African socio-political and economic difficulties from colonialism to decolonization, from decolonization to a search for independence and complete emancipation from the shackles of imperialism. It is used to reflect on the works of some African thinkers and to relate them to the problem of the development of African-centered Philosophical Consciencism and its use for internal social instrumentation, and the management of commands and controls of the African transformation path without external interference. The solution or the lack thereof to the problem of the development of African-centered Philosophical Consciencism has a direct connection to generating *controllability conditions* that will be transformed to *sufficient conditions* for Africa's internal transformation in accordance to Africa's collective preference and will on the basis of the foundations of African morality. This problem and the required solution are defined by Africa's cognitive activities in the epistemological space of knowing, decision and control.

I The Cognitive Damage and the African Collective Behavior

Let us keep in mind Nkrumah's reflection on the relationship between thought space and the practice space. *Practice without thought is blind; thought without practice is empty* [R1.202. p.78]. It must be added that the wrong thought system is not only blind but dangerous to the practice of decision-choice activities in the action space which, under the principle of opposites, is composed of negative and positive actions. An action is said to be positive if it promotes increasing benefits to the transformative motion of the progress of the individual-collective behavior otherwise, it is said to be negative in the sense of being inimical to the individual-collective contribution to internal transformation. A behavior in a wrong thought system leads to wrong action and an undesirable result; similarly, a behavior in a right thought system leads to a right action and a desirable result. It was the recognition of the relational structure of the wrong thought system and the danger of practice by the individual and the collective that lead Carter G. Woodson to reflect on the general cognitive conditions of Africans in America, which under the Pan-African extensions applies to all Africans in the Americas, Caribbean and Africa itself, in terms of the damage that Eurocentric education framework has produced. In that regard, he states:

> The so-called modern education, with all its defects, however, does others so much more good than it does the Negro, because it has been worked out in conformity to the needs of those who have enslaved [colonized] and oppressed weaker people. For example, the philosophy and ethics resulting from our educational system have justified slavery, peonage, segregation, and lynching. The oppressor has the right to exploit, to handicap, and to kill the oppressed. Negroes [Africans] daily educated in the tenets of such a religion of the strong have accepted the status of the weak as divinely ordained, and during the last three generations of their nominal freedom [decolonization] they have done nothing to change it. Their pouting and resolutions indulged in by a few of the race have been of little avail [R1.294, pp. xxxii-xxxiii] (see also in [R1.1] [R1.12][R1.20],[R1.34][R1.35], [R1.83a] [R1.87] [R1.213][R1214]] [R1.95][R1.96] [R1.92] [R1.218] and others.

With these reflections on the historical context of the cognitive problem of the Negro [African] in the thought space, Cater G Woodson reflected on the contents of education and cognition within the development of the mind and thinking of Negroes (Africans).

> No systematic effort toward change has been possible, for, taught the same economics, history, philosophy, literature and religion which have established the present code of moral, the Negro's [African] mind has been brought under the control of his oppressor. The problem of holding the Negro [African] down, therefore, is easily solved. When you control a man's thinking you do not have to worry about his actions. You do not have to tell him not to stand here or go yonder. He will find his "proper place" and will stay in it. You do not need to send him to the back door. He will go without being told. In fact, if there is no back, he will cut one for his special benefit. His education makes it necessary [R1.294, p.xxxiii].

Given the relational structure of educational contents and learning, he further reflected on the negative effects of mis-education of the Negro [African] in the action and thinking spaces of human progress. He continued:

> The same educational process which inspires and stimulate the oppressor with the thought that he is everything and has accomplished everything worth while, depresses and crushes at the same time the spark of genius in the Negro [African] by making him feel that his race does not amount to much and never will measure up to the standards of other peoples. The Negro [African] thus educated is a hopeless liability of the race [R1.294, p. xxxiii].

This hopeless liability expresses itself in multiple social fronts. The traditional self-motivation of the African in taking bold initiatives, creating instrumentation and designing management of command and control systems for his progress was lost for a long period of imperial and neocolonial destruction and continual interference of neocolonialists and imperialists. This hopeless liability is also expressed in the destruction of the true African Personality where the colonial and neocolonial education and mindset have estranged the educated from their African roots. The educated African is a stranger in his own land. The cognitive frame of this educated African is his/her oppressor. Educated Africans do not even know or admit that they are managing client states for the benefit of the imperial and neocolonial order making African independence a mockery of their intellects. Their minds are under a system of external controllers from education, the curricula of which are designed by anti-African Philosophical Consciencism to produce subservient personalities and not African personality contrary to Africentric tradition as it were.

The African education under the curricula derived from Eurocentric Philosophical Consciencism is nothing more or less than production of neo-colonial mindset controlled from without but as a continuity of the colonial mindset where the emphasis is on memory and not on thinking. This educational process has brought into the contemporary African a continuity of a colonized mind into the contemporary African social existence forms of governing class with enhanced colonial and neocolonial mindset that is under remote control. The members of these forms of governing class has taken the Eurocentric Philosophical

Individualism to absolute absurdity, where the ambitions of the members of this governing class and the supporting intelligentsia are individual and personal trivialities under competitive waste. Their ambitions do not ascend to the level of nation building from the inherited colonial damage and Africa's destruction. They are controlled by their irrelevant education of mimicry and blinded by trivialities under competitive waste of individualism without understanding that the problem of a nation building is a collective phenomenon and after decolonization the ambitions of the members of the governing must be directed the reconstruction and nation building of Africa. The solutions to the problems of absurd individualism, Africa's reconstruction and nation building are creatively found in the Africentric principle of opposites and individual-community duality at the guidance of African-centered Philosophical Consciencism. If the African mind is sculptured by conditions of African-centered Philosophical Consciencism then the minds of the members of the governing class will by design be controlled from within Africa and their decision-choice actions will be positive to Africa and not to plunder Africa's resources to other lands.

The problem of African people is a serious one that has created unbearable lives for them. The problem being experienced is the same for the individual and the collective. The generality of the problem is that it affects all Africans in Africa and all its diaspora in the Americas, Europe, the Caribbean and other places. The problem is not material poverty and hence cannot be solved by poverty reduction. It is not health and physical wellbeing and hence cannot be solved by attending physical bodies. The problem is not even that our people are dying and systematically being reduced to non-existence just like the Caribe Indians. The problem is simply that the mindset that Africans currently hold is a destructive force against the African personality and African existence. The curricula of African education and the materials that are used for instructions and mimicry are directed to study the African as a problem and the solution to this African problem lies outside Africa and the struggle to solve the problem of African nation building from within Africa is hopeless. The African education has not produced a mindset that appreciates the *Sankofa-anoma* principle of time trinity of past, present and future. Today's Africans seem to forget their traditions where our ancestors were clear of this simple statement by Nkrumah.

Progress does not come by itself, neither desire nor time can alone ensure progress. Progress is not a gift, but a victory. To make progress, man has to work, strive and toil, tame the elements, combat environment, recast institutions, subdue circumstances, and at all times be ideologically alert and awake [R1.207, p.113].

It is not the meaning of progress that is important here, neither is the meaning of the actions towards victory. The most important framework to achieve progress irrespective of how it is defined and measured is within the phrase ideologically alert and awake. These ideological starts of awareness and awakening are the products of the mind set and its conceptual system that motivate decision choice activities and actions to manufacture instrumentation and design command and control system for management, where the instrumentation, command and control systems are related to work, toil, element-taming, institution recasting and combating of the environment that will shape the path of progress. For the path of progress to be sustainable, the instrumentation and the design of the management of the command and control system must be internally induced from positive actions to promote progress and to fight enemies that generate negative resistance.

II Progress, Information, Knowledge, Thinking and Decisions

The instrumentation of commands and controls, and the design of the management of commands and controls depend on the type of a thinking system that the members of the society individually and collectively hold. The thinking system and its development are at the mercy of the social ideology and the supporting philosophy that create intentionalities over the decision-choice and action spaces. The development of the thinking system is encapsulated within the boundaries of a paradigm with laws of thought where such a paradigm is at the mercy of socio-political philosophy. The paradigm development is at the mercy of the ideals of law, order, goodness and truth within the principle of opposites with relational continuum and unity. It is here that *Maat*, viewed within the African-centered principle of opposites, presents itself not simply as the *Goddess of Truth* but as a philosophical system of ethical principles of African traditions in relation to intentionalities of decision-choice activities and social actions to guide the individual and collective behaviors in the individual-community duality with relational

359

continuum and unity. In other words, *Maat* is a collection of social algorithms to solve the *Anoma-kokone-kone problem* and *funtummereku-denkyemmereku problem* under *Asantrofi-anoma rationality* in the social set-up where the experiential information structure obeys the general principle of *Sankofa-anoma* (the time trinity of past, present and future). It is useful to keep in mind that knowledge production, economic production, social formations and other such endeavors are decision-choice activities and human actions in the social space. These decision-choice activities, social actions and intentionalities are undertaken through individual and collective thinking guided by the principles implied in the ethical philosophy, and in the African case by those implied by *Maat*. It is this thinking under the principles implied in *Goddess of Truth*, (*Maat*) and defined within the principle of opposites, that *African consciousness* can find meaning and *African conscience* can find an unshakable stand. In these times, there are deficits of African creative thinking and consciousness that have produced social deficits and given rise to crises in the African conscience from colonization to the present. Reflecting on such social deficits and crises, Attoh Ahuma produced the following statement:

> Is the soul of our nation losing glamour and romance inseparably associated with primitive conditions of life? Is it making intelligent and vigorous effort to deserve a seat amongst the master-souls of the age? What are our credentials and passports? What are our assets as a nation? Can we, do we, stand before other races and people with heads erect and with a free independent spirit? What is our mental, moral, and social equipment worth? These are pointed, dominant notes of interrogations that demand sensible, direct and practical rejoinders. They are questions of unique importance and immediate urgency; for they affect the honour, prosperity, and security – the very life of our people and country [African people and Africa]. The blood of our nation [Africa] requires enrichment, and the freest possible circulation; it calls for invigoration; it needs recuperation, that the Body Politic may be quickened, strengthened, and purified. When altruism or passionate devotion to humanity permeates every pore, and when true patriotism [Africanness] or the love of service and sacrifice for the Homeland pours nutrition to all parts of our national [African] system, then shall we acknowledge with joyful pride the existence of our nation [Africa] and the destiny that lies before us as members of the Negro race [R1.171, p.164]. (The non-italicized words are those of the author)

The conditions of the honor, prosperity and security, the strengthening, and the purification of African energy flows are the activities of the creative thinking system which must be developed on the basis of Philosophical Consciencism. In this respect, the African problem in all its multiple dimensions is defined in the epistemic space and can only be solved in the same space when a way out is specified. Reflecting on the way out of this consciousness and the many crises Attoh Ahuma, continued with his previous reflection.

> When we become conscious of the place we occupy in nature, and our eyes are opened, all selfish individualism will sink into oblivion, and with the expansion of the soul shall come the yearning, burning zeal and love for country [Africa] and race. Among the virtues necessary to the development of the nation [Africa] must be the assiduous cultivation of public spirit, that animating principle that belongs to and enthuses all collective bodies – an "esprit de corps." The new element, with its foreign attributes introduced by the dominant power in its government and protection of our interest, has unpremeditatedly made us self-suppressive. It is an axiomatic fact that 'where a dominant race rules another, the mildest form of government is a despotism. It has been so at all times and among all nations in every part of the world.' As a people, we are not educated to the point of appreciating the finical forms and methods of government, which at present must necessarily spell oppression and wanton waste [R1.171, pp.164-165].

When the African consciousness and the crises in conscience are referred to the nature, structure and form of Africa's progress, we must deal with the general meaning of social progress, the specific meaning of social progress, the corresponding measures and the criteria of progress that is relevant to Africa with full reference to African tradition as seen within the individual-collective duality and Africentric principle of opposites. From the foundations of African epistemic tradition for social formation, it becomes necessary to balance the relational structure of the individual to the collective and under the ethical principles of Maat, the moral good, epistemic truth where social progress is such that all selfish individualism will sink into oblivion, be taken over by the yearning, burning zeal and love for Africa collectivity and race. Under the supreme guidance of the *principles of Maat* social progress is of the people, by the people and for the people who own it. Social progress is not of an individual, by an individual and for an individual. The social progress is within the framework where the operating principle is each for all and all for each under a balance of collective-individual efforts within the individual-collective duality to solve the *anoma-kokone-kone* problem in favor of the collective existence, as well as to solve the *funtummereku-denkyemmereke problem* in favor of social unity. The measure and criterion for African social progress under the fundamental ethical principle of collectivism must be seen in terms of continual general uplifting of African people in all areas of human endeavor. In reflecting on the meaning and path of social progress, Esuman-Gwira Sekyi provided the following statement to conceptualize.

> Two wrongs do not make a right; there is no escape from the truth of this law. Europe has misunderstood social progress as other groups of people misunderstood it in the past. Like those decayed Empires, Europe has evolved, and perhaps is capable of evolving, nothing but Imperialism and not the Brotherhood of nations. Imperialism is self-contradictory in idea and consequently, self-destructive in realization. "Let us not induce ourselves to think and believe that the only way to 'survive' Europe's aggression is by organizing on European (including America) lines" for that must involve conflict with Europe, and in time we must become like Europe, ever creating new wants to supply an insatiable desire for conquest, ever oppressing others to further this conquest, and bound to end by consuming all that has been acquired by conquest in universal holocaust kindled by the demon of Greed. "If we are to formulate any really sound and practicable scheme for our future, let us set before us, and try to understand, the ideal of living as men, and not seek the compromise of surviving as persecuted persons" [R1.171, pp.244-245]

In a further reflection on European imperialist arrogance and crudeness, Esuman-Gwira Sekyi stated:

> The success of the white man in establishing dominion over the black man, in my opinion, instead of proving, as the white man naively assumes, the superiority of the white man to the black man, in my opinion, instead of proving, as the white man naively assumes superiority of the white man to the black man, on the contrary is further evidence of the crudeness of the white man's sense of respect; for it is clear that in any transaction between a polite man and a rude man, if brute force is of the essence of that transaction, the rude man

is very likely to gain the upper hand; and any man is rude who often confuses respect with servility.

This then, is the position to which I have been trying to lead. The unity which our worldly-wise advisors are urging us to effect will be possible only if we adopt the absolutely anti-social methods of Europe;...[R1.171, p. 250].

With critical reflection on the African essence and African historical unity, epistemic unity and cultural unity as defining the path of progress in terms of the problem-solution process of *Sankofa-anoma, Anoma-kokone-kone, Funtummireku-denkyem-mireku* and *Asantrofi-anoma* [R1.92]phenomena within the fundamental moral principles of Maat, the Goddess of good, truth, law and order [R1.55] [R1.57],[R1.176a], Esuman-Gwira Sekyi instructed Africans with a further statement on the essence of Africanness by stating:

In particular, I wish to speak to the African. Let the native African not heed anything that is said by non-African peoples against him. Let him be proud of his African soul, his black soul, the soul that evolved all that has tamed a good deal of the aggressiveness that has heretofore characterized the lighter people. Let him seek always to remain African, for that which makes him African, that which for convenience I will call Africanity, always has been, and always will be , the leaven with which the crude meal of humanity can be leavened. In spite of all that prejudiced archaeology and anthropology have sought to establish, the real reason why negroes are darkest in colour will soon enough be clear to all who can think. Africa was old when Europe was young: reflection therefore is African, while impulse is European. Let us, the children of Africa, by remaining true to our Africanity, help to raise the children of Europe, our juniors, in the art of living socially, above the impulse of their unreflective social youth [R1.171, p. 251].

III Thinking, Decisions and Progressive Social Transformation

Philosophical Consciencism defines an epistemic framework of control instrumentation of the necessary controllers for the transformation of the social formation by setting the potential pole against the actual pole in the social actual-potential polarity. The success of the design of this epistemic framework depends on knowledge, thinking and decision-choice actions. Knowledge is acquired by thinking and decision-choice actions; thinking and decision-decision choice actions are enhanced by knowledge while knowledge and decision-choice actions are the products of thinking in constructionism-reductionism duality which affects the behavior of the ignorance-knowledge polarity in relation to the epistemic distance as well as the enveloping of the social transformation process for any given social experiential information structure. In this logical frame, the enveloping of the social transformation process, in the last analysis, is the work of the art and science of human thinking that simultaneously controls knowledge production and decision-choice actions. Social progress is a transformation and the path of that social transformation in the creation of an internal thinking system is not simply the imitation of someone else thinking. Knowledge of other people's thinking and the path of the knowledge-production process are useful to the extent to which they enhance one's internal thinking. This internal thinking that must be produced by cognitive actions of the African society is indeed wanting. The African epistemic process that is required to deal with the elements in the problem-solution dualities has been hijacked and reduced to art and science of mimicry contrary to the African tradition as we know it from antiquity in the villages and rural areas. This mimicry has been perfected to the point of Africa's cognitive destruction, where thinking on the part of the educated African has become a lost art and the science of thinking has become a difficult African problem in the search of solutions to the elements in the problem-solution duality.

It is on the reflection on this cognitive destruction and mimicry of other peoples' epistemic process without particular reference to African conditions that lead Atto Ahuma to the following statement.

> We often confuse memory and recollection with the whole of that mental operation which produces thought. The capacity for, or the exercise of the very highest intellectual functions is not characteristic of the man in the street, nor indeed of the average man [person] of intelligence the wide world over. Thinking is an Art; it is the greatest blessing in the gift of Heaven and may not even be found in some talented man [person] who could box the compass of the whole circle of academical education.
>
> As a people [Africans], we have ceased to be a THINKING NATION. Our forebears, with all their limitations and disadvantages, had occasion to originate ideas and to contrive in their own order. They sowed incorruptible thought-seeds, and we are reaping a rich harvest to-day, though. For the most part, we are scarcely conscious of the debt

we owe them. Western education or civilization undiluted, unsifted, has more or less enervated our minds and made them passive and catholic. Our national [African] life is semi-paralysed; our mental machinery dislocated, the inevitable consequence being, speaking generally, the resultant production of a Race of men and women 'who think too little and talk too much. But neither garrulity nor loquacity forms an indispensable element in the constitution of a state or nation [R1.171, p.166].

A reflective understanding of this statement by Attoh Ahuma is instructive on the role of cognitive operations in human existence. There is a component of the zone of *passive cognitive operations* composed of memory and recollection of acquaintances, the experiential information structure and the existing content of the availability of the claimed knowledge structure. The passive cognitive operations are memorization, retention and simple reproduction of information and knowledge. There is the zone of *active cognitive operations* where acquaintances, the experiential information structure and the claimed knowledge structure are mentally processed to produce thought and new forms of knowledge. This active cognitive operation is thinking which the highest form of the human intellect is. It is this active cognitive operation as the highest form of intellectual activity that Philosophical Consciencism is about. From the point of view of Philosophical Consciencism in the instrumentation of controllable elements and the management of command and control under continual changing information structure, the important component of education is not the passive cognitive operation but the active cognitive operation that establishes a thinking society in creating new forms. In other words, transformative education must emphasize thinking which is active and the highest form of the cognitive operation for which much discussions have been undertaken in [R1.95] [R1.294] [R3.10] [R3.13]. The concept and the practice of the development of a thinking society through appropriate methodological processes have always formed the epistemic centrality of my learning, teaching, research and writing within the academic work. The cognitive foundation of the above statement lies in the idea that new knowledge and refutation of bankrupt ideas are at the mercy of thinking and thinking system which place one in a *conflict zone* of ignorance-knowledge duality for continual knowledge production. An epistemic mimicry of existing knowledge, devoice of the principle of doubt, leads to a condition where one lives in a zone of familiarity which is encapsulated in the *zone of cognitive imbecility*.

Education is composed of teaching, learning and research. The greatest effective output of education is to enhance the cognitive capacity of the learner to think by practicing the *principle of doubt* to escape the epistemic enslavement of the oppressive chambers of cognitive familiarity which is produced by memory and recollection under the principle of credulity. The passive cognitive operation is simply an intermediary product that is to serve as an input into the active cognitive operation responsible for generating new outcomes to expand the domain of the passive cognitive operation and the uses to which it may be demanded within the problem-solution duality. Any society stuck in the zone of passive collective cognitive activity is creating its own cognitive extinction and the destruction of progressive internal transformation of its social form. Education with more emphasis on passive cognitive operations and on the principle of intellectual credulity leads to a process of generating an individual and collective increase of cognitive decay the cure of which is simply reversing the process with emphasis on education in thinking and encouragement of unrestricted intellectual inquiry that is relevant to the social set-up.

Since experiential information structure varies over societies, so also is the nature of solution-problem duality and hence a solution to the problem of one society is not directly transferable to other societies without epistemic restructuring through thinking. The nature of the problem-solution duality of different societies offers external conditions that may indicate the need to change and intensify dualistic activities associated with social actual-potential polarities of other societies. The developmental structure of African-centered Philosophical Consciencism recognizes that there is a problem within the African educational process to impart information in the experiential information structure without also imparting the art and science of thinking. To prevent the cultivation of falsehood in the experiential information structure composed of defective and deceptive information sub-structures in the African transformation process African-centered Philosophical Consciencism provides a framework of guidance to implement the elements in the zone of active cognitive operation that will help to create the sufficient conditions of social transformation of Africa.

With the understanding of the distinction between passive and active cognitive operations and their relationships to education and social transformation in the social decision-choice space, Atto Ahuma provided us with further reflection using the following statement.

Existence is a mere parody unless embroidered with the flowers of the intellect and the fruitage of the soul. When the executive forces of a man's [person's] life are wholly enlisted in the daily gratification of selfish pursuits and individual aggrandizement – when emergent novelties of a foreign strand absorb the energies of mind and soul and strength. Ideas cannot germinate though disseminated by Cherubim and Seraphim. We shall always miss the pulsation, vibrancy and full volume of life as a nation [African Nation] until we have understood what it is to think nationally – to spend and be spent for the highest good of our country [Africa] and our race. Until we have discovered for ourselves this 'missing link,' we do well to despair of the collective realization of the ancient prophecy, "Ethiopia shall soon stretch out her hands." 'Africa shall rise,' but only when we begin to think continentally and nationally. This want of real, vital and solid thinking has its moral dangers; for 'as a man thinketh in his heart, so is he.' In prehistoric days, Europe looked to Africa for new ideas, for fresh inspirations, and the saying was perpetuated and handed down from generation to generation, 'Semper aliquid novi ex Africa' – There is always something new from Africa.... Now lies she there. And none so poor to do her reverence. All because thinking in our age has become a lost Art [R1.171, p.167].

Human activities in the zone of active cognitive operation encompassing the science and art of thinking are the most difficult to practice, learn, and teach in epistemic activities over epistemological space that must be related and linked to ontological space. The nature of the difficulty rests on the fact that it is a complex system involving activities of assessment of paradigms and laws of thought, knowledge production, understanding with comprehension of a system of codes, evaluation of truth and falsity of validity of claimed knowledge, analysis of epistemic structures and sub-structures, synthesis of epistemic structures and sub-structures and applications to correct and create new forms of socio-natural elements wherever possible. It is because of these activities in the active cognitive operation with creative thinking why old and non-useful ideas are revised, phantom ideas are discarded and new ideas are generated. The art and science of thinking are subject to social intentionalities both of which are under the control of Philosophical Consciencism, and in the African case, the individual and collective thinking must draw its strength from the guidance of African-centered Philosophical Consciencism to make the thinking relevant in all African problem-solution dualities, as well as to avoid spending African time on phantom social problems. It is the complexity of the active cognitive operation in all its dimensions that motivated the following statement by Atto Ahuma.

> The most difficult problem of our times is how to think so that Africa may regain her lost Paradise. How to think the thoughts that galvanize and electrify into life souls that are asleep unconscious of their destiny; How to think the thoughts that produce, multiply, divide and circulate for the general good – the thoughts that make crooked places straight, that pulverize gates of brass and cut in sunder all bars of iron – the power that gives friends and foes alike the treasuries of darkness and hidden riches of secret places – the Art that brings National Evangels, binding up broken and despairing hearts, proclaiming liberty and freedom to captives, and the opening of the Prison to them that are bound or have bound themselves. To effect such an end, we must leave severely alone the empty pageantries of triflers, the eccentricities of pedants, the inanities of agitators, and the ingenuities of sycophants. These are conditions more abiding and worth contending for, achieving and overcoming; in this sign we shall conquer, if we learn to think our hardest and strive to transmute our innermost thoughts into action for THE SAFETY OF THE PUBLIC AND THE WELFARE OF THE RACE [R1.171, p.168].

IV Credulity, Mimicry, Mendacity and Philosophical Consciencism

The solution to the problem of thinking in African cognitive space is made more difficult by Africa's credulity in the information structure and irrelevant knowledge produced by colonialists and imperialists, with further intensification by neo-colonial maneuvers in the information-knowledge space backed by Western imperial terror and ruthless destruction of social fabrics. Education in the process of producing a neo-colonial mindset through laws of thought and the classical paradigm, under the principle of excluded middle in dualism, where logical contradiction is not accepted as a truth value, is incompatible with the African ways of a traditional paradigm of thought. This Eurocentric way of thought creates social formation that separates the individual from the community without reunification. The emphasis is on the fundamental ethical principle of individualism and individual freedom as an ethical measure of a good society where the question is not how well is the society doing but how well is the individual doing. In this respect, individual or a group of individuals is encouraged to exploit others in the social set-up at the expense of collective social progress. Critical African thinking on this Eurocentric paradigm will immediately reveal that this is incompatible with traditional African social formation.

The African paradigm and laws of thought, on the other hand, work with the principle of opposites under duality and polarity with relational continuum and unity, where there is no excluded middle. It is this concept of relational continuum and unity projected onto the space of conflicts and contradictions with resolutions and reconciliations which forms the logical foundation of African social formation of collectivity and humanism, where the individual and community are organizationally separated but reconnected in an inseparable oneness. The emphasis is on the fundamental ethical principle of collectivism, humanism and collective freedom as an ethical measure of a good society, where the question is not how well is an individual doing at any time but how well is the society doing. In the African educational process with emphasis on learning, the members of an educated class are created with a personality and mindset that are induced by the claimed information-knowledge structure. To the extent to which this information-knowledge structure is imperialist and neocolonialist derived with substantial social attachment of credulity, the cultivated mindset will be neocolonial and the resulting personality will be non-African.

As such, there is always a fundamental question as to what is the learner learning and to what extent is the degree of credulity. The principle of credulity in the information-knowledge structure violates the principle of doubt that requires thinking and not crude mimicry. In the African cognitive space, credulity and mimicry lead to the crude copying of incompatible behavior to the internal transformation of African societies. It is useful to keep in mind that the imperial information-knowledge structure is composed of defective and deceptive information-knowledge sub-structures from which anti-African propaganda and imperial ideology with their supporting Western-centered Philosophical Consciencism with it imperial intentionalityare constructed. It is the understanding of the social effects of the relational structure of credulity, mendacity and education that lead Attoh Ahuma on reflecting while the Whiteman and Africans, to produce the following statement:

Thanks to the letters C.O.D., facilities are afforded the young upstart to gratify his unworthy ambition. What the Whiteman eats, he eats, what he drinks and smokes, he drinks and smokes, thereby securing what , in his deluded opinion, is considered the Hall-mark of respectability, civilization and refinement. If his lord and master holds a cigar in a peculiar manner, it is copied; his gait, mode of expression, his expletives, smiles, laughter and other mannerisms with the fidelity of an Edisonian Phonograph. These are the things the black wretch in his Boeotian ignorance and folly, regards as signs of perfect manhood – this thin veneer of polish – and there the lesson ends. The thoughtful, judicious and discreet Young African, naturally versed in the principles of Selection – who differentiates and discriminates between essentials and unessentials, who studiously rejects and selects, skips what does not concern him or does not correspond with his environments, who recognizes limitations, and is independent of foreign ways, customs and manners, is accordingly ridiculed and reprobated as 'de trop' an unclassed. He is Hottentot or a Bushman who does not successfully compete with the Whiteman in his sartorial equipment [R1.171, p.170-171].

This is an imitation and mimicry perfected to the art and science of insanity towards the destruction of African cognitive unity, cultural unity, collective creativity and socio-political unity. In this mode of imitation and mimicry African collective thinking becomes a casualty in the epistemological space where Africa becomes a consumer of some information-knowledge structures which are not only irrelevant to her existence, but extremely destructive of the spirit of her social transformation from within, and where qualitative movement is externally imposed. In this cognitive imitative process, the educated African has no social transformative character and has become a hopeless liability to his/her people and exploiter of his/her supporting material and spiritual foundation. This behavior is duplicate in all walks of African life where creative thinking has been lost and the education process has lost its African intentionality. Working with Western-centered Philosophical Consciencism, educated Africans are in cahoots with the neo-colonialists and imperialists to hasten and encourage the destruction of Africa. These members of the African educated class as well as the leadership have surrendered like master slaves leading their flock of sheep into poverty and social destruction. They fail to understand that social transformation is about nation-building; and this nation-building is of the people, by the people and for the people. It is an act of collective existence and not individual existence. The credulity attached to the Western epistemic structure and Philosophical Consciencism by unthinking educated Africans and members of the leadership remotely control them to look for solutions to African problems from without rather than from within. Refusing to think, they look for new ideas for consumption outside of Africa, rather than remembering the old European saying: *ex Africa semper aliquid novi* (from Africa comes always something new). The theory of Philosophical Consciencism, as a guide to social decision-choice system for transformation of social actual-potential polarities affirms the position that thinking mind is the standard of a transformative person and that just and lawful behavior is the standard of civilized nation. It is this thinking mind that creates great persons who then then define social vision, pathway of social progress and construct great institutions for the management of control-command systems of transformations in the nation-building process, the social intentionality of which finds expressions in Philosophical Consciencism. This is the audacity of thinking.

V The Audacity of Thinking, Vision and Pathways under African-Centered Philosophical Consciencism in Social Systemicity

The design of African resurgence, vision and pathway of social transformation are the products of cognitive operations. These operations must be active and not passive, if a desirable pathway of motion in quality-time space is to be created to move the social system over the desirable and acceptable transformation enveloping path in terms of the welfare of the African people. The social transformation requires work in creating the sufficient conditions for instruments of control and their management to move the social system from one categorial equilibrium to another on the basis of knowledge obtained from the African experiential information structured. These cognitive operations are the work of the audacity of thinking. The audacity of thinking is critical examination of the African experiential information, creating cognitive capacity of its processing into knowledge structure, analyzing and synthesizing the various components, comprehending its meaning in unity and continuum and putting it to use for qualitative change relevant to the nation-building social program. All these require policy constructs by thinking to select the pathway of African progress. The strategic thinking to destroy the conditions of colonialism in order to transform the African territories to decolonized territories is particular to the colonialism-decolonization polarity which then gives rise to a new actual-potential social polarity. This new actual-potential social polarity requires a new and more critical strategic thinking to select a pathway of African progress and security from within. It was on this recognition by Nkrumah that brought about the following statement:

> When independence has been gained, Positive Action requires a new orientation away from the sheer destruction of colonialism and toward national reconstruction [R1.207, p.112].

The importance of the audacity of thinking cannot be underestimated. Everything that happens in a society is within the thinking-unthinking duality. In the African social set-ups, the audacity of the strategic thinking must be guided by African centered Philosophical Consciencism to induce the pathway selection and manufacture the decision-choice activities over the selected pathway. After the transformation of the colonialism-decolonization polarity, the new actual-potential polarity presents itself to two organic pathways of decision-choice action of critical importance. One pathway leads to Africa's attachment to an imperial system of exploitation and oppression under a thinking system guided by Western-centered Philosophical Consciencism. This pathway leads to the establishment of the decolonization-neo-colonization polarity with a complete or partial focus on international begging and hopeless development strategies shredded in humiliation and hopelessness. This is not the African pathway to resurgence, Africa's tradition nor her golden years to come. The second pathway leads to Africa's detachment from the imperial system of exploitation and oppression under a thinking system guided by African-centered Philosophical Consciencism. This pathway leads to the establishment of the independent-emancipation polarity with complete focus on construction of Africa's tomorrow from within Africa's internal strength on the basis of African will and preferences with dignity, no begging, no humiliation and no hopelessness, where the transformational activities in the underdevelopment-development polarity is Africa's own thinking, will and preference under Africentric conceptual system. However, this is Africa's cognitive dilemma. Should Africa's cognitive operation and audacity of thinking be guided by the Western-centered Philosophical Consciencism under the imperial iron walls of logical familiarity and historical exploitation or be guided by African-centered Philosophical Consciencism under the walls of creativity and vibrancy that connect Africa cognitive today to its traditional roots?

It is on the recognition of this African cognitive dilemma and critical thinking with the conflicts and contradiction in choosing the pathway to African progress that Ayi Kwei Armah's reflection on Africa's present pathway to destruction, and a change of continuity to the past pathway to current *regeneration* becomes an important input into the development of contemporary African thinking.

> What remains? To sing regret, curse ancestors and throughout stagnant lives pass down the malediction on those yet to come? Easy that lazy existence, sweetly dragged the life spent waiting upon death. Easy the falling slide, even for rememberers.
>
> We who hear the call not to forget what is in our nature, have we not betrayed it in this blazing noonday of the killer's? Around us they have placed a plethora of things screaming denial of our nature, things welcoming us against ourselves, things luring us into the whiteness of destruction. We too have drunk oblivion, and overflowing with it, have joined the exhilarated chase after death.
>
> We cannot continue so. For a refusal to change direction, for the abandonment of the way, for such perverse persistence there are no reasons, only hollow, unconvincing lies.
>
> And the seers, the hearers, the utterers? What sufficiency is there in our hearing only this season's noise, seeing only the confusion around us here, uttering, like cavernous mirrors, a wild echo only of howling cacophony engulfing us? That is not the nature of our seeing, that is not our hearing, not our uttering. Only our drugged weariness, unjustified, unjustified, keeps us bound to the present.[R1.16, p. xiii].

Ayi Kwei Armah's comparative reflections on the pathways of contemporary Africa and historic traditions of Africa's social transformation continue with analytical instruction. By noting that the contemporary decision-choice pathway of Africa is destructive and leads to Africa's demise, he asks a series of questions around which to think. He continues:

> How have we come to be mere mirrors to annihilation? For whom do we aspire to reflect our people's death? For whose entertainment

shall we sing our agony? In what hopes? That the destroyers, aspiring to extinguish us, will suffer conciliatory remorse at the sight of their own fantastic success? The last imbecile to dream such dreams is dead, killed by the savior of his dreams. Such idiot hopes come from a territory far beyond rebirth. Those utterly dead, never again to wake, such is their muttering. Leave them in their graves. Whatever waking form they wear, the stench of death pours ceaseless from their mouths. From every opening of their possessed carcases comes death's excremental pus. Their soul itself is dead and long since purified. Would you have your intercourse with these creatures from the graveyard? Go to them then, and speak your message to long rotten ash.

This sight, this hearing, this our uttering: these are not for dumb recording of the senseless present, unless the vocation we too have fallen into wanting is merely to be part of the cadaverous stampede, hurrying on the rush to destruction[R1.16, p. xiii].

This pathway to Africa's destruction is under the guidance of Western-centered Philosophical Consciencism that has created the colonial and neocolonial mindset in the African decision-choice space where African leadership and the educated class are always looking outside of Africa to solve Africa's problems instead of realizing that sustainable progressive social transformation is internally generated, and not from without. A critical question the arises in the space of African cognition and decision-choice activities as to how to change from the pathway of destruction and death guided by Western-centered Philosophical Consciencism to the pathway of life and progress guided by African-centered Philosophical Consciencism. Here, Ayi Kwei Armah reflecting on the resolutions to the conflicts and contradictions within the actual-potential polarity of Western-centered Philosophical Consciencism and African-centered Philosophical Consciencism appeals to the tradition of the *Sankofa-Anoma principle* [the time trinity of past-present-future]:

The linking of those gone, ourselves here, those coming; our continuation, our flowing not along any meretricious channel but along our living way, the way: it is that remembrance that calls us. The eyes of seers should range far into purposes. The ears of the hearers should listen far towards origins. The utterers' voice should make knowledge of the way, of heard sounds and visions seen, the voice of the utterers should make this knowledge inevitable, impossible to lose.

A people losing sight of origins are dead. A people deaf to purposes are lost. Under fertile rain, in scorching sunshine there is no difference: their bodies are mere corpses, awaiting final burial. [R1.16, p. xiv].

...For those returning, salvaging blistered selves from death, and those advancing still hypnotized by death, in the absence of the utterers' work what will they be but beasts devouring beasts, zombis fighting zombis, a continuation along the road of death in place of regeneration, the rediscovery of our way, the way?

Leave the killers' spokesmen, the predators' spokesmen, leave the destroyers spokesmen to cast contemptuous despair abroad. That is not our vocation. That will not be our utterance [R1.16, p.xvii].

Let us keep in mind that Western-centered Philosophical Consciencism projects philosophy and ideology to process Western experiential information structure which is infested with anti-African viruses for colonial and neocolonial advantage, in order to abstract benefits and transfer costs to Africa contrary to *Asantrofi-Anoma* rationality. It does not offer a logical pathway to sustainable transformational solutions in the African actual-potential social polarity with objectivity, and indeed the use of the mindset from it will always produce the opposite of African progressive transformation with dependency syndromes, indebtedness, chaotic social environments, exploitative intentionality and a loss of African personality and its human essence. This is the modus operandi of Western imperialism which produces social decay for Africa. The nature of this modus operandi of the predators has been explained in [R1.91]. The use of Western-centered Philosophical Consciencism to process and evaluate African experiential information structure into knowledge input of African decision-choice actions merely offers a set of prejudices and distorted conclusions under an increasing degree of logical African irrationality. The outcome is governed by a cost-benefit process in dealing with conflicting interests, social vision and goal-objective formation in managing different societies under different political economies. These conflicts are enhanced by a principal-agent phenomenon in contractual and dependency relations with the Western imperial system which is extensively analyzed and explained in [R13.8] [R13.9].

The Western-centered Philosophical Consciencism of the principle of opposites with excluded middle separate the individual from the community emphasizing the fundamental ethical principle of individualism which paves the way for human time extinction, where the principle of each for himself without regard for the welfare of all becomes the operational guide in the organizing and management of the social formation. The emphasis on individualism neglects the cardinal fact of the individual finality in the time structure and the time infinitude of the social set-up. The preservation of human existence in the infinite time domain is sacrificed for the simple and immediate preservation of the individual, thus paving the way for the complete destruction of the African Ancestral Tree as well as the elimination of its unlimited growth and continuity. From this organizational perspective, the principle of opposites under the Western conceptual system promotes low culture and the destruction of human continuity for species evolution, which is inconsistent with the Africentric universal concept of continual creation without end. Here, the individual is separated from the collective by the principle of excluded middle with no logical connectivity as abstracted from continual creation and evolution under the actual-potential polarity with a well-defined primary category and derived categories from the principle of opposite with relational continuum and unity. The emphasis is simply on individual liberty where the social foundation within the principle of individual liberty, law and order moves over increasing low culture with wars and human suffering and destruction. It is on the basis of this Africa's transformation problem and the choice of the appropriate pathway that Cheikh Anta Diop provided us with the following reflection:

> Our generation is out of luck, so to speak, in what we will not be able to avoid the intellectual storm; willy-nilly, we will be forced to take the bull by the horns, to rid our minds of intellectual formulas and tidbits of thought in order to enter resolutely upon the only truly dialectical path toward the solution of the problem that history forces upon us. This connotes active research, in the most authentic meaning of that term, by clear and fertile minds capable of producing effective solutions and realizing them without intellectual guardianship.
>
> Historical circumstances now demand of our generation that it solve in a felicitous manner the vital problems that face Africa, most especially the cultural problem. If we do not succeed in this , we will appear in the history of the development of our people as the watershed generation that was unable to insure the unified cultural survival of the African continent; the generation which, out of political and intellectual blindness, committed the error fatal to our national

future. We will have been the unworthy generation "par excellence"[R1.87, pp. 13-14].

The concept *without intellectual guardianship* is powerfully instructive and challenging. It means that Africa must light her way through tunnels of epistemic darkness with African-centered Philosophical Consciencism to the openness of progress in defense of herself and her children. It is here that when an awakened African intelligentsia emerges from the womb of Africa it will excavate the great intellectual treasure hidden in the bosom of her belly. This awakened African intelligentsia will no longer seek compromise with the intellectual world of Africa's imperial occupiers and neocolonialists by searching for crumbs for ungraceful eating with the hope of receiving accolades in the shamelessness of their positional servitude within the imperial kingdom of Africa's oppressors. This awakened intelligentsia will realize Nkrumah's advice with a full and uncompromising awareness and dignity without apology that:

> What is called for as a first is a body of connected thought which will determine the general nature of our action in unifying the society which we have inherited, this unification to take account, at all times, of the elevated ideals underlying the traditional African society. Social revolution must therefore have, standing firmly behind it, an intellectual revolution, a revolution in which our thinking and philosophy are directed towards the redemption of our society. Our philosophy must find its weapons in the environment and living conditions of the African people. It is from those conditions that the intellectual content of our philosophy must be created [R1.202, p. 78].

These are the challenges of the members of the African intelligentsia that some great African thinkers like Abraham, Amo Afer, Ben-Jochannan, Anta Diop, DuBois, Nkrumah, Woodson and many others have articulated. Given these challenges, there are two cohorts of unenlightened and enlightened members which constitute a duality with relational continuum and unity. The composite challenge requires that some members of the African intelligentsia graduate into the cohort of the members of the enlightened African intelligentsia. The emancipation of the African continent, therefore, is the emancipation of the African traditional conceptual system from corruption and enslavement. This firstly requires a separation from Western-centered Philosophical Consciencism and then into a pursuance of a progressive development of

377

the traditional African thought system under African-centered Philosophical Consciencism. The first requirement is then reinforced by a second requirement of intense mobilization and dissemination of all elements of the African experiential information structure as well as the members of the African intelligentsia.

On this pathway to African epistemic redemption, these interdependent monographs have added a voice toward *rediviva Africana* in the development of high culture where freedom and justice are the foundation of the changing positions within the individual-community duality with relational continuum and unity, and where the principle of each for all and all for each establishes the identities of individuals and the community. The principle of opposites with relational continuum and unity emphasizes the fundamental ethical principle of collectivity which paves the way to human time continuum where the principle of each for all and all for each becomes operational guide in organizing and management of the social formation. The individual has finality in the time structure while the social setup has infinite time domain. The preservation of human existence is the preservation of the human collectivity represented by the African Ancestral Tree that has an unlimited growth. The principle of opposite under the Africentric conceptual system promotes high culture of human continuity for species evolution which is consistent with the Africentric universal concept of continual creation without end. Here is the logical connectivity as abstracted from the following conceptual statement on creation and evolution under the actual-potential polarity with a well-defined primary category and derived categories from the principle of opposites with relational continuum and unity, where the social foundation within the principle of collective freedom and justices moves over increasing high culture

I am he, who evolved himself under the form of the god Khepera, I, the evolver of the evolutions evolved myself, the evolver of all evolutions, after many evolutions and developments which came forth from my mouth. (I developed myself from primeval matter, which I had made).

No heaven existed, and no earth, and no terrestrial animals or reptiles had come into being. I formed them out of the inert mass of watery matter, I found no place whereon to stand...I was alone, and the gods Shu and Tefnut had not gone forth from me; there existed none other who worked with me. I laid the foundation of all things by my will; All things evolved themselves therefrom. I united myself to my shadow, and I sent forth Shu and Tefnut out from myself; thus being one god I became three, and Shu and Tefnut gave birth to Nut and Seb, and Nut gave birth to Osiris, Horus-Khent-an-maa. Sut, Isis, and Nephthys, at one birth, one after the other, and their children multiplied upon this earth [1.56, pp xcix-c] [Note Tefnut=water, Shu= air, Geb= earth]

The problem of transformations and evolutions is resolved in Africentic philosophical traditions through the interplay of opposing patterns of multiplicity of rhythms and their organizationally behavioral forms. Nothing is quantitatively lost in transformation and nothing is quantitatively gained in transformation, only qualitative transformations of states, categories and processes creating varieties in particular and universal unity.

THE THEORY OF PHILOSOPHICAL CONSCIENCISM: PRACTICE FOUNDATIONS OF NKRUMAISM IN SOCIAL SYSTEMICY

MULTIDISCIPLINARY REFERENCES

R1: ON AFRICA AND FOUNDATIONS OF AFRICAN THOUGHT SYSTEM

[R1.1] Abraham, W.E., *The Mind of Africa*, London, Weidenfeld and Nicholson, 1962.

[R1.2] Ahuma, Attoh S.R.B., *The Gold Cost Nation and National Consciousness*, in Langley, Ayo J. (ed.), *Ideologies of Liberation in Black Africa: 1856-1970*, London, Rex Collings, 979. pp. 161-172.

[R1.3] Ajala, Adekunle, *Pan-Africanism: Evolution, Progress and Prospects*, London, Andre Deutsch, 1973.

[R1.4] Allen, J. *et al., Without Sanctuary: The Lynching Photography in America*, Santa Fe, N.M., Tween Palms Publishers, 2000.

[R1.5] Allen, James P., *Genesis in Egypt: The Philosophy of Ancient Egyptian Creation* Accounts (Yale Egyptological Studies, #2), New Haven, Yale University, Department of Near Eastern Languages and Civilizations, 1988.

[R1.6] Alston, W. and R.B. Brandt (eds.), *The Problems of Philosophy*, Boston, MA, Allyn and Bacon, 1967.

[R1.7] Amate, C.O.C., *Inside OAU: Pan-Africanism in Practice*, New York, St. Martin's Press, 1986.

[R1.8] Amen, Ra Un Nefer, *Metu Neter* Vol. 1, Bronx, New York, Khamit Co. Pub., 1977.

[R1.9] Amin, Samir, *Neo-Colonialism in West Africa*, New York, Monthly Review Press, 1973.

[R1.10] Amin, Samir, *Eurocentrism*, New York, Monthly Review Press, 1989.

[R1.11] Amo Afer, A.G., *The Absence of Sensation and the Faculty of Sense in the Human Mind and Their Presence in our Organic and Living Body, PhD Dissertation and other Essays1727-1749*, Halle, Wittenberg, Jena, Martin Luther University Translation, 1968.

[R1.12] Ani, M., *Yurugu: An African-Centered Critique of European Cultural Thought and Behavior*, Trenton, N.J., Africa World Press, 1994.

[R1.13] Apostel, Leo, *African Philosophy: Myth or Reality*, Ghent, Belgium, Scientific Publishers, 1981.

[R1.14] Appolus, Emil (ed.), *The Resurgence of Pan-Africanism*, London, Freedman Brothers, 1974.

[R1.15] Aristote, N., *Politique*, Books I and II, Paris, Les Belles Lettres, 1960.

[R1.16] Armah, Ayi Kwei, *Two Thousand Seasons*, Oxford, Heineman, 1973.

[R1.17] Armah, Ayi Kwei, *Osiris Rising*, Dakar, Senegal, Africa Per Ankh, 1995.

[R1.18] Asante, Molefi K. *et al.* (eds.), *African Culture: The Rhythms of Unity*, Trenton, New Jersey, African World Press Inc., 1990.

[R1.19] Asante, Molefi K., *Afrocentricity*, Trenton, New Jersey, Africa World Press, Inc., 1989.

[R1.20] Asante, Molefi K. *et al.* (eds.), *African Intellectual Heritage: Book of Sources*, Philadelphia, PA, Temple University Press, 19

[R1.21] Asante, Molefi K., *Kement, Afrocenticity and Knowledge*, Trenton, New Jersey, Africa World Press, Inc., 1990.

[R1.22] Asante, Molefi K and Ama Mazama (Eds.), Encyclopedia of African Religion Thousand Oaks, California, Sage, 2009

[R1.23] Asante, S.K.B., *Pan-African Protest: West Africa and the Italo-Ethiopian Crisis*, 1934-1941, Legon History Series, London, Longmans, 1977.

[R1.24] Ashby, M.A., *The African Origins of Civilization: Book I, Mystical Religion and Yoga Philosophy*, Miami, FL, Cruzan Mystic Books, 1995.

[R1.25] Ashby, M.A., *The African Origins of Civilization, Book II: African Origins of Western Civilization, Religion and Philosophy*, Miami, FL, Cruzan Mystic Books, 2001.

[R1.26] Ashby, M.A., *The African Origins of Civilization Book III: The African Origins of Eastern Civilization, Religion, Yoga Mysticism and Philosophy*, Miami, FL, Cruzan Mystic Books, 2001

[R1.27] Ashby, M.A., Egyptian *Yoga Vol. I,: The Philosophy of Enlightenment*, Miami, FL, Cruzan Mystic Books, 1997   ,

[R1.28a] Auma-Osolo, A., *Cause-Effects of Modern African Nationalism on the World Market*, University Press of America, 1983.

[R1.28b] Austen, R. A., *In Search of Sunjata: The Mande Oral Epic as History, Literature and Performance*, Bloomington, Indiana University Press, 1999

[R1.29] Axinn, Sidney, "Kant, Logic and Concept of Mankind," *Ethics*, Vol. 48 (XLVIII), 1958, pp. 286-291.

[R1.30] Bakewell, Charles M., *Source Book in Ancient Philosophy*, New York, Charles Scribner's Sons, 1909.

[R1.30a] Bangura, Abdul K., *African Mathematics: From Bones to Computers*, Lanhan, MD, USA, University Press of America, 20012.

[R1.31] Bascom, William, *African Art in Cultural Perspective: Introduction*, New York, Norton, 1973.

[R1.32] Bebey, Francis, *African Music: A People's Art*, Westport, CT, Lawrence Hill and Co., 1980.

[R1.33] Bell, Richard H., *Understanding African Philosophy: A Cross-cultural Approach to Classical and Contemporary Issues*, New York, Routledge, 2002.

[R1.34] Ben-Jochannan, Joseph A.A., *Cultural Genocide in the Black and African Studies Curriculum*, New York, River Nile Universal Books, 1972.

[R1.35] Ben-Jochannan, Joseph A.A., *Africa: Mother of Western Civilization*, Baltimore, MD, Black Classic Press, 1988.

[R1.36] Ben-Jochannan, Joseph A.A. *et al., African Origins of Major Western Religions*, New York, Alkebulan Books, 1970.

[R1.37] Ben-Jochannan, Joseph A.A., *Black Man of the Nile*, Baltimore, MD, Black Classic Press, 1989.

[R1.38] Ben-Jochannan, Joseph A.A., *We the Black Jews*, Baltimore, MD, Black Classic Press,1983.

[R1.39] Ben-Jochannan, Joseph A.A., Hugh Brooks and Kempton Webb, *Africa: Land, People and Cultures of the World*, New York, W.H. Sadlier, 1971.

[R1.40] Berkeley, George, "Material Things are Experiences of Men or God" in [R1.5], 1967, pp. 658-668.

[R1.41] Blyden, E.W., "African Life and Customs" in Langley, Ayo J. (ed.), *Ideologies of Liberation in Black Africa:1856-1970*, London, Rex Collings, 1979 pp. 78-87..

[R1.42] Blyden, Edward W., *Christianity, Islam and the Negro Race*, Baltimore, MD, Black Classic Press, 1994.

[R1.43] Boahen, A. Adu, *African Perspectives on Colonialism*, Baltimore, Johns Hopkins University Press, 1987.

[R1.44] Bonnefoy, Yves, *Mythologies*, Vols 1-2, Chicago, University of Chicago Press, 1991.

[R1.45] Bovill, E.W., *The Golden Trade of the Moors*, London, Oxford University Press, 1958.

[R1.46] Bovill, E.W., *Caravans of the Old Sahara*, London, Oxford University Press, 1933.

[R1.47] Breasted, James H., *Ancient Records of Egypt*, Vols. 1-5, Chicago, The University of Chicago Press, 1906-1907.

[R1.48] Breasted, James Henry, *Development of Religion and Thought in Ancient Egypt*, New York, Charles Scribner's Sons, 1912.

[R1.49] Browdes, Anthony T., *Exploiting the Myths, Vol. 1: Nile Valley Contribution to Civilization*, Washington, D.C., The Institute of Karmic Guidance, 1992.

[R1.50] Browdes, Anthony T., *Egypt on the Potomac*, Washington, D.C., IKG Publishers, 2004.

[R1.51] Brown, Lee (ed.), *African Philosophy: New and Traditional Perspectives*, New York, Oxford University Press, 2004.

[R1.52] Budge, Willis E.A., *Osiris and the Egyptian Resurrection*, Vols. 1 and 2, New York, Dover, 1973.

[R1.53] Budge, Willis E.A., *The Gods of Egyptians*, Vols. 1 and 2, New York, Dover, 1969.

[R1.54] Budge, Willis E.A., *Amulets and Talismans*, New Hyde Park, University Books, 1961.

[R1.55] Budge, Willis E.A., *The Bandlet of Righteousness, an Ethiopian Book of the Dead*, London, Luzac and Co., 1929.

[R1.56] Budge, Willis E.A., *The Egyptian Book of the Dead*, New York, Dover, 1967.

[R1.57] Budge, Willis E.A., *The Papyrus of Ani*, Vols. 1, 2 and 3, New York, G. P. Putman's sons, 1913.

[R1.58] Budge, Willis E.A., *The Negative Confession*, New York, Bell Publishing Co., 1960.

[R1.59] Budge, Wallis, *The Egyptian Sudan*. Vols. I and II, London, Kegan, Trench & Co. 1907.

[R1.60a] Cabral, A., *Return to the Source*, New York, Monthly Review Press, 1973.

[R1.60b] Calame-GRIAULE, G., *Words and the Dogon World*, Trans D. LaPin, Philadelphia, Institute for the Study of Human Issues, 1986.

[R1.61] Cameron, J., *The African Revolution*, New York, Random House, 1961.

[R1.62] Chinweizu, The *West and the Rest of Us: White Predators, Black Slaves and the African Elite*, New York, Vintage Books, 1975.

[R1.63] Chomsky, Noam, *Pirates and Emperors: International Terrorism in the Real World*, New York, Claremont Research Publication, 1986.

[R1.64] Chomsky, Noam, *Profit Over People*, New York, Seven Stories Press, 1999.

[R1.65] Chomsky, Noam and E. S. Herman, *The Washington Connection and Third World Fascism*, New York, South End Press, 1979.

[R1.66] Clark, Gordon Haddon, *Thales to Dewey: a History of Philosophy*, Boston, MA, Houghton Mifflin, 1957.

[R1.67] Clark, Rundle R.T., *Myth and Symbol in Ancient Egypt*, New York, Thames and Hudson, 1978.

[R1.68] Clarke, John H., *Notes on African World Revolution: Africa at the Crossroads*, Trenton, N.J. African World Press, 1991.

[R1.69a] Clarke, John H., *Marcus Garvey and the Vision of Africa*, New York, Random House, 1974.

[R1.69b] Comaroff, J. and J. Comaroff, *Of Revelation and Revolution: Christianity Colonization and Consciousness in South Africa*, Chicago, University of Chicago Press, 1991

[R1.70a] Cowan, L.G., *The Dilemmas of African Independence*, New York, Warker and Co., 1964.

[R1.70b] Craton, Michael, *Sinews of Empire*, New York, Anchor Press, 1974.

[R1.71] Cromwell, Adelaide M. (ed.), *Dynamics of the African Afro-American Connection: From Dependency to Self-Reliance*, Washington, D.C., Howard University Press, 1987.

[R1.72] Cruse, Harold, *The Crisis of the Negro Intellectual: A History and Analysis of the Failure of Black Leadership*, New York, Quill Press, 1967.

[R1.73] Danquah, J.B., *Friendship and Empire*, London, Fabian Colonial Bureau, 1949.

[R1.74] Danquah J.B., *The Akan Doctrine of God: A Fragment of Gold Coast Ethics and Religion*, London, Frank Cass and Co., 1968.

[R1.75] Danquah, J.B., *Ancestors, Heroes and God*, Kibi, Ghana, George Boakie Pub. Co., 1938.

[R1.76] Davidson, Basil, *The Lost Cities of Africa*, Boston, Little, Brown & Co., 1959.

[R1.77] Davis, Kortright, *Emancipation Still Comin': Explorations in Caribbean Emancipatory Theology*, New York, Orbis, 1990.

[R1.78] Dawson, Christopher, *The Making of Europe, Part 1: The Foundations*, New York, The World Pub. Co., 1956.

[R1.79] De Buck, Adriaan and Alan H. Gardiner (eds.), *The Egyptian Coffin Texts*, Vols. 1-7, Chicago, University of Chicago Press, 1935-1961.

[R1.80] Descartes, René, *The Philosophical Works of Descartes*, Cambridge, Cambridge University Press, 1931.

[R1.81] Descartes, René, "Man as Two Substances" in [R1.5], 1962, pp. 386-402.

[R1.82] Descartes, René, *Meditations on First Philosophy*, New York, Boobs-Merrill and Co, 1960.

[R1.83a] Diop, Cheikh A., *Towards the African Renaissance: Essays in Culture and Development, 1946-1960*, Berwick upon Tweed .Great Britain, Martins the Printers, *1996*

[R1.83b] Diop, Cheikh A., *The African Origins of Civilization: Myth or Reality*, Brooklyn, New York, Lawrence Hill, 1974

[R1.84] Diop, Cheikh A., *The Cultural Unity of Black Africa*, Chicago, Third World Press, 1978

[R1.85] Diop, Cheikh A., *Pre-colonial Black Africa*, Brooklyn, New York, Lawrence Hill, 1987.

[R1.86] Diop, Cheikh A., *Civilization or Barbarism*, Brooklyn, New York, Lawrence Hill, 1991.

[R1.87] Diop, Cheikh A., *Black Africa: The Economic and Cultural Base for a Federated State*, Brooklyn, New York, Lawrence Hill, 197.

[R1.88] Doane, Thomas W., *Bible Myths and Their Parallels in Other Religions*, New Hyde Park, New York, University Book, 1971.

[R1.89] Dodson, H. *et al.*, *Jubilee: The Emergence of African-American Culture*, Schomburg Center for Research in Black Culture, New York, New York Public Library, 2002.

[R1.90a] Dompere, Kofi K., *Africentricity and African Nationalism*, Langley Park, MD, IAAS Publishers, 1992.

[R1.90b] Dompere, Kofi K., *The Theory of Categorial Conversion: Rational Foundations of Nkrumaism*, Working Monograph, Washington, D.C., Howard University Economics Department, 2013

[R1.91] Dompere, Kofi K., *African Union: Pan-African Analytical Foundations*, London, Adonis- Abbey Pub., 2006.

[R1.92] Dompere, Kofi K., *Polyrhythmicity: Foundations of African Philosophy*, London, Adonis-Abbey Pub., 2006.

[R1.93] Douglass, Frederick, "Fourth of July Oration," in Asante, Molefi K. *et al.* (eds.), *African Intellectual Heritage: Book of Sources*, Philadelphia, PA, Temple University Press, 19, pp. 637-640.

[R1.94] Dray, W.H., "Historical Understanding as Re-thinking," in Baruch Brody (ed.) Readings in the Philosophy of Science, Englewood Cliffs, NJ, Prentice-Hall Inc, 1970, pp. 167-179.

[R1.95] DuBois, W.E.B., *W. E. B. Du Bois: Reader* [edited by Eric J. Sundquist], New York, Oxford University Press 1996.

[R1.96] DuBois, W.E.B., *Dust of Dawn: An Essay Toward an Autobiography of a Race Concept*, New York, Harcourt, Brace

[R1.97] DuBois, W.E.B., *The World and Africa*, New York, International Publishers, 1987.

[R1.98] DuBois, W.E.B., *On Sociology and Black Community*, Chicago, The University of Chicago Press, 1978.

[R1.99] DuBois, W.E.B., *W.E.B. DuBois Speaks: Speeches and Addresses 1890-1919*, New York, Pathfinder, 1970.

[R1.100] DuBois, W.E.B., *The Education of Black People; Ten Critiques, 1906-1960* (Ed. H. Aptherker), New York, Monthly Review Press, 1975.

[R1.101] Duchein, R.N., *The Pan-African Manifesto*, Accra, Guinea Press Ltd., 1957.

[R1.102] Emerson, Rupert, *From Empire to Nation*, Boston, MA, Beacon Press, 1963.

[R1.103] Engels, Frederick, *Dialectics of Nature*, New York, International Pub., 1940.

[R1.104] Esedebe, P.O., *Pan-Africanism: The Idea and the Movement 1963-1976*, Washington, D.C., Howard University Press, 1982.

[R1.105] Fagg, William B., *Nigerian Images, The Splendor of African Sculpture*, New York, Praeger, 1963.

[R1.106] Fanon, Frantz, *The Wretched of the Earth*, New York, Grove Press Inc., 1963.

[R1.107] Fanon, Frantz, *Toward the African Revolution*, New York, Grove Press Inc., 1964.

[R1.108] Fanon, Frantz, *Black Skin, White Masks*, New York, Dover Press Inc., 1967.

[R1.109] Faulkner, R.C., *The Ancient Egyptian Coffin Texts*, Warminster, England, Aris and Philips, 1973.

[R1.110] Faulkner, R.C., *The Ancient Egyptian Pyramid Texts*, Oxford, Clarendon, 1969.

[R1.111] Felder, C.H., *Troubling Biblical Waters: Race. Class and Family*, New York, Orbis Press, 1989.

[R1.112a] Foote, George W. and W.P. Bell (eds.), *The Bible Handbook for Freethinkers and Inquiring Christians*, London, The Pioneer Press, 1921.

[R1.112b] Forde, D. (ed.) *African Worlds: Studies in the Cosmological Ideas and Social Values of African Peoples*, Oxford, Oxford University Press, 1954.

[R1.113] Frankford, Henri, *The Intellectual Adventure of Ancient Man*, Chicago, University of Chicago Press, 1957.

[R1.114] Frankford, Henri, *Ancient Egyptian Religion: An Interpretation*, New York, Columbia University Press, 1948.

[R1.115] Fraser, Douglas (ed.), *African Art as Philosophy*, New York, Interbook, 1974.

[R1.116] Frazer, James G., *The Golden Bough; a Study in Magic and Religion*, Vols. 1 – 13, London, Macmillan, 1911-1936.

[R1.117] Frobenius, Leo, *The Voice of Africa*, Vols. 1 and 2, London, Hutchinson and Company, 1913.

[R1.118] Gabre-Medhim, Tsegaye, "The Origin of the Trinity in Art and Religion," in Ben-Jochannan, Joseph A.A. *et al., African Origins of Major Western Religions*, New York, Alkebulan Books, 1970, pp. 99-120.

[R1.119] Gadamer, Hans Georg, *The Beginning of Knowledge* (Translated by Rod Coltman), New York, Continuum, 2001.

[R1.120] Garvey, Marcus, *The Philosophy and Opinions of Marcus Garvey*, Dover, MA, The Majority Press, 1986.

[R1.121] Geiss, I., *The Pan-African Movement*, London, Methuen Press, 1974.

[R1.122] Ghana Ministry of Information and Broadcasting, *Nkrumah's Subversion in Africa*, Accra, Ghana, Government Printing Press, 1966.

[R1.123] Gillings, Richard J., *Mathematics in the Times of the Pharaohs*, London, M.I.T. Press, 1972.

[R1.124] Glover, Ablade E. (ed.), *Adinkra Symbolism*, Accra, Liberty Press Ltd., 1971.

[R1.125] Glover, Ablade E., *Linguist Staff Symbolism*, Accra, Liberty Press Ltd., 1971.

[R1.126] Glover, Ablade E., *Stools Symbolism*, Accra, Liberty Press, 1971.

[R1.127] Goma, L. K. H., *The Hard Road to the Transformation of Africa*, Aggrey-Fraser Guggisberg Memorial Lecture 1991, University of Ghana, Legon-Accra, Communication Studies Press, 1991.

[R1.128] Graves, Robert and Raphael Patai, *Hebrew Myths: The Book of Genesis*, New York, Doubleday and Company, 1964.

[R1.129] Green, R.H. *et al., Unity or Poverty: the Economics of Pan-Africanism*, Baltimore, MD, Penguin Books, 1968.

[R1.130] Griaule, M., *Conversations with Ogotemmele*, London, Oxford University Press, 1969.

[R1.131] Groves, Charles P., *Planting of Christianity in Africa*, Vols.1-4, London, Lutterworth Press, 1948-1958.

[R1.132] Guyer, David, *Ghana and the Ivory Coast*, New York, An Exploration Press, 1970.

[R1.133] Gyekye, K., *An Essay on African Philosophical Thought: The Akan Conceptual Schemes*, New York, Cambridge University Press, 1987.

[R1.134] Gyekye, Kwame, *Tradition and Modernity: Philosophical Reflections on African Experience*, New York, Oxford University Press, 1997.

[R1.135] Hargreaves, J. D., *Prelude to the Partition of West Africa*, London, Macmillan, 1963.

[R1.136] Harris, Joseph E. (ed.), *Global Dimensions of the African Diaspora*, Washington, D.C., \ Howard University Press, 1982.

[R1.137] Hayford, Casely J.E., "African Nationality," in Langley, Ayo J. (ed.), *Ideologies of Liberation in Black Africa:1856-1970*, London, Rex Collings, 1979, pp. 203-219.

[R1.138] Herman, Edward N. and N. Chomsky, *Manufacturing Consent*, New York, Pantheon Books, 1988.

[R1.139] Hess, R., "Travels of Benjamin of Tudela," *Journal of African History*, Vol. 6, 1965, p. 17 (also in [R1.35] p. 5).

[R1.140] Hill, Cromwell A., and Martin Kilson (eds.), *Apropos of Africa: Sentiments of Negro American Leaders on Africa From 1800 to the 1950*, London, Frank Cass and Co., 1969.

[R1.141] Hillard, Asa G. (eds.), *The Teachings of Ptahhotop: The Oldest Book in the World*, Atlanta, Blackwood, 1987.

[R1.142] Hirschman, A. O., *National Power and the Structure of Foreign Trade*, Berkeley, University of California Press, 1945

[R1.143] Hochschild, Adam, *King Leopold's Ghost*, New York, Houghton Mifflin Co., 1998.

[R1.144] Hodgkin, Thomas, "National Movements in West Africa", *The Highway*, February, 1952, pp. 169-175.

[R1.145] Hoskins, L Halford (ed.) *Aiding Underdeveloped Areas Abroad, The Annals*, Vol. 268, March 1950.

[R1.146] Hountondji, Paulin J., *African Philosophy: Myth and Reality*, Bloomington, Indiana, Indiana University Press, 1983.

[R1.147] Hufbauer, G.C. *et al.*, *Economic Sanctions in Support of Foreign Policy Goals*, Washington, D.C., Institute for international Economics, 1983.

[R1.148] Hughes, L. and M. Meltzer, *A Pictorial History of the Negro (African American)*, New York, Crown, 1963.

[R1.149] Ilyenkov, E.V., *Dialectical Logic: Essays on its History and Theory*, Moscow, Progress Pub. 1977

[R1.150] Jahn, Janheinz, *Muntu, The New African Culture* (Trans. Marjorie Grene), New York, Grove, 1961.

[R1.151] James, C.L.R., "Kwame Nkrumah: Founder of African Emancipation", *Black World*, Vol. XXI (9), July, 1972.

[R1.152] James, George G.M., *Stolen Legacy*, Newport News, Virginia, United Brothers Communication System, 1989.

[R1.153] Jeffreys, M.D.W., " The Negro Enigma," *West African Review*, September, 1951.

[R1.154] Johnson, De Graft J.C., *African Glory*, New York, Praeger, 1955.

[R1.155] July, R.W., *The Origins of Modern African Thought*, London, Faber, 1967.

[R1.156] Kant, Immanuel, *On History*, New York, Bobb-Merrill and Co. Inc, 1963.

[R1.157] Karenga, N., *Maat: The moral Ideal in Ancient Egypt; A Study in Classical African Ethics*, Los Angeles, University of Sancore Press, 2006.

[R1.158a] Karenga, N., *Kwanzaa: Origin, Concepts, Practices*, Los Angeles, Kawaida Publishers, 1988.

[R1.158b] Karenga, N., *The African American Holiday of Kwanzaa*, Los Angeles, University of Sancore Press, 1988.

[R1.159] Kaunda, K., "Ideology and Humanism", *Pan-African Journal*, Vol.1 (1) 1968, pp. 5-6.

[R1.160] Kaunda, K., *Humanism in Africa*, London, Longmans, 1966.

[R1.161] Keita, L., " African Philosophical Systems: A Rational Reconstruction," *The* Philosophical *Forum*, Vol. 9 (23) Winter-Spring, 1977.

[R1.162] Keita, L., " Two Philosophies of African History: Hegel and Diop," *Présence* Africaine, No. 91, Third Quarter, 1974.

[R1.163] Keltie, J.S., *The Partition of Africa*, London, E. Stanford Press, 1893.

[R1.164] Kenyatta, Jomo, *Harambee!: The Prime Minister of Kenya's Speeches 1963-1964*, New York, Oxford University Press, 1964.

[R1.165] Kenyatta, Jomo, *Facing Mount Kenya*, New York, Vintage Books, 1965.

[R1.166] Knight, Richard P., *The Symbolic Language of Ancient Art and Mythology*, New York, J.W. Bouton, 1876.

[R1.167] Kohn, H., *African Nationalism in the Twentieth Century*, Princeton, New Jersey, Van Nostrand, 1965.

[R1.168] Krafona, Kwesi (ed.), *Organization of African Unity: Essays in Honour of Kwame Nkrumah*, London, Afroword Publishing Co., 1988.

[R1.169a] Lane, Ambrose I. Sr., *For Whites Only? How and Why America Became a Racist Nation*, Washington, DC., 21ˢᵗ Century Pub. Inc., 1999

[R1.169b] Langley, J. Ayodele, *Pan-Africanism and Nationalism in West Africa 1900-1945: A study in Ideology and Social Classes*, Oxford, Clarendon Press, 1973.

[R1.170] Langley, J.A., "Garveyism and African Nationalism," *Race*, Vol. 11 (2), pp. 157-172.

[R1.171] Langley, Ayo J. (ed.), *Ideologies of Liberation in Black Africa: 1856-1970*, London, Rex Collings, 1979.

[R1.172a] Leach, E. R.] *Culture and Communication: The Logic by which Symbols are Connected*, Cambridge, Cambridge University Press, 1976

[R1.172b] Legum, C., *Pan-Africanism*, New York, Praeger Pub., 1962.

[R1.173] Lynch, H.R., *Edward Wilmot Blyden: Pan-Negro Patriot*, London, Oxford University Press, 1967.

[R1.174] MacEwan, A., *International Economic Instability and U.S. Imperial Decline*, New York, Monthly Review Press, 1990.

[R1.175] Mandela, Nelson, *The Struggle for My Life*, New York, Pathfinder Press, 1986.

[R1.176a] Martin, D., Maat and Order in African Cosmology: A Conceptual Tool for Understanding Indigenous Knowledge, *Jour. of Black Studies*, Vol. 38, #6, Jul 2008, pp.951-967.

[R1.176b] Martin, Tony, *The Pan-African Connection: From Slavery to Garvey and Beyond*, Dover, MA, The Majority Press, 1983.

[R1.177] Mark Cohen, S. *et al.* (eds.), *Readings in Ancient Greek Philosophy: From Thales to Aristotle*, Indianapolis, Indiana, Hackett, 2000.

[R1.178] Massey, Gerald, *Ancient Egypt: The Light of the World*, Baltimore, MD, Black Classic Press, 1992.

[R1.179] Massey, Gerald, *A Book of the Beginnings*, Vols. 1-2, London, William and Norgate,   1881.

[R1.180] Massey, Gerald, *Pyramid Text*, Vols. 1-4, New York, Longmans Green, 1952.

[R1.181] Massey, Gerald, *The Natural Genesis*, Vols. 1 and 2, Baltimore, MD, Black Classic Press, 1998 (First published 1883).

[R1.182] Mathews, Wendell, *Basic Symbols and Terms of the Church*, New York, Fortress Press, 1971.

[R1.183] Mazrui, Ali A.A., *Nationalism and New States in Africa from About 1935*, Nairobi, Kenya, Heinemann, 1984

[R1.184] Mazrui, Ali A.A., toward *a Pax Africana*, Chicago, The University of Chicago Press,   1967.

[R1.185] Mbiti, J.S., *African Religions and Philosophy*, New York, Anchor, 1970.

[R1.186] M'buyinga, E., *Pan-Africanism or Neo-Colonialism: The Bankruptcy of the O.A.U.*, London, Zed Press,1975

[R1.187] McCray, W.A., *The Black Presence in the Bible*, vols. 1 and 2, Black Light Fellowship, 1990.

[R1.188] Mendoza, M.G. *et al.* (eds), *Trade Rules in the Making: Challenges in Regional and Multilateral Negotiations*, Washington, D.C., Brookings Institutions Press, 1999.

[R1.189] Meyerowitz, Eva L.R., *The Divine Kingship in Ghana and Ancient Egypt*, London, Faber and Faber, 1960.

[R1.190a] Moody, R.A., *Life After Life*, New York, Bantam, 1976.

[R1.190b] Mudimbe, V. Y., *The Inventions of Africa; Gnosis, Philosophy, and Other   Knowledge*, Womington, Indiana University Press, 1988.

[R1.191] Murapa, R., "Nkrumah and Beyond: Osagyefo, Pan-Africanist Leader," *Black World*,   Vol. XXI (9), July, 1972.

[R1.192] Niane, D.T., *Sundiata: An Epic of Old Mali in Transactions*, G. D. Pickett (ed.), London, Longmans, 1965

[R1.193] Nketia, J.H., *Ethnomusicology in Ghana,* Legon, Ghana, Ghana University Press, 1969.

[R1.194] Nketia, J.H., *Ethnomusicology and African Music: Collected Papers Vol.1: Models of Inquiry and Interpretation, Accra, Afram Publication, 2005*

[R1.195] Nketia, J.H., *The Music of African, New York, Norton, 1974*

[R1.196] Nketia, J.H. and J. C. Dje Dje, (eds.) *Selected Reports in Ethnomusicology Vol. V Studies in African Music,* Los Angeles, Univ. of California Press 1984.

[R1.197] Nkrumah, Kwame, *Toward Colonial Freedom,* London, Heinemann, 1962 (First Published 1946).

[R1.198] Nkrumah, Kwame, *Ghana, The Autobiography of Kwame Nkrumah,* London, Thomas Nelson and Sons Ltd., 1957.

[R1.199] Nkrumah, Kwame, Speech at the Conference of Independent African States, Accra, April 15, 1958.

[R1.200] Nkrumah, Kwame, *I Speak of Freedom,* London, PANAF Press, 1965.

[R1.201] Nkrumah, Kwame, Speech at the Closing Session of Casablanca Conference, Casablanca, January 7, 1961.

[R1.202] Nkrumah, Kwame, *Africa Must Unite*, New York, International Publishers, 1963.

[R1.203] Nkrumah, Kwame, *Consciencism*, London, Heinemann, 1964.

[R1.204] Nkrumah, Kwame, *Neo-Colonialism*, New York,

[R1.205] Nkrumah, Kwame, Speech at the Fourth Afro-Asian Solidarity Conference, Winneba, Ghana, May 10, 1965.

[R1.206] Nkrumah, Kwame, Challenge of the Congo, London, PANAF Press, 1967.

[R1.207] Nkrumah, Kwame, *Axioms*, New York, International Publishers, 1967.

[R1.208] Nkrumah, Kwame, Dark Days of Ghana, London, PA-NAF Press, 1968.

[R1.209] Nkrumah, Kwame, *Handbook of Revolutionary Warfare*, London, PANAF Press, 1968.

[R1.210] Nkrumah, Kwame, *Class Struggle in Africa*, London, PANAF Press, 1970.

[R1.211] Nkrumah, Kwame, *Revolutionary Path*, London, PANAF Press, 1973.

[R1.212] Nkrumah, Kwame, *Rhodesian File*, London, PANAF Press, 1976.

[R1.213] Nkrumah, Kwame, "Principles of African Studies," Address Delivered at the Time of Official Opening of the Institute for African Studies, University of Ghana, Legon, *Voice of Africa*, Vol. 3, (3) December, 1963.

[R1.214] Nkrumah, Kwame, *Selected Speeches, Vols. 1-6,* Compiled by Samuel Obeng, Accra, Afram Pub. 1997

[R1.215] Nsanze, Terence, "In Search of an African Ideology," *Pan-African Journal,* Vol.1 (1), 1968, pp. 27-30.

[R1.216] Nyerere, Julius K., *Freedom and Unity (Uhuru Na Umoja),* Nairobi, Oxford University Press, 1966.

[R1.217] Nyerere, Julius K., *Freedom and Socialism (Uhuru Na Ujamaa),* Nairobi, Oxford University Press, 1968.

[R1.218a] Obenga, Theophile T, *African Philosophy During The Period Of The Pharaohs, 2800- 300BC,* Popenguine, Senegal, W.A., Per Ankh Pub., 2006

[R1.218b] Obenga, Theophile T, *African Philosophy in World History,* Popenguine, Senegal, W.A., Per Ankh Pub., 1998

[R1.218c] Obenga, Theophile T, *Ancient Egypt and Black Africa,* London, Karnak Books, Africa World Press, 1992

[R1.218d] Odinga, Oginga, *Not Yet Uhuru,* New York, Hill and Wang, 1967.

[R1.219] Ofori-Ansah, Kwaku P., *Symbols of Adinkra Cloth,* Washington, D.C., 1999.

[R1.220] Ofosu-Appiah, L.H., *Encyclopaedia Africana* Vols. 1 and 2, New York, Reference Publication, Inc., 1977.

[R1.221] Okwonko, R. L., "The Garvey Movement in British West Africa," *Journal of African History,* Vol. 21, 1980, pp. 105-117.

[R1.222] Olugboji, D., *The United States of Africa and Realpolitik,* Lagos, Nigeria, CMS Press, 1959.

[R1.223] Omari, Peter T., *Kwame Nkrumah: The Anatomy of African Dictatorship,* New York, Africana Publishers, 1970.

[R1.224] Owusu-Ansah, J.V., *New Versions of the Traditional Motifs,* Kumasi, Ghana, Degraft Graphics and Publications, 1992.

[R1.225] Organization of African Unity, *Lagos Plan of Action for Economic Development of Africa, 1980-2000,* Geneva, IILS, 1981.

[R1.226] Ovason, David, *The Secret Architecture of Our National Capital,* New York, HarperCollins, 2000.

[R1.227] Padmore, George, *Pan-Africanism or Communism,* London, Dennis Dobson, 1956.

[R1.228] Padmore, George, *Africa: Britain's Third Empire*, London, Dennis Dobson, 1949.

[R1.229] Paterson, Thomas G. (ed.), *Major Problems in American Foreign Policy: Documents and Essays*, Lexington, MA, Heath and Co., 1978.

[R1.230] Payer, Cheryl, *The Debt Trap: The International Monetary Fund and the Third World*, New York, Monthly Review Press, 1974.

[R1.231] Perham, M., "Psychology of African Nationalism," *Optima*, Vol. 10, 1960, pp. 27-36.

[R1.232] Perkins, John, *Confessions of Economic Hit Man*, San Francisco, Berrett-Koehler Pub. Inc., 2004.

[R1.233] Peters, Jonathan A., *A Dance of Masks: Senghor, Achebe, Soyinka*, Washington, D.C., Three Continents Press, 1978.

[R1.234] Petrie, William M.F., *The Pyramids and Temples of Gizeh*, London, Field and Tuer, 1885.

[R1.2351] Rattray, Robert S., *Religion and Art in Ashanti*, New York, AMS Press, 1979.

[R1.236] Resnick, Idrian N., "The University and Development in Africa," *Pan-African Journal*, Vol. 1 (1), 1968, pp. 30-34.

[R1.237] Robeson, P., "Power of Negro Action," in [R1.20], pp. 522-532.

[R1.238] Robinson, E.A.G. (ed.), *Economic Consequences of the Size of Nations*, New York, Macmillan, 1963.

[R1.239] Rodney, Walter, *How Europe Underdeveloped Africa*, London, Bogle-L'Ouvertune Pub., 1972.

[R1.240] Roger, Joel A., *Africa's Gift to America: The Afro-American in the Making and Saving of the United States*, New York, J.A. Rogers Publications, 1959.

[R1.241] Rotberg, R.I., *The Rise of Nationalism in Central Africa: The Making of Malawi and Zambia, 1873-1964*, Cambridge, MA, Harvard University Press, 1965.

[R1.242] Sampson, G.P. (ed.), *The Role of the World Trade Organization in Global Governance*, New York, United Nations University Press, 2001.

[R1.243] Schwaller de Lubicz, R.A., *The Egyptian Miracle: An Introduction to the Wisdom of the Temple*, Rochester (Vermont) Inner Traditions International, 1985

[R1.244] Schwaller de Lubicz, R.A., *A Study of Numbers: A Guide to The constant creation of The Universe*, Rochester, (Vermont) Inner Traditions International, 1986.

[R1.245] Schwaller de Lubicz, R.A., *The Temple in Man: The Secrets of Ancient Egypt* Brookline, (Massachusetts), Autumn Press, 1977.

[R1.246] Schaller de Lubicz, R.A., *The Temple In Man: Sacred Architecture and The Perfect Man.*, Rochester, (Vermont), Inner Traditions Published, 1981

[R1.247] Schwaller de Lubicz, R.A., *Symbol and the Symbolic: Egypt, Science, and The Evolution of Consciousness*, Brookline, (Massachusetts) Autumn Press, 1978.

[R1.248] Schwaller de Lubicz, R.A., *The Temple of Man: Apet of The South at Luxor*, Rochester, (Vermont) Inner Traditions, 1998.

[R1.249] Schwaller de Lubicz, R.A., *Sacred Science: The King of Pharaonic Theocracy*, New York, Inner Traditions International, 1982.

[R1.250] Schwaller de Lubicz, R.A., *The Temples of Karnak*, London, Thames & Hudson, 1999).

[R1.251a] Scranton, Laird, *The Science of the Dogon: Decoding the African Mystery Tradition*, New York, Inner Traditions,

[R1.251b] Seligman, Charles G., *Egypt and Negro Africa: A Study in Divine Kingship*, London, George Routledge and Sons, 1934.

[R1.252] Seme, Pixley Isaka, "The Regeneration of Africa," in Langley, Ayo J. (ed.), *Ideologies of Liberation in Black Africa: 1856-1970*, London, Rex Collings, 1979, pp. 261-265.

[R1.253] Serequeberhan, Tsenay, *The Hermeneutics of African Philosophy: Horizon and Discourse*, New York, Routledge, 1994.

[R1.254] Shorter, A., *African Christian Theology—Adaptation or Incarnation*, New York, Orbis, 1977.

[R1.255] Shorter, Aylward, W. ., *African Culture and Christian Church: An Introduction to Social and Pastoral Anthropology*, New York, Orbis Books, 1974.

[R1.256] Sithole, N., *African Nationalism*, New York, Oxford University Press, 1969.

[R1.257] Smertin, Y., *Kwame Nkrumah*, New York, International Publishers, 1987.

[R1.258] Smith, A., *The Geopolitics of Information, How Western Culture Dominates the World*, New York, Oxford University Press, 1980.

[R1.259] Snowden Jr., Frank M., *Blacks in Antiquity*, Cambridge, MA, Harvard University Press, 1970.

[R1.260] Steindorff, G. and K.C. Seele, *When Egypt Ruled the East*, Chicago, Chicago University Press, 1957.

[R1.261] Snowden Jr., Frank M., *Blacks in Antiquity: Ethiopians in the Greco-Roman Experience*, Cambridge, MA, The Belknap Press of Harvard University Press, 1970.

[R1.262] Sundkler, B.G.M., *Bantu Prophets in South Africa*, London, Oxford University Press, 1961.

[R1.263] Tambo, Oliver, *Preparing for Power: Oliver Tambo Speaks*, New York, George Braziller Inc., 1988.

[R1.264] Tekyi, K., "Racial Unity," in Langley, Ayo J. (ed.), *Ideologies of Liberation in Black Africa:1856-1970*, London, Rex Collings, 1979 pp. 402-404.

[R1.265a] Temples, Father Placide, *Bantu Philosophy*, Paris, Présence Africaine, 1959.

[R1.265b] Temple, R. G., *The Sirius Mystry*, London, Sidwick and Jackson, 1976.

[R1.266] Tennemann, Wilhelm G., *A Manual of the History of Philosophy*, London, H. G. Bohn, 1852.

[R1.267] Tetteh, M.N., *The Ghana Young Pioneer Movement*, The Institute of African Studies, University of Ghana, Tema, Optimum Design and Publishing Service 1985.

[R1.268] The Editors, "One Afrikan-Centric Continental Afrikan Government Now", *The Afrikan Truth Magazine* Vol. 1 (2) 1995.

[R1.269] *The Lost Books of the Bible and Forgotten Book of Eden*, New York, The World Pub Co., 1963.

[R1.270] *The Spark* Editor, *Some Essential Features of Nkrumaism*, London, PANAF Press, 1975.

[R1.271] *The Torah: The Five Books of Moses*, Philadelphia, The Jewish Publication of America, 1962.

[R1.272] Thompson, V. Bakpetu, *Africa and Unity*, London, Longman, 1969.

[R1.273] Thompson, W. S., *Ghana's Foreign Policy 1957-1966*, Princeton, New Jersey, Princeton University Press, 1969.

[R1.274] Toure, A. Sekou, *Africa on the Move*, London, PANAF Press, 1977.

[R1.275] Tuafo, Kofi Y., *Kwame Nkrumah*, Accra, Elaine Book, 2012.

[R1.276] United States National Security Council, *Memorandum 200 Study: Implications of World Wide Population Growth for U.S. Security and*

Overseas Interest, December 10, 1974, Classified by h.c. Blaney III, Declassified July 3, 1989, Executive Order 12358.

[R1.277] United States of America, Department of States, *Foreign Relations of the United States, Vol. XXIV: 190*, Africa, 1964-1968.

[R1.278] Van Sertima, I., *African Presence in Early Europe*, New Brunswick, New Jersey, Transaction, 1986.

[R1.279] Van Sertima, I., *They Came Before Columbus: The African Presence in Ancient America*, New York, Random House, 1977.

[R1.280] Van Sertima, I. (ed.), *Great African Thinkers, Vol. 1, Cheikh Anta Diop*, New Brunswick, New Jersey, Transaction, 1987.

[R1.281] Van Sertima, I. (ed.), *Blacks in Science*, New Brunswick, N.J., Transaction Publishers, 1998.

[R1.282] Wallerstein, I., *Africa: The Politics of Unity*, New York, Random House, 1967.

[R1.283] Wallerstein, I., *Africa: The Politics of Independence*, New York, Random House, 1961.

[R1.284] Walters, Ronald W., "The Afrocentic Concept at Howard University", *New Directions*, Fall, 1990.

[R1.285] Walters, Ronald W., *Pan Africanism in the African Diaspora: An Analysis of Modern Afrocentric Political Movements*, Detroit, Michigan, Wayne State University Press, 1997.

[R1.286] Watson, R. L., *The Slave Question: Liberty and Property in South Africa*, Hanover, New England, University of New England Press, 1990.

[R1.287] Weber, Alfred, *History of Philosophy*, New York, Scribner's Sons, 1896.

[R1.288] Welsing, Cress F., *The Isis Papers: The Keys to the Colors*, Chicago, Third World Press, 1991.

[R1.289] Whitman, Daniel, "The Dual Soul and Double in Africa and West," *Chrysalis: Aspects of African Spirit*, Vol. 3 (1) Spring, 1988, pp. 22-27.

[R1.290] Williams, Chancellor, *The Destruction of Black Civilization: Great Issues of Race From 4500BC- 2000AD*, Detroit, Harlo Press, 1974

[R1.291] Williamson, J., *The Lending Policies of the International Monetary Fund*, Washington, D.C., Institute for International Economics, 1982.

[R1.292] Wilson, H. S. (ed.), *Origins of West African Nationalism*, London, Macmillan, 1969.

[R1.293] Witt, R. E., *Isis in the Graeco-Roman World,* Ithaca, N.Y., Cornell University Press, 1971

[R1.294] Woodson, Carter G., *Mis-Education of the Negro,* Washington, D.C., The Associated Publication, 1933.

[R1.295] Woodson, Carter G., *The African Background,* Washington, D.C., The Associated Press, 1936.

[R1.296] World Health Organization, *Interrelationships between Health Programs and Socio- economic Development,* Public Health Paper #49, Geneva, WHO, 1973.

[R1.297] Wright, Richard A. (ed.), *African Philosophy: An Introduction,* New York, University Press of America, 1984

[R1.298] Yesufu, T.M. (ed.), *Creating the African University: Emerging Issues of the 1970s,* Ibadan, Oxford University Press, 1973.

[R1.299] Zartman, William I., *International Relations in the New Africa,* Englewood Cliffs, New Jersey, Prentice-Hall, 1966.

[R1.300] Zeller, Edward, *Outline of the History of Greek Philosophy,* London, Longmans, Green and Co., 1914.

[R1.301] Zeller, Edward, *A History of Greek Philosophy From the Earliest Period to the Time of Socrates,* Vols 1 – 2, London, Longmans, Green and Co., 1881.

R 2. CATEGORY THEORY IN MATHEMATICS, LOGIC AND SCIENCES

[R2.1] Awodey, S., "Structure in Mathematics and Logic: A Categorical Perspective," Philosophia *Mathematica*, Vol. 3. 1996, pp209-237.

[R2.2] Bell, J. L., "Category Theory and the Foundations of Mathematics," British *Journal of Science*, Vol. 32, 1981, pp349-358.

[R2.3] Bell, J. L., "Categories, Toposes and Sets," *Synthese*, Vol. 51, 1982, pp. 393-337.

[R2.4] Black, M., *The Nature of Mathematics*, Totowa, N.J., Littlefield, Adams and Co., 1965.

[R2.5] Blass, A., "The Interaction between Category and Set Theory," *Mathematical Applications of Category Theory*, Vol. 30, 1984, pp. 5-29.

[R2.6] Brown, B. and J Woods (eds.), *Logical Consequence; Rival Approaches and New Studies in exact Philosophy: Logic, Mathematics and Science*, Vol. II Oxford, Hermes, 2000.

[R2.8] Domany, J. L.,et al., *Models of Neural Networks III*, New York Springer, 1996

[R2.12] Feferman, S., "Categorical Foundations and Foundations of Category Theory," in R. Butts (ed.), Logic, *Foundations of Mathematics and Computability*, Boston, Mass., Reidel, 1977, pp149-169.

[R2.13] Gray, J.W. (ed.) *Mathematical Applications of Category Theory* (American Mathematical Society Meeting 89th Denver Colo. 1983)., Providence, R.I., American Mathematical Society, 1984.

[R.2.14] Johansson, Ingvar, *Ontological Investigations: An Inquiry into the Categories of Nature, Man, and Society*, New York, Routledge, 1989.

[R2.15] Kamps, K. H., D. Pumplun, and W. Tholen (eds.) *Category Theory: Proceedings of the International Conference*, Gummersbach, July 6-10, New York, Springer, 1982.

[R2.18] Landry, E., Category Theory: the Language of Mathematics," *Philosophy of Science*, Vol. 66, (Supplement), S14-S27.

[R2.19] Landry E. and J.P Marquis, "Categories in Context: Historical, Foundational and Philosophical," *Philiosophia Mathematica*, Vol. 13, 2005, pp. 1-43.

[R2.20] Marquis, J. –P., "Three Kinds of Universals in Mathematics," in B. Brown, and J. Woods (eds.), *Logical Consequence; Rival Approaches and New Studies in exact Philosophy: Logic, Mathematics and Science*, Vol. II Oxford, Hermes, 2000, pp 191-212.

[R2.21] McLarty, C., "Category Theory in Real Time," *Philosophia Mathematica*, Vol. 2, 1994, pp. 36-44.

[R2.22] McLarty, C., "Learning from Questions on Categorical Foundations," *Philosophia Mathematica*, Vol.13, 2005, pp44-60.

[R2.23] Rodabaugh, S. et. al., (eds.), *Application of Category Theory to Fuzzy Subsets*, Boston, Mass., Kluwer 1992.

[R2.24] Taylor, J.G. (ed.), *Mathematical Approaches to Neural Networks*, New York North-Holland, 1993.

[R2.25] Van Benthem, J. et al.(eds.), *The Age of Alternative Logics: Assessing Philosophy of Logic and Mathematics Today*, New York, Springer, 2006.

R3. FUZZY LOGIC IN KNOWLEDGE PRODUCTION

[R3.1] Baldwin, J.F., "A New Approach to Approximate Reasoning Using a Fuzzy Logic," *Fuzzy Sets and Systems*, Vol. 2, #4, 1979, pp. 309-325.

[R3.2] Baldwin, J.F., "Fuzzy Logic and Fuzzy Reasoning," *Intern. J. Man-Machine Stud.*, Vol. 11, 1979, pp. 465-480.

[R3.3] Baldwin, J.F., "Fuzzy Logic and Its Application to Fuzzy Reasoning," in. M. M. Gupta et al. (eds.), *Advances in Fuzzy Set Theory and Applications*, New York, North-Holland, 1979, pp.96-115.

[R3.4] Baldwin, J.F. et al., "Fuzzy Relational Inference Language," *Fuzzy Sets and Systems*, Vol. 14, #2, 1984, pp. 155-174.

[R3.5] Baldsin, J. and B.W. Pilsworth, "Axiomatic Approach to Implication For Approximate Reasoning With Fuzzy Logic," *Fuzzy Sets and Systems*, Vol. 3, #2, 1980, pp. 193-219.

[R3.6] Baldwin, J.F. et. al., "The Resolution of Two Paradoxes by Approximate Reasoning Using A Fuzzy Logic," *Synthese*, Vol. 44, 1980, pp. 397-420.

[R3.7] Dompere, K. K., *Fuzzy Rationality: Methodological Critique and Unity of Classical, Bounded and Other Rationalities*, (Studies in Fuzziness and Soft Computing, vol.235) New York, Springer, 2009.

[R3.8] Dompere Kofi K., *Epistemic Foundations of Fuzziness*, (Studies in Fuzziness and Soft Computing, vol. 236) New York, Springer, 2009.

[R3.9] Dompere Kofi K., *Fuzziness and Approximate Reasoning: Epistemics on Uncertainty, Expectation and Risk in Rational Behavior*, (Studies in Fuzziness and Soft Computing, vol. 237) New York, Springer, 2009.

[R3.10] Dompere, Kofi K., *The Theory of the Knowledge Square: The Fuzzy Rational Foundations of Knowledge-Production Systems*, New York, Springer, 2013.

[R3.11] Dompere, Kofi K., "Cost-Benefit Analysis, Benefit Accounting and Fuzzy Decisions: Part I, Theory", *Fuzzy Sets and Systems*, Vol. 92, 1997, pp. 275-287.

[R3.12] Dompere, Kofi K., "The Theory of Social Cost and Costing For Cost-Benefit Analysis in a Fuzzy Decision Space", Fuzzy Sets and Systems. Vol. 76, 1995, pp. 1-24.

[R3.13] Dompere, Kofi K., *Fuzzy Rational Foundations of Exact and Inexact Sciences,* New York, Springer, 2013.

[R3.14] Gaines, B.R., "Foundations of Fuzzy Reasoning," *Inter. Jour. of Man-Machine Studies,* Vol. 8, 1976, pp. 623-668.

[R3.15] Gaines, B.R., "Foundations of Fuzzy Reasoning," in Gupta, M.M. et. al.(eds.), *Fuzzy Information and Decision Processes,* New York North-Holland, 1982, pp. 19-75.

[R3.16] Gaines, B. R., "Precise Past, Fuzzy Future," *International Journal. Of Man-Machine Studies.,* Vol. 19, #1,1983, pp.117-134.

[R3.17] Giles, R., "Lukasiewics Logic and Fuzzy Set Theory," *Intern. J. Man-Machine Stud.,* Vol. 8, 1976, pp. 313-327.

[R3.18] Giles, R., "Formal System for Fuzzy Reasoning," *Fuzzy Sets and Systems,* Vol. 2, #3, 1979, pp. 233-257.

[R3.19] Ginsberg, M. L.(ed.), *Readings in Non-monotonic Reason,* Los Altos, Ca., Morgan Kaufman, 1987.

[R3.20] Goguen, J.A., "The Logic of Inexact Concepts," *Synthese,* Vol. 19, 1969, pp. 325-373.

[R3.21] Gottinger, H.W., "Towards a Fuzzy Reasoning in the Behavioral Science," *Cybernetica,* Vol. 16, #2, 1973, pp. 113-135.

[R3.22] Gottinger, H.W., "Some Basic Issues Connected With Fuzzy Analysis", in H. Klaczro and N. Muller (eds.), *Systems Theory in Social Sciences,* Birkhauser Verlag, Basel, 1976, pp. 323-325.

[R3.23] Gupta, M.M. et. al., (eds.), *Approximate Reasoning In Decision Analysis,* North Holland, New York, 1982.

[R3.24] Höhle Ulrich. and E.P. Klement, *Non-Clasical Logics and their Applications to Fuzzy Subsets: A Handbook of the Mathematical Foundations of Fuzzy Set Theory,* Boston, Mass. Kluwer, 1995.

[R3.25] Kaipov, V., Kh. et. al., "Classification in Fuzzy Environments," in M. M. Gupta et al. (eds.), *Advances in Fuzzy Set Theory and Applications,* New York, North-Holland, 1979, pp. 119-124,

[R3.26] Kaufman, A., "Progress in Modeling of Human Reasoning of Fuzzy Logic" in M. M. Gupta et al. (eds.), *Fuzzy Information and Decision Process,* New York, North-Holland, 1982, pp. 11-17.

[R3.27] Lakoff, G., "Hedges: A Study in Meaning Criteria and the Logic of Fuzzy Concepts," *Jour. Philos. Logic,* Vol. 2, 1973, pp. 458-508.

[R3.28] Lee, R.C.T., "Fuzzy Logic and the Resolution Principle," *Jour. of Assoc. Comput.* Mach., Vol. 19, 1972, pp. 109-119.

[R3.29] LeFaivre, R.A., "The Representation of Fuzzy Knowledge", *Jour. of Cybernetics,* Vol. 4, 1974, pp. 57-66.

[R3.30] Negoita, C.V., "Representation Theorems for Fuzzy Concepts," *Kybernetes*, Vol. 4, 1975, pp. 169-174.

[R3.31] Nowakowska, M., "Methodological Problems of Measurements of Fuzzy Concepts in Social Sciences", *Behavioral Sciences*, Vol. 22, #2, 1977, pp. 107-115.

[R3.32] Skala, H.J., *Non-Archimedean Utility Theory*, D. Reidel Dordrecht, 1975.

[R3.33] Skala, H.J., "On Many-Valued Logics, Fuzzy Sets, Fuzzy Logics and Their Applications," *Fuzzy Sets and Systems*, Vol. 1, #2, 1978, pp. 129-149.

[R3.34] Tan, S K.et al., "Fuzzy Inference Relation Based on the Theory of Falling Shadows," *Fuzzy Sets and Systems*, Vol. 53, #2, 1993, pp. 179-188.

[R3.35] Van Fraassen, B.C., "Comments: Lakoff's Fuzzy Propositional Logic," in D. Hockney et. al., (Eds.), *Contemporary Research in Philosophical Logic and Linguistic Semantics* Holland, Reild, 1975, pp. 273-277.

[R3.36] Yager, R.R. et. al. (eds.), An Introduction to Fuzzy Logic Applications in *Intelligent Systems*, Boston, Mass., Kluwer 1992.

[R3.37] Ying, M.S., "Some Notes on Multidimensional Fuzzy Reasoning," *Cybernetics and Systems*, Vol. 19, #4, 1988, pp. 281-293.

[R3.38] Zadeh, L.A., "Quantitative Fuzzy Semantics," *Inform. Science*, Vol. 3, 1971, pp. 159-176.

[R3.39] Zadeh, L.A., "A Fuzzy Set Interpretation of Linguistic Hedges," *Jour. Cybernetics*, Vol. 2, 1972, pp. 4-34.

[R3.40] Zadeh, L.A., "Fuzzy Logic and Its Application to Approximate Reasoning," *Information Processing 74, Proc. IFIP Congress 74*, #3, North Holland, New York, pp. 591-594, 1974.

[R3.41] Zadeh, L.A, "The Concept of a Linguistic Variable and Its Application to Approximate Reasoning," in K.S. Fu et. al. (eds.), *Learning Systems and Intelligent Robots*, Plenum Press, New York, 1974, pp. 1-10.

[R3.42] Zadeh, L.A., et. al., (eds.), *Fuzzy Sets and Their Applications to Cognitive and Decision Processes*, New York, Academic Press, 1974.

[R3.43] Zadeh, L.A., "The Birth and Evolution of Fuzzy Logic," *Intern. Jour. of General Systems*, Vol. 17, #(2-3) 1990, pp. 95-105.

R4. FUZZY MATHEMATICS IN APPROXIMATE REASONING UNDER CONDITIONS OF INEXACTNESS AND VAGUENESS

[R4.1] Bellman, R.E., "Mathematics and Human Sciences," in J. Wilkinson et. al. (eds.), *The Dynamic Programming of Human Systems*, New York, MSS Information Corp., 1973, pp. 11-18.

[R4.2] Bellman, R.E and Glertz, M., "On the Analytic Formalism of the Theory of Fuzzy Sets," Information *Science*, Vol. 5, 1973, pp. 149-156.

[R4.3] Butnariu, D., Fixed Points for Fuzzy Mapping," *Fuzzy Sets and Systems*, Vol. 7, #2, pp. 191-207, 1982.

[R4.4] Butnariu, D., "Decompositions and Range for Additive Fuzzy Measures", *Fuzzy Sets and Syst ems*, Vol. 10, #2, pp. 135-155, 1983.

[R4.5] Cerruti, U.,"Graphs and Fuzzy Graphs" in *Fuzzy Information and Decision Processes*, New York, North-Holland, 1982, pp.123-131.

[R4.6] Chakraborty, M. K. et al., "Studies in Fuzzy Relations Over Fuzzy Subsets," *Fuzzy Sets and Systems*, Vol. 9, #1, pp. 79-89, 1983.

[R4.8] Chang, C.L., "Fuzzy Topological Spaces," *J. Math. Anal. and Applications*, Vol. 24, 1968,pp. 182-190.

[R4.9] Chang, S.S.L., "Fuzzy Mathematics, Man and His Environment", *IEEE Transactions on Systems, Man and Cybernetics*, SMC-2 1972, pp. 92-93,

[R4.10] Chang, S.S.L. et. al., "On Fuzzy Mathematics and Control," *IEEE Transactions, System, Man and Cybernetics*, SMC-2, 1972, pp. 30-34.

[R4.11] Chang, S.S., "Fixed Point Theorems for Fuzzy Mappings," Fuzzy *Sets and Systems*, Vol. 17, 1985, pp. 181-187.

[R4.12] Chapin, E.W., "An Axiomatization of the Set Theory of Zadeh," *Notices, American Math. Society*, 687-02-4 754, 1971.

[R4.13] Chaudhury, A. K. and P. Das, "Some Results on Fuzzy Topology on Fuzzy Sets," *Fuzzy Sets and Systems*, Vol. 56, 1993, pp. 331-336.

[R4.14] Chitra, H., and P.V. Subrahmanyam, "Fuzzy Sets and Fixed Points," *Jour. of Mathematical Analysis and Application*, Vol. 124, 1987, pp. 584-590.

[R4.15] Czogala, J. et. al., Fuzzy Relation Equations On a Finite Set," *Fuzzy Sets and Systems,* Vol. 7, #1,1982. pp. 89-101.

[R4.16] DiNola, A. et. al., (eds.), *The Mathematics of Fuzzy Systems*, Koln, Verlag TUV Rheinland, 1986.

[R4.17] Dompere, Kofi K., *Cost-Benefit Analysis and the Theory of Fuzzy Decisions: Identification and Measurement Theory* (Series: Studies in Fuzziness and Soft Computing, Vol. 158), Berlin, Heidelberg, Springer, 2004.

[R4.18] Dompere, Kofi K., *Cost-Benefit Analysis and the Theory of Fuzzy Decisions: Fuzzy Value Theory* (Series: Studies in Fuzziness and Soft Computing, Vol.160), Berling, Heidelberg, Springer, 2004.

[R4.19] Dubois, D. and H. Prade, *Fuzzy Sets and Systems*, New York, Academic Press, 1980.

[R4.20] Dubois, "Fuzzy Real Algebra: Some Results," *Fuzzy Sets and Systems*, Vol. 2, #4, pp. 327-348, 1979.

[R4.21] Dubois, D. and H. Prade, "Gradual Inference Rules in Approximate Reasoning," Information *Sciences*, Vol. 61(1-2), 1992, pp. 103-122.

[R4.22] Dubois, D. and H. Prade, "On the Combination of Evidence in various Mathematical Frameworks,." In: Flamm. J. and T. Luisi, (eds.), *Reliability Data Collection and Analysis.* Kluwer, Boston, 1992, pp. 213-241.

[R4.23] Dubois, D. and H. Prade, "Fuzzy Sets and Probability: Misunderstanding, Bridges and Gaps." *Proc. Second IEEE Intern. Conf. on Fuzzy Systems*, San Francisco, 1993, pp. 1059-1068.

[R4.24] Dubois, D. and H. Prade [1994], "A Survey of Belief Revision and Updating Rules in Various Uncertainty Models," *Intern. J. of Intelligent Systems*, Vol. 9, #1, pp. 61-100.

[R4.25] Filev, D.P. et. al., "A Generalized Defuzzification Method via Bag Distributions," *Intern. Jour. of Intelligent Systems*, Vol. 6, #7, 1991, pp. 687-697.

[R4.26] Goetschel, R. Jr., et. al., "Topological Properties of Fuzzy Number," *Fuzzy Sets and Systems*, Vol. 10, #1, pp. 87-99, 1983.

[R4.27] Goodman, I.R., "Fuzzy Sets As Equivalence Classes of Random Sets" inYager, R.R. (ed.), *Fuzzy Set and Possibility Theory: Recent Development*, New York, Pergamon Press 1992. pp. 327-343.

[R4.28] Gupta, M.M. et. al., (eds), *Fuzzy Antomata and Decision Processes*, New York, North-Holland, 1977.

[R4.29] Gupta, M.M. and E. Sanchez (eds.), *Fuzzy Information and Decision Processes*, New York, North-Holland, 1982.

[R4.30] Higashi, M. and G.J. Klir, "On measure of fuzziness and fuzzy complements," Intern. *J. of General Systems*, Vol. 8 #3, 1982, pp. 169-180.

[R4.31] Higashi, M. and G.J. Klir, "Measures of uncertainty and information based on possibility distributions," *International Journal of General Systems*, Vol. 9 #1, 1983, pp. 43-58.

[R4.32] Higashi, M. and G.J. Klir, "On the notion of distance representing information closeness: Possibility and probability distributions," *Intern. J. of General Systems*, Vol. 9 #2, 1983, pp. 103-115.

[R4.33] Higashi, M. and G.J. Klir, "Resolution of finite fuzzy relation equations," *Fuzzy Sets and Systems*, Vol. 13, #1,1984, pp. 65-82.

[R4.34] Higashi, M. and G.J. Klir, "Identification of fuzzy relation systems," IEEE *Trans. on Systems, Man, and Cybernetics*, Vol.14 #2, 1984, pp. 349-355.

[R4.35] Jin-wen, Z., "A Unified Treatment of Fuzzy Set Theory and Boolean Valued Set theory: Fuzzy Set Structures and Normal Fuzzy Set Structures," *Jour. Math. Anal. and Applications*, Vol. 76, #1, 1980, pp. 197-301.

[R4.36] Kandel, A. and W.J. Byatt, "Fuzzy Processes," *Fuzzy Sets and Systems*, Vol. 4, #2, 1980, pp. 117-152.

[R4.37] Kaufmann, A. and M.M. Gupta, Introduction *to fuzzy arithmetic: Theory and applications*, New York Van Nostrand Rheinhold,. 1991.

[R4.38] Kaufmann, A., *Introduction to the Theory of Fuzzy Subsets*, Vol. 1, New York, Academic Press, 1975.

[R4.39] Kaufmann,A., *Theory of Fuzzy Sets*, Paris, Merson Press, 1972.

[R4.40] Klement, E.P. and W. Schwyhla, "Correspondence Between Fuzzy Measures and Classical Measures," *Fuzzy Sets and Systems*, Vol. 7, #1, 1982.pp. 57-70.

[R4.41] Klir, George and Bo Yuan, *Fuzzy Sets and Fuzzy Logic*, Upper Saddle River, NJ Prentice Hall, 1995.

[R4.42] Kokawa, M. et. al. "Fuzzy-Theoretical Dimensionality Reduction Method of Multi-Dimensional Quality," in Gupta, M.M. and E. Sanchez (eds.),*Fuzzy Information and Decision Processes*, New York, North-Holland, 1982, pp. 235-250.

[R4.43] Kruse, R. et al., *Foundations of Fuzzy Systems*, New York, John Wiley and Sons, 1994.

[R4.44] Lasker, G.E. (ed.), *Applied Systems and Cybernetics, Vol. VI: Fuzzy Sets and Systems*, Pergamon Press, New York, 1981.

[R4.45] Lientz, B.P., "On Time Dependent Fuzzy Sets", *Inform, Science*, Vol. 4, 1972, pp. 367-376.

[R4.46] Lowen, R., "Fuzzy Uniform Spaces," *Jour. Math. Anal. Appl.*, Vol. 82, #21981,pp. 367-376.

[R4.47] Michalek, J., "Fuzzy Topologies," Kybernetika, Vol. 11, 1975, pp. 345-354.

[R4.48] Negoita, C.V. et. al., *Applications of Fuzzy Sets to Systems Analysis*, Wiley and Sons, New York, 1975.

[R4.49] Negoita, C.V., "Representation Theorems for Fuzzy Concepts," *Kybernetes*, Vol. 4, 1975, pp. 169-174.

[R4.50] Negoita, C.V. et. al., "On the State Equation of Fuzzy Systems," *Kybernetes*, Vol. 4, 1975, pp. 231-214.

[R4.51] Netto, A.B., "Fuzzy Classes," *Notices, American Mathematical Society,* Vol.68T-H28, 1968, pp.945.

[R4.52] Pedrycz, W., "Fuzzy Relational Equations with Generalized Connectives and Their Applications," *Fuzzy Sets and Systems*, Vol. 10, #2, 1983, pp. 185-201.

[R4.53] Raha, S. et. al., "Analogy Between Approximate Reasoning and the Method of Interpolation," *Fuzzy Sets and Systems*, Vol. 51, #3, 1992, pp. 259-266.

[R4.54] Ralescu, D., "Toward a General Theory of Fuzzy Variables," *Jour. of Math. Analysis and Applications*, Vol. 86, #1, 1982, pp. 176-193.

[R4.55] Rodabaugh, S.E., "Fuzzy Arithmetic and Fuzzy Topology," in G.E. Lasker, (ed.), *Applied Systems and Cybernetics, Vol. VI: Fuzzy Sets and Systems*, Pergamon Press, New York, 1981 pp. 2803-2807.

[R4.56] Rosenfeld, A., "Fuzzy Groups," *Jour. Math. Anal. Appln.*, Vol. 35, 1971, pp. 512-517.

[R4.57] Ruspini, E.H., "Recent Developments In Mathematical Classification Using Fuzzy Sets," in G.E. Lasker, (ed.), *Applied Systems and Cybernetics, Vol. VI: Fuzzy Sets and Systems*, Pergamon Press, New York, 1981. pp. 2785-2790.

[R4.58] Santos, E.S., "Fuzzy Algorithms," *Inform. and Control*, Vol. 17, 1970, pp. 326-339.

[R4.59] Stein, N.E. and K Talaki, "Convex Fuzzy Random Variables," *Fuzzy Sets and Systems*, Vol. 6, #3, 1981, pp. 271-284.

[R4.60] Triantaphyllon, E. et. al., "The Problem of Determining Membership Values in Fuzzy Sets in Real World Situations," in D.E. Brown et. al. (eds), *Operations Research and Artificial Intelligence: The Integration of Problem-solving Strategies*, Boston, Mass., Kluwer, 1990, pp. 197-214.

[R4.61] Tsichritzis, D., "Participation Measures," *Jour. Math. Anal. and Appln.*, Vol. 36, 1971, pp. 60-72.

[R4.62] Turksens, I.B., "Four Methods of Approximate Reasoning with Interval-Valued Fuzzy Sets," *Intern. Journ. of Approximate Reasoning*, Vol. 3, #2, 1989, pp. 121-142.

[R4.63] Turksen, I.B., "Measurement of Membership Functions and Their Acquisition," *Fuzzy Sets and Systems*, Vol. 40, #1, 1991, pp. 5-38.

[R4.64] Wang, P.P. (ed.), *Advances in Fuzzy Sets, Possibility Theory, and Applications*, New York, Plenum Press, 1983.

[R4.65] Wang, Zhenyuan, and George Klir, *Fuzzy Measure Theory*, New York, Plenum Press, 1992.

[R4.66] Wang, P.Z. et. al. (eds.), *Between Mind and Computer: Fuzzy Science and Engineering*, Singapore, World Scientific Press, 1993.

[R4.67] Wang, S., "Generating Fuzzy Membership Functions: A Monotonic Neural Network Model," *Fuzzy Sets and Systems*, Vol. 61, #1, 1994, pp.71-82.

[R4.68] Wong, C.K., "Fuzzy Points and Local Properties of Fuzzy Topology," *Jour. Math. Anal. and Appln.*, Vol. 46, 19874, pp. 316-328.

[R4.69] Wong, C.K., "Categories of Fuzzy Sets and Fuzzy Topological Spaces," *Jour. Math. Anal. and Appln.*, Vol. 53, 1976, pp. 704-714.

[R4.70] Yager, Ronald R., and Dimitor P Filver, *Essentials of Fuzzy Modeling and Control*, New York, John Wiley and Sons, 1994.

[R4.71] Yager, R.R. et. al., (Eds.), *Fuzzy Sets, Neural Networks, and Soft Computing*, New York, Nostrand Reinhold, 1994.

[R4.72] Yager, R.R., "On the Theory of Fuzzy Bags," *Intern. Jour. of General Systems*, Vol. 13, #1, 1986, pp. 23-37.

[R4.73] Zadeh, L.A., "A Computational Theory of Decompositions," *Intern. Jour. of Intelligent Systems*, Vol. 2, #1, 1987, pp. 39-63.

[R4.74] Zimmerman, H.J., *Fuzzy Set Theory and Its Applications,* Boston, Mass, Kluwer, 1985.

R5. FUZZY OPTIMIZATION, DECISION-CHOICES AND APPROXIMATE REASONING IN SCIENCES

[R5.1] Bose, R.K. and Sahani D, "Fuzzy Mappings and Fixed Point Theorems," *Fuzzy Sets and Systems*, Vol. 21, 1987, pp. 53-58.

[R5.2] Butnariu D. "Fixed Points for Fuzzy Mappings," *Fuzzy Sets and Systems*, Vol. 7, 1982, pp.191-207.

[R5.3] Dompere, Kofi K., "Fuzziness, Rationality, Optimality and Equilibrium in Decision and Economic Theories" in Weldon A. Lodwick and Janusz Kacprzyk (Eds.), *Fuzzy Optimization: Recent Advances and Applications* (Series: Studies in Fuzziness and Soft Computing, Vol. 254), Berlin, Heidelberg, Springer, 2010.

[R5.4] Eaves, B.C., "Computing Kakutani Fixed Points," *Journal of Applied Mathematics*, Vol. 21, 1971, pp. 236-244.

[R5.5] Heilpern, S. "Fuzzy Mappings and Fixed Point Theorem," *Journal of Mathematical Analysis and Applications*, Vol. 83, 1981, pp.566-569.

[R5.6] Kacprzyk, J. et. al., (eds.), *Optimization Models Using Fuzzy Sets and Possibility Theory*, Boston, Mass., D. Reidel, 1987.

[R5.7] Kakutani, S., "A Generalization of Brouwer's Fixed Point Theorem," *Duke Mathematical Journal*, Vol. 8, 1941, pp. 416-427.

[R5.8] Kaleva, O. "A Note on Fixed Points for Fuzzy Mappings", *Fuzzy Sets and Systems*, Vol. 15, 1985, pp. 99-100.

[R5.9] Lodwick, Weldon A and Janusz Kacprzyk (eds.), *Fuzzy Optimization: Recent Advances and Applications*, (Studies in Fuzziness and Soft Computing, Vol. 254), Berlin Heidelberg, Springer, 2010,

[R5.10] Negoita, C.V. et. al., "Fuzzy Linear Programming and Tolerances in Planning," *Econ. Group Cybernetic Studies*, Vol. 1, 1976, pp. 3-15.

[R5.11] Negoita, C.V., and A.C. Stefanescu, "On Fuzzy Optimization," in Gupta, M.M. et. al., (eds.) *Approximate Reasoning In Decision Analysis*, North Holland, New York, 1982. pp. 247-250.

[R5.12] Negoita, C.V., "The Current Interest in Fuzzy Optimization," *Fuzzy Sets and Systems*, Vol. 6, #3, 1981, pp. 261-270.

[R5.13] Negoita, C.V., et. al., "On Fuzzy Environment in Optimization Problems," in J. Rose et. al., (eds.), *Modern Trends in Cybernetics and Systems*, Springer, Berlin, 1977, pp.13-24.

[R5.14] Ralescu, D., "0ptimization in a Fuzzy Environment," in M. M. Gupta et al. (eds.), *Advances in Fuzzy Set Theory and Applications,* New York, North-Holland, 1979, pp.77-91.

[R5.15] Warren, R.H., "Optimality in Fuzzy Topological Polysystems," *Jour. Math. Anal.,* Vol. 54, 1976, pp. 309-315.

[R5.16] Zimmerman, H.-J., "Description and Optimization of Fuzzy Systems," *Intern. Jour. Gen. Syst.* Vol. 2, #4, 1975, pp. 209-215.

[R5.17] Zimmerman, H.J., "Applications of Fuzzy Set Theory to Mathematical Programming," *Information Science*, Vol. 36, #1, 1985, pp. 29-58.

R6. FUZZY PROBABILITY, FUZZY RANDOM VARIABLE AND RANDOM FUZZY VARIABLE

[R6.1] Bandemer, H., "From Fuzzy Data to Functional Relations," *Mathematical Modelling*, Vol. 6, 1987, pp. 419-426.

[R6.2] Bandemer, H. et. al., *Fuzzy Data Analysis*, Boston, Mass, Kluwer, 1992.

[R6.3] Kruse, R. et. al., *Statistics with Vague Data*, Dordrecht, D. Reidel Pub. Co., 1987.

[R6.4] Chang, R.L.P., et. al., "Applications of Fuzzy Sets in Curve Fitting," *Fuzzy Sets and Systems*, Vol. 2, #1, pp. 67-74.

[R6.5] Chen, S.Q., "Analysis for Multiple Fuzzy Regression," *Fuzzy Sets and Systems*, Vol. 25, #1, pp. 56-65.

[R6.6] Celmins, A., "Multidimensional Least-Squares Fitting of Fuzzy Model," *Mathematical Modelling*, Vol. 9, #9, pp. 669-690.

[R6.7] El Rayes, A.B. et. al., "Generalized Possibility Measures," *Information Sciences*, Vol. 79, 1994, pp. 201-222.

[R6.8] Dumitrescu, D., "Entropy of a Fuzzy Process," *Fuzzy Sets and Systems*, Vol. 55, #2, 1993, pp. 169-177.

[R6.9] Delgado, M. et. al., "On the Concept of Possibility-Probability Consistency," *Fuzzy Sets and Systems*, Vol. 21, #3, 1987, pp. 311-318.

[R6.10] Devi, B.B. et. al., "Estimation of Fuzzy Memberships from Histograms," *Information Sciences*, Vol. 35, #1, 1985, pp. 43-59.

[R6.11] Diamond, P., "Fuzzy Least Squares", *Information Sciences*, Vol. 46, #3, 1988, pp. 141-157.

[R6.12] Dubois, D. et. al., "Fuzzy Sets, Probability and Measurement," *European Jour. of Operational Research*, Vol. 40, #2, 1989, pp. 135-154.

[R6.13] Fruhwirth-Schnatter, S., "On Statistical Inference for Fuzzy Data with Applications to Descriptive Statistics," *Fuzzy Sets and Systems*, Vol. 50, #2, 1992, pp. 143-165.

[R6.14] Fruhwirth-Schnatter, S., "On Fuzzy Bayesian Inference," *Fuzzy Sets and Systems*, Vol. 60, #1, 1993, pp. 41-58.

[R6.15] Gaines, B.R., "Fuzzy and Probability Uncertainty logics," *Information and Control*, Vol. 38, #2, 1978, pp. 154-169.

[R6.16] Geer, J.F. et. al., "Discord in Possibility Theory," *International Jour. Of General Systems*, Vol. 19, 1991, pp. 119-132.

[R6.17] Geer, J.F. et. al., "A Mathematical Analysis of Information-Processing Transformation Between Probabilistic and Possibilistic Formulation of Uncertainty," *International Jour. of General Systems*, Vol. 20, #2, 1992,
pp. 14-176.

[R6.18] Goodman, I.R. et. al., *Uncertainty Models for Knowledge Based Systems*, New York, North-Holland, 1985.

[R6.19] Grabish, M. et. al., *Fundamentals of Uncertainty Calculi with Application to Fuzzy Systems*, Boston, Mass., Kluwer, 1994.

[R6.20] Guan, J.W. et. al., *Evidence Theory and Its Applications*, Vol. 1, New York, North-Holland, 1991.

[R6.21] Guan, J.W. et. al., *Evidence Theory and Its Applications*, Vol. 2, New York, North-Holland, 1992.

[R6.22] Hisdal, E., Are Grades of Membership Probabilities?," *Fuzzy Sets and Systems*, Vol. 25, #3, 1988, pp. 349-356.

[R6.23] Höhle Ulrich , "A Mathematical Theory of Uncertainty," in R.R. Yager (ed.) *Fuzzy Set and Possibility Theory: Recent Developments*, New York, Pergamon, 1982, pp. 344 – 355.

[R6.24] Kacprzyk, Janusz and Mario Fedrizzi (eds.) *Combining Fuzzy Imprecision with Probabilistic Uncertainty in Decision Making*, New York, Plenum Press, 1992.

[R6.25] Kacprzyk, J. et. al., *Combining Fuzzy Imprecision with Probabilistic Uncertainty in Decision Making*, New York, Springer-Verlag, 1988.

[R6.26] Klir, G.J., "Where Do we Stand on Measures of Uncertainty, Ambignity, Fuzziness and the like?" *Fuzzy Sets and Systems*, Vol. 24, #2, 1987, pp. 141-160.

[R6.27] Klir, G.J. et. al., *Fuzzy Sets, Uncertainty and Information*, Englewood Cliff, Prentice Hll, 1988.

[R6.28] Klir, G. J. et. al., "Probability-Possibility Transformations: A Comparison," *Intern. Jour. of General Systems*, Vol. 21, #3, 1992, pp. 291-310.

[R6.29] Kosko, B., "Fuzziness vs Probability," *Intern. Jour. of General Systems*, Vol. 17, #(1-3) 1990, pp. 211-240.

[R6.30] Manton, K.G. et. al., *Statistical Applications Using Fuzzy Sets*, New York, John Wiley, 1994.

[R6.31] Meier, W., et. al., "Fuzzy Data Analysis: Methods and Indistrial Applications," *Fuzzy Sets and Systems*, Vol. 61, #1, 1994, pp. 19-28.

[R6.32] Nakamura, A., et. al., "A logic for Fuzzy Data Analysis," *Fuzzy Sets and Systems,* vol. 39, #2, 1991, pp. 127-132.

[R6.33] Negoita, C.V. et. al., *Simulation, Knowledge-Based Compting and Fuzzy Statistics,* New York, Van Nostrand Reinhold, 1987.

[R6.34] Nguyen, H.T., "Random Sets and Belief Functions," *Jour. of Math. Analysis and Applications,* Vol. 65, #3, 1978, pp. 531-542.

[R6.35] Prade, H. et. al., "Representation and Combination of Uncertainty with belief Functions and Possibility Measures," *Comput. Intell.* ,Vol. 4, 1988, pp. 244-264.

[R6.36] Puri, M.L. et. al., "Fuzzy Random Variables," *Jour. of Mathematical Analysis and Applications,* Vol. 114, #2, 1986, pp. 409-422.

[R6.37] Rao, N.B. and A. Rashed, "Some Comments on Fuzzy Random Variables," *Fuzzy Sets and Systems,* Vol. 6, # 3, 1981, pp.285-292.

[R6.38] Sakawa, M. et. al., "Multiobjective Fuzzy linear Regression Analysis for Fuzzy Input-Output Data," *Fuzzy Sets and Systems,* Vol. 47, #2, 1992, pp. 173-182.

[R6.39] Schneider, M. et. al., "Properties of the Fuzzy Expected Values and the Fuzzy Expected Interval," *Fuzzy Sets and Systems,* Vol. 26, #3, 1988, pp. 373-385.

[R6.40] Slowinski, Roman and Jacques Teghem (eds) *Stochastic versus Fuzzy Approaches to Multiobjective Mathematical Programming under Uncertainty,* Dordrecht, Kluwer, 1990.

[R6.41] Stein, N.E. and K Talaki, "Convex Fuzzy Random Variables," *Fuzzy Sets and Systems,* Vol. 6, #3, 1981, pp. 271-284.

[R6.42] Sudkamp, T., "On Probability-Possibility Transformations," *Fuzzy Sets and Systems,* Vol. 51, #1, 1992, pp. 73-82.

[R6.43] Tanaka, H. et. al., "Possibilistic Linear Regression Analysis for Fuzzy Data," *European Jour. of Operational Research,* Vol. 40, #3, 1989, pp. 389-396.

[R6.44] Walley, P., *Statistical Reasoning with Imprecise Probabilities,* London Chapman and Hall, 1991.

[R6.45] Wang, G.Y. et. al., "The Theory of Fuzzy Stochastic Processes," *Fuzzy Sets and Systems,* Vol. 51, #2 1992, pp. 161-178.

[R6.46] Wang, X. et. al., "Fuzzy Linear Regression Analysis of Fuzzy Valued Variable," *Fuzzy Sets and Systems,* Vol. 36, #1, 19.

[R6.47] Zadeh, L. A., "Probability Measure of Fuzzy Event," *Jour. of Math Analysis and Applications,* Vol. 23, 1968, pp. 421 – 427.

R7. IDEOLOGY AND THE KNOWLEDGE CONSTRUCTION PROCESS

[R7.1] Abercrombie, Nicholas et al., *The Dominant Ideology Thesis*, London, Allen and Unwin, 1980.

[R7.2] Abercrombie, Nicholas, *Class, Structure, and Knowledge: Problems in the Sociology of Knowledge*, New York, New York University Press, 1980.

[R7.3] Aron, Raymond, *The Opium of the Intellectuals*, Lanham, MD, University Press of America, 1985.

[R7.4] Aronowitz, Stanley, *Science as Power: Discourse and Ideology in Modern Society*, Minneapolis, University of Minnesota Press, 1988.

[R7.5] Barinaga, M. and E. Marshall, *Confusion on the Cutting Edge*, Science, Vol. 257, July 1992, pp. 616-625.

[R7.6] Barnett, Ronald, *Beyond All Reason: Living with Ideology in the University*, Philadelphia, PA., Society for Research into Higher Education and Open University Press, 2003.

[R7.7] Barth, Hans, *Truth and Ideology*, Berkeley, University of California Press, 1976.

[R7.8] Basin, Alberto, and Thierry Verdie, "The Economics of Cultural Transmission and the Dynamics of Preferences," *Journal of Economic Theory*, Vol. 97, 2001, pp. 298-319.

[R7.9] Beardsley, Philip L. *Redefining Rigor: Ideology and Statistics in Political Inquiry*, Bevery Hills, Sage Publications, 1980.

[R7.10] Bikhchandani, Sushil et al., "A Theory of Fads, Fashion, Custom, and Cultural Change," *Journal of political Economy*, Vol. 100 1992, pp 992-1026.

[R7.11] Boyd Robert and Peter J Richerson, *Culture and Evolutionary Process*, Chicago, University of Chicago Press, 1985.

[R7.12] Buczkowski, Piotr and Andrzej Klawiter, *Theories of Ideology and Ideology of Theories*, Amsterdam, Rodopi, 1986.

[R7.13] Chomsky, Norm, *Manufacturing Consent*, New York, Pantheo Pess, 1988.

[R7.14] Chomsky, N., *Problem of Knowledge and Freedom*, Glasgow, Collins, 1972.

[R7.15] Cole, Jonathan, R., "Patterns of Intellectual influence in Scientific Research," *Sociology of Education*, Vol. 43, 1968, pp. 377-403.

[R7.16] Cole Jonathan, R. and Stephen Cole, *Social Stratification in Science*, Chicago, University of Chicago Press, 1973.

[R7.17] Debackere, Koenraad and Michael A. Rappa, "Institutioal Varations in Problem Choice and Persistence among Scientists in an Emerging Fields," *Research Policy*, Vol. 23, 1994, pp425-441.

[R7.18] Fraser, Colin and George Gaskell (eds.), *The Social Psychological Study of Widespread Beliefs*, Oxford, Clarendon Press, 1990.

[R7.19] Gieryn, Thomas, F. "Problem Retention and Problem Change in Science," *Sociological Inquiry*, Vol. 48, 1978, pp. 96-115.

[R7.20] Harrington, Joseph E. Jr, "The Rigidity of social Systems," *Journal of Political Economy*, Vol. 107, pp. 40-64.

[R7.21] Hinich, Melvin and Michael Munger, *Ideology and the Theory of Political Choice*, Ann Arbor University of Michigan Press, 1994.

[R7.22] Hull, D. L., *Science as a Process: An Evolutionary Account of the Social and Conceptual Development of Science*, Chicago, University of Chicago Press, 1988.

[R7.23] Marx, Karl and Friedrich Engels, The German Ideology, New York, International Pub, 1970

[R7.24] Mészáros, István , *Philosophy, Ideology and Social Science*: Essay in Negation and Affirmation, Brighton, Sussex, Wheatsheaf, 1986.

[R7.25] Mészáros, István *The Power of Ideology*, New York, New York University Press, 1989.

[R7.26] Newcomb, Theodore M. et. al., *Persistence and Change*, New York, John Wiley, 1967.

[R7.27] Pickering, Andrew, *Science as Practice and Culture*, Chicago, University of Chicago Press, 1992.

[R7.28] Therborn, Göran , *The Ideology of Power and the Power of Ideology*, London, NLB Publications, 1980.

[R7.29] Thompson, Kenneth, *Beliefs and Ideology*, New York, Tavistock Publication, 1986.

[R7.30] Ziman, John, "The Problem of 'Problem Choice'," *Minerva*, Vol. 25, 1987, pp. 92-105.

[R7.31] Ziman, John, *Public Knowledge: An Essay Concerning the Social Dimension of Science*, Cambridge, Cambridge University Press, 1968.

[R7.32] Zuckerman, Hrriet, "Theory Choice and Problem Choice in Science," *Sociological Inquiry*, Vol.48, 1978, pp. 65-95.

R 8. INFORMATION, THOUGHT AND KNOWLEDGE

[R8.1] Aczel, J. and Z. Daroczy, *On Measures of Information and their Characterizations*, New York, Academic Press, 1975.

[R8.2] Anderson, J. R., *The Architecture of Cognition*, Cambridge, Mass., Harvard University Press, 1983.

[R8.3] Angelov, Stefan and Dimitr Georgiev, "The Problem of Human Being in Contemporary Scientic Knowledge," *Soviet Studies in Philosophy*, summer, 1974, pp. 49-66.

[R8.4] Ash, Robert, *Information Theory*, New York, John Wiley and Sons, 1965.

[R8.5] Barlas, Y. and S. Carpenter, "Philosophical Roots of Model Validation: Two Paradigms," *System Dynamic Review*, Vol. 6, 1990, pp148-166.

[R8.6] Bergin, J., "Common Knowledge with Monotone Statistics," *Econometrica*, Vol. 69, 2001, pp. 1315-1332.

[R8.7] Bestougeff, Hélène and Gerard Ligozat, *Logical Tools for Temporal Knowledge Representation*, New York, Ellis Horwood, 1992.

[R8.8] Brillouin, L., *Science and information Theory*, New York, Academic Press, 1962.

[R8.9] Bruner, J. S., et. al., *A Study of Thinking*, New York, Wiley, 1956.

[R8.10] Brunner, K. and A. H. Meltzer (eds.), *Three Aspects of Policy and Policy Making: Knowledge, Data and Institutions*, Carnegie-Rochester Conference Series, Vol. 10, Amsterdam, North-Holland, 1979.

[R8.11] Burks, A. W., *Chance, Cause, Reason: An Inquiry into the Nature of Scientific Evidence*, Chicago, University of Chicago Press, 1977.

[R8.12] Calvert, Randall, *Models of Imperfect Information in Politics*, New York, Hardwood Academic Publishers, 1986.

[R8.13] Cornforth, Maurice, *The Theory of Knowledge*, New York, International Pub. 1972

[R8.14] Cornforth, Maurice, *The Open Philosophy and the Open Society*, New York, International Pub. 1970

[R8.15] Coombs, C. H., *A Theory of Data*, New York, Wiley, 1964.

[R8.16] Dretske, Fred. I., *Knowledge and the Flow of Information*, Cambridge, Mass., MIT Press 1981.

[R8.17] Dreyfus, Hubert L., "A Framework for Misrepresenting Knowledge," in Martin Ringle (ed.) *Philosophical Perspectives in*

Artificial Intelligence, Atlantic Highlands, N.J., Humanities press, 1979.

[R8.18] Fagin R. et al., *Reasoning About Knowledge,* Cambridge, Mass, MIT Press, 1995.

[R8.19] Geanakoplos, J., "Common Knowledge," *Journal of Economic Perspectives,"* Vol. 6, 1992, pp53-82.

[R8.20] George, F. H., *Models of Thinking,* London, Allen and Unwin, 1970.

[R8.21] George, F. H., "Epistemology and the problem of perception," *Mind,* Vol.66, 1957, pp.491-506.

[R8.22] Harwood, E. C., *Reconstruction of Economics,* Great Barrington, Mass, American Institute for Economic Research, 1955.

[R8.23] Hintikka, J., *Knowledge and Belief,* Ithaca, N. Y., Cornell University Press, 1962.

[R8.24] Hirshleifer, Jack., "The Private and Social Value of Information and Reward to inventive activity," *American Economic Review,* Vol. 61, 1971, pp.561-574.

[R8.25] Kapitsa, P. L., "The Influence of Scientific Ideas on Society," *Soviet Studies in Philosophy,* Fall, 1979, pp.52-71.

[R8.26] Kedrov, B. M., "The Road to Truth," *Soviet Studies in Philosophy,* Vol. 4, 1965, pp 3 – 53.

[R8.27] Klatzky, R. L., *Human Memory: Structure and Processes,* San Francisco, Ca., W. H. Freeman Pub., 1975.

[R8.28] Kreps, David and Robert Wilson, "Reputation and Imperfect Information," *Journal of Economic Theory,* Vol. 27. 1982, pp253-279.

[R8.29] Kubát, Libor and J. Zeman (eds.), *Entropy and Information,* Amsterdam, Elsevier, 1975.

[R8.30] Kurcz, G. and W. Shugar et al (eds.), *Knowledge and Language,* Amsterdom, North-Holland, 1986.

[R8.31] Lakemeyer, Gerhard,and Bernhard Nobel (eds.), *Foundations of Knowledge Representation and Reasoning,* Berlin, Springer-Verlag, 1994.

[R8.32] Lektorskii, V. A., "Principles involved in the Reproduction of Objective in Knowledge,", *Soviet Studies in Philosophy,* Vol. 4, #4, 1967, pp. 11-21.

[R8.33] Levi, I., *The Enterprise of Knowledge,* Cambridge, Mass. MIT Press 1980.

[R8.34] Levi, Isaac, "Ignorance, Probability and Rational Choice", *Synthese,* Vol. 53, 1982, pp. 387-417.

[R8.35] Levi, Isaac, "Four Types of Ignorance," *Social Science,* Vol. 44, pp745-756.

[R8.36] Marschak, Jacob, *Economic Information, Decision and Prediction: Selected Essays,* Vol. II, Part II, Boston, Mass. Dordrecnt-Holland, 1974.

[R8.37] Menges, G. (ed.), *Information, Inference and Decision,* D. Reidel Pub., Dordrecht, Holland, 1974.

[R8.38] Michael Masuch and László Pólos (eds.), *Knowledge Representation and Reasoning Under Uncertainty,* New York, Springer-Verlag, 1994.

[R8.39] Moses, Y. (ed.), *Proceedings of the Fourth Conference of Theoretical Aspects of Reasoning about Knowledge,* San Mateo, Morgan Kaufmann, 1992.

[R8.40] Nielsen, L.T. et al., "Common Knowledge of Aggregation Expectations," *Econometrica,* Vol. 58, 1990, pp. 1235-1239.

[R8.41] Newell, A., *Unified Theories of Cognition,* Cambridge, Mass. Harvard University Press, 1990.

[R8.42] Newell, A., *Human Problem Solving,* Englewood Cliff, N.J. Prentice-Hall, 1972.

[R8.43] Ogden, G. K. and I. A., *The Meaning of Meaning,* New York, Harcourt-Brace Jovanovich, 1923.

[R8.44] Planck, Max, Scientific Autobiography and Other Papers, Westport, Conn., Greenwood, 1968.

[R8.45] Pollock, J., *Knowledge and Justification,* Princeton, Princeton University Press, 1974.

[R8.46] Polanyi, M., *Personal Knowledge,* London, Routledge and Kegan Paul, 1958.

[R8.47] Popper, K. R., *Objective Knowledge,* London, Macmillan, 1949.

[R8.48] Popper, K. R., *Open Society and it Enemies, Vols. 1 and 2* Princeton, Princeton Univ.Press, 2013

[R8.49] Popper, K. R., *The Poverty of Historicism* New York, Taylor and Francis, 2002

[R8.50] Price, H. H., *Thinking and Experience,* London, Hutchinson, 1953.

[R8.51] Putman, H., *Reason, Truth and History,* Cambridge, Cambridge University Press, 1981.

[R8.52] Putman. H., *Realism and Reason*, Cambridge, Cambridge University Press, 1983.

[R8.53] Putman, H., *The Many Faces of Realism*, La Salle, Open Court Publishing Co., 1987.

[R8.54] Russell, B., *Human Knowledge, its Scope and Limits*, London, Allen and Unwin, 1948.

[R8.55] Russell, B., *Our Knowledge of the External World*, New York, Norton, 1929.

[R8.56] Samet, D., "Ignoring Ignorance and Agreeing to Disagree," *Journal of Economic Theory*, Vol. 52, 1990, pp. 190-207.

[R8.57] Schroder, Harold, M. and Peter Suedfeld (eds.), *Personality Theory and Information Processing*, New York, Ronald Pub. 1971.

[R8.58] Searle J., *Minds, Brains and Science*, Cambridge, Mass., Harvard University Press, 1985.

[R8.59] Shin, H., "Logical Structure of Common Knowledge," *Journal of Economic Theory*, Vol. 60, 1993, pp. 1-13.

[R8.60] Simon, H. A., *Models of Thought*, New Haven, Conn., Yale University Press, 1979.

[R8.61] Smithson, M., *Ignorance and Uncertainty, Emerging Paradigms*, New York, Springer-Verlag, 1989.

[R8.62] Sowa, John F., *Knowledge Representation: Logical, Philosophical, and Computational Foundations*, Pacific Grove, Brooks Pub., 2000.

[R8.63] Stigler, G. J., The Economics of Information," *Journal of Political Economy*, Vol.69, 1961, pp.213-225.

[R8.64] Tiukhtin, V. S., "How Reality Can be Reflected in Cognition: Reflection as a Property of All Matter," *Soviet Studies in Philosophy*, Vol.3 #1, 1964, pp.3-12.

[R8.65] Tsypkin, Ya Z., *Foundations of the Theory of Learning Systems*, New York, Academic Press, 1973.

[R8.66] Ursul, A. D., "The Problem of the Objectivity of Information," in Libor Kubát, and J. Zeman (eds.), *Entropy and Information*, Amsterdam, Elsevier, 1975. pp. 187 – 230.

[R8.67] Vardi, M. (ed.), *Proceedings of Second Conference on Theoretical Aspects of Reasoning about Knowledge*, Asiloman, Ca., Los Altos, Ca, Morgan Kaufman, 1988.

[R8.68] Vazquez, Mararita, et al., "Knowledge and Reality: Some Conceptual Issues in System Dynamics Modeling," *Systems Dynamics Review*, Vol. 12, 1996, pp. 21-37.

[R8.69] Zadeh, L. A., "A Theory of Commonsense Knowledge," in Skala, Heinz J. et al., (eds.), *Aspects of Vagueness*, Dordrecht, D. Reidel Co. 1984. pp 257 – 295.

[R8.70] Zadeh, L. A., "The Concept of Linguistic Variable and its Application to Approximate reasoning," *Information Science*, Vol. 8, 1975, pp. 199 – 249(Also in Vol. 9, pp. 40 – 80).

R9. LANGUAGE AND THE KNOWLEDGE-PRODUCTION PROCESS

[R9.1] Aho, A. V. "Indexed Grammar - An Extension of Context-Free Grammars" *Journal of the Association for Computing Machinery*, Vol. 15, 1968, pp. 647-671.

[R9.2] Black, Max (ed.), *The Importance of Language*, Englewood Cliffs, N.J, Prentice- Hall, 1962.

[R9.3] Carnap, Rudolff, Meaning and Necessity: A Study in Semantics and Modal Logic, Chicago, University of Chicago Press, 1956.

[R9.4] Chomsky, Norm, "Linguistics and Philosophy" in S. Hook (ed.) *Language and Philosophy*, New York, New York University Press, 1968, pp. 51-94.

[R9.5] Chomsky, Norm, *Language and Mind*, New York, Harcourt Brace Jovanovich, 1972.

[R9.8] Cooper, William S., *Foundations of Logico-Linguistics: A Unified Theory of Information, Language and Logic*, Dordrecht, D. Reidel, 1978.

[R9.9] Cresswell, M.J.., *Logics and Languages*, London, Methuen Pub. 1973.

[R9.10] Dilman, IIham, *Studies in Language and Reason*, Totowa, N.J., Barnes and Nobles, Books, 1981.

[R9.11] Dompere, Kofi K., A General Theory of Information: Definitional Foundations and a Critique of the Tradition, A Working Paper on Economics, Decision, Philosophy and Mathematics, Department of Economics Howard University, Washington, D.C., 2015.

[R9.12] Dompere, Kofi K., A Theory of Infodynamics: An Introduction, A Working Paper on Economics, Decision, Philosophy and Mathematics, Department of Economics Howard University, Washington, D.C., 2016

[R9.13] Fodor, Jerry A., *The Language and Thought*, New York, Thom as Y. Crowell Co, 1975

[R9.14] Givon, Talmy, *On Understanding Grammar*, New York, Academic Press, 1979

[R9.15] Gorsky, D.R., *Definition,* Moscow, Progress Publishers, 1974.

[R9.16] Hintikka, Jaakko, The Game of Language, Dordrecht, D. Reidel Pub. 1983.

[R9.17] Johnson-Lair, Philip N. *Mental Models: Toward Cognitive Science of*

Language, Inference and Consciousness, Cambridge, Mass, Harvard University Pres, 1983.

[R9.19] Kandel, A., "Codes Over Languages," *IEEE Transactions on Systems Man and Cybernetics*, Vol. 4, 1975, pp. 135-138.

[R9.20] Keenan, Edward L. and Leonard M. Faltz, *Boolean Semantics for Natural Languages*, Dordrecht, D. Reidel Pub.,1985

[R9.21] Lakoff, G. Linguistics and Natural Logic, *Synthese*, Vol. 22, 1970, pp. 151-271.

[R9.22] Lee, E.T., et. al., "Notes On Fuzzy Languages," *Information Science*, Vol. 1, 1969, pp. 421-434.

[R9.23] Mackey, A. and D. Merrill (eds.) *Issues in the Philosophy of Language*, New Haven, CT. Yale University Press, 1976.

[R9.24] Nagel, T., "Linguistics and Epistemology" in S. Hook(ed.) *Language and Philosophy*, New York, New York University Press, 1969, pp. 180-184.

[R9.25] Pike, Kenneth, *Language in Relation to a Unified Theory of Structure of Human Behavior*, The Hague, Mouton Pub., 1969.

[R9.26] Quine, W.V. O. *Word and object*, Cambridge, Mass, MIT Press, 1960.

[R9.27] Russell, Bernard, *An Inquiry into Meaning and Truth*, Penguin Books, 1970.

[R9.28] Tarski, Alfred, *Logic, Semantics and Matamathematics*, Oxford, Clarendon Press, 1956.

[R9.29] Whorf, B.L. (ed.), *Language, Thought and Reality*, New York, Humanities Press, 1956.

R.10. POSSIBLE WORLDS AND THE KNOWLEDGE PRODUCTION PROCESS

[R10.1] Adams, Robert M., "Theories of Actuality," *Noûs*, Vol. 8, 1974, pp211-231.

[R10.2] Allen, Sture (ed.) *Possible Worlds in Humanities, Arts and Sciences,* Proceedings of Nobel Symposium, Vol. 65, New York, Walter de Gruyter Pub. , 1989.

[R10.3] Armstrong, D. M., *A Combinatorial Theory of Possibility.* Cambridge University Press, 1989.

[R10.4] Armstrong, D.M *A World of States of Affairs*, Cambridge, Cambridge University Press 1997.

[R10.5] Bell, J.S., "Six Possible Worlds of Quantum Mechanics" in Allen, Sture (Ed.) *Possible Worlds in Humanities, Arts and Sciences*, Proceedings of Nobel Symposium, Vol. 65, New York, Walter de Gruyter Pub.,1989.pp. 359-373....

[R10.6] Bigelow, John. "Possible Worlds Foundations for Probability", *Journal of Philosophical Logic,* 5 (1976), pp. 299-320.

[R10.7] Bradley, Reymond and Norman Swartz, *Possible World: An Introduction to Logic and its Philosophy*, Oxford, Bail Blackwell, 1997.

[R10.8] Castañeda, H.-N. "Thinking and the Structure of the World", *Philosophia*, 4 (1974), pp. 3-40.

[R10.9]Chihara, Charles S. *The Worlds of Possibility: Modal Realism and the Semantics of Modal Logic,* Clarendon, 1998.

[R10.10]Chisholm, Roderick. "Identity through Possible Worlds: Some Questions", *Noûs,* 1 (1967), pp. 1-8; reprinted in Loux, *The Possible and the Actual.*

[R10.11] Divers, John, Possible *Worlds*, London: Routledge, 2002.

[R10.12] Forrest, Peter. "Occam's Razor and Possible Worlds", *Monist,* 65 (1982), pp. 456-64.

[R10.13] Forrest, Peter. and Armstrong, D. M. "An Argument Against David Lewis' Theory of Possible Worlds", *Australasian Journal of Philosophy,* 62 (1984), pp. 164-168.

[R10.14] Grim, Patrick, "There is No Set of All Truths", *Analysis*, Vol. 46, 1986, pp. 186-191.

[R10.15] Heller, Mark. "Five Layers of Interpretation for Possible Worlds", *Philosophical Studies*, 90 (1998), pp. 205-214.

[R10.16] Herrick, Paul , *The Many Worlds of Logic,*. Oxford: Oxford University Press1999.

[R10.17] Krips, H. "Irreducible Probabilities and Indeterminism", *Journal of Philosophical Logic,* Vol. 18, 1989, pp. 155-172.

[R10.18] Kuhn, Thomas S., "Possible Worlds in History of Science" in Allen, Sture (ed.) *Possible Worlds in Humanities, Arts and Sciences,* Proceedings of Nobel Symposium, Vol. 65, New York, Walter de Gruyter Pub. , 1989. pp. 9-41.

[R10.19] Kuratowski, K. and Mostowski, A. *Set Theory: With an Introduction to Descriptive Set Theory,* New York: North-Holland, 1976.

[R10.20] Lewis, David, *On the Plurality of Worlds,* Oxford, Basil Blackwell, 1986.

[R10.21] Loux, Michael J. (ed.) *The Possible and the Actual: Readings in the Metaphysics of Modality,* Ithaca & London: Cornell University Press, 1979.

[R10.22] Parsons, Terence, *Nonexistent Objects,* New Haven, Yale University Press, 1980.

[R10.23] Perry, John, "From Worlds to Situations", Journal of Philosophical Logic, Vol. 15, 1986, pp. 83-107.

[R10.24] Rescher, Nicholas and Brandom, Robert. *The Logic of Inconsistency: A Study* in *Non-Standard Possible-World Semantics And Ontology,* Rowman and Littlefield, 1979.

[R10.25] Skyrms, Brian. "Possible Worlds, Physics and Metaphysics", *Philosophical Studies,* Vol. 30, 1976, pp. 323-32.

[R10.26] Stalmaker, Robert C. "Possible World", *Noûs,* Vol. 10, 1976, pp. 65-75

[R10.27] Quine, W.V.O. *Word and Object,* M.I.T. Press, 1960.

[R10.28] Quine, W.V.O "Ontological Relativity", *Journal of Philosophy,* 65 (1968), pp. 185-212.

R11. RATIONALITY, INFORMATION, GAMES, CONFLICTS AND EXACT REASONING

[R11.2] Border, Kim, *Fixed Point Theorems with Applications to Economics and Game Theory*, Cambridge, Cambridge University Press 1985.

[R11.3] Brandenburger, Adam, "Knowledge and Equilibrium Games," *Journal of Economic Perspectives*, Vol.6, 1992, pp. 83-102.

[R13.4] Campbell, Richmond and Lanning Sowden, *Paradoxes of Rationality and Cooperation: Prisoner's Dilemma and Newcomb's Problem*, Vancouver, University of British Columbia Press, 1985.

[R11.6] Gates Scott and Brian Humes, *Games, Information, and Politics: Applying Game Theoretic Models to Political Science*, Ann Arbor, University of Michigan Press, 1996.

[R11.7] Gjesdal, Froystein, "Information and Incentives: The Agency Information Problem," *Review of Economic Studies*, Vol.49, 1982, pp373-390.

[R11.8] Harsanyi, John, "Games with Incomplete Information Played by 'Bayesian' Players I: The Basic Model," *Management Science*, Vol.14, 1967, pp.159-182.

[R11.9] Harsanyi, John, "Games with Incomplete Information Played by 'Bayesian' Players II: Bayesian Equilibrium Points," *Management Science*, Vol.14, 1968, pp.320-334.

[R11.10] Harsanyi, John, "Games with Incomplete Information Played by 'Bayesian' Players III: The Basic Probability Distribution of the Game," *Management Science*, Vol.14, 1968, pp.486-502.

[R11.11] Harsanyi, John, *Rational Behavior and Bargaining Equilibrium in Games and Social Situations*, New York Cambridge University Press, 1977.

[R11.14] Krasovskii, N.N. and A.I. Subbotin, *Game-theoretical Control Problems*, New York, Springer-Verlag, 1988.

[R11.15] Kuhn, Harold (ed.) *Classics in Game Theory*, Princeton, Princeton University Press, 1997.

[R11.16] Lagunov, V. N., *Introduction to Differential Games and Control Theory*, Berlin, Heldermann Verlag, 1985.

[R11.17] Luce, D. R. and H. Raiffa, *Games and Decisions*, New York, John Wiley and Sons, 1957.

[R11.18] Maynard Smith, John, *Evolution and the Theory of Games*, Cambridge, Cambridge University Press, 1982.

[R11.20] Myerson, Roger, *Game Theory: Analysis of Conflict*, Cambridge, Mass. Harvard University Press, 1991.

[R11.21] Rapoport, Anatol and Albert Chammah, *Prisoner's Dilemma: A Study in Conflict and Cooperation*, Ann Arbor, University of Michigan Press, 1965.

[R11.22] Roth, Alvin E., "The Economist as Engineer: Game Theory, Experimentation, and Computation as Tools for Design Economics," *Econometrica*, Vol.70, 2002, pp1341-1378.

[R11.23] Shubik, Martin, *Game Theory in the Social Sciences: Concepts and Solutions*, Cambridge, Mass., MIT Press, 1982.

[R11.24] Smart, D.R., *Fixes point Theorems*, Cambridge, Cambridge University Press, 1980.

[R11.26] Von Neumann, John and Oskar Morgenstern, *The Theory of Games in Economic Behavior*, New York, John Wiley and Sons, 1944.

R12. RATIONALITY AND PHILOSOPHY OF EXACT AND INEXACT SCIENCES IN THE KNOELEDGE PRODUCTION

[R12.1] Achinstein, P., "The Problem of Theoretical Terms," in Brody, Baruch A. (Ed.) *Reading in the Philosophy of Science*, Englewood Cliffs, NJ., Prentice Hall, 1970.

[R12.2] Amo Afer, A. G., *The Absence of Sensation and the Faculty of Sense in the Human Mind and Their Presence in our Organic and Living Body,Dissertation and Other essays 1727-1749,* Halle Wittenberg, Jena, Martin Luther Universioty Translation, 1968.

[R12.3] Beeson, M. J., *Foundations of Constructive Mathematics*, Berlin/New York, Springer, 1985.

[R12.4] Benacerraf, P., "God, the Devil and Gödel," *Monist,* Vol. 51, 1967, pp.9-32.

[R12.5] Benecerraf, P and H. Putnam (eds.), *Philosophy of Mathematics: Selected Readings*, Cambridge, Cambridge University Press, 1983.

[R12.6] Black, Max, *The Nature of Mathematics,* Totowa, Littlefield, Adams and Co. 1965.

[R12.7] Blanche, R., *Contemporary Science and Rationalism*, Edinburgh, Oliver and Boyd, 1968.

[R12.8] Blanshard, Brand, *The Nature of Thought*, London Allen and Unwin, 1939.

[R12.9] Blauberg, I. V., V.N. Sadovsky and E.G. Yudin, Systems Theory: Philosophical and Methodological Problems, Moscow, Progress Publishers, 1977.

[R12.10] Braithwaite, R. B., *Scientific Explanation*, Cambridge, Cambridge University Press. 1955.

[R12.11] Brody, Baruch A. (ed.), *Reading in the Philosophy of Science*, Englewood Cliffs, N.J., Prentice Hall1970.

[R12.12] Brody, Baruch A., "Confirmation and Explanation," in Brody, Baruch A. (ed.) *Reading in the Philosophy of Science*, Englewood Cliffs, N.J., Prentice-Hall, 1970, pp. 410-426.

[R12.13] Brouwer, L.E.J., "Intuitionism and Formalism", *Bull of American Math.Soc.* Vol. 20, 1913, pp81-96.; Also in Benecerraf, P. and H. Putnam (eds.), *Philosophy of Mathematics: Selected Readings*, Cambridge, Cambridge University Press, 1983. pp. 77-89.

[R12.14] Brouwer, L.E.J., "Consciousness, Philosophy, and Mathematics," in Benecerraf, P. and H. Putnam (eds.),

Philosophy of Mathematics: Selected Readings, Cambridge, Cambridge University Press, 1983. pp90-96.

[R12.15] Brouwer, L. E. J., Collected *Works, Vol. 1: Philosophy and Foundations of Mathematics* [A Heyting (ed.)], New York, Elsevier, 1975.

[R12.16] Campbell, Norman R., *What is Science?*, New York, Dover, 1952.

[R12.17] Carnap, R., "Foundations of Logic and Mathematics," in *International Encyclopedia of Unified Science*, Chicago, Univ. of Chicago, 1939, pp.143-211.

[R12.18] Carnap, Rudolf, "On Inductive Logic," *Philosophy of Science*, Vol. 12, 1945, pp. 72-97.

[R12.19] Carnap, Rudolf, "The Methodological Character of Theoretical Concepts," in Herbert Feigl and M. Scriven (eds.) *Minnesota Studies in the Philosophy of Science, Vol. I*, 1956, pp. 38-76.

[R12.20] Charles, David and Kathleen Lennon (eds.), *Reduction, Explanation, and Realism*, Oxford, Oxford Unive3rsity Press, 1992.

[R12.21] Cohen, Robert S. and Marx W. Wartofsky (eds.), *Methodological and Historical Essays in the Natural and Social Sciences*, Dordrecht, D. Reidel Publishing Co. 1974.

[R12.22] Dalen van, D. (ed.), *Brouwer's Cambridge Lectures on Intuitionism*, Cambridge, Cambridge University Press, 1981.

[R12.23] Davidson, Donald, *Truth and Meaning: Inquiries into Truth and Interpretation*, Oxford, Oxford University Press, 1984.

[R12.24] Davis, M., *Computability and Unsolvability*, New York, McGraw-Hill, 1958.

[R12.24b] Denonn. Lester E. (ed.), The Wit and Wisdom of Bertrand Russell, Boston, MA., The Beacon Press, 1951.

[R12.25] Dummett, M., "The Philosophical Basis of Intuitionistic Logic," in Benecerraf, P. and H. Putnam (eds.), *Philosophy of Mathematics: Selected Readings*, Cambridge, Cambridge University Press, 1983. pp97-129

[R12.26] Feigl, Herbert and M. Scriven (eds.), Minnesota *Studies in the Philosophy of Science*, Vol. I, 1956.

[R1.27] Feigl, Herbert and M. Scriven (eds.), *Minnesota Studies in the Philosophy of Science*, Vol. II, 1958.

[R12.28] Garfinkel, Alan, *Forms of Explanation:Structures of Inquiry in Social Science*, New Haven, Conn., Yale University Press, 1981.

[R12.29] George, F. H., *Philosophical Foundations of Cybernetics*, Tunbridge Well, Great Britain, 1979.

[R12.30] Gillam, B., "Geometrical Illusions," *Scientific American* , January, 1980, pp.102-111.

[R12.31] Gödel, Kurt., "What is Cantor's Continuum Problem?" in Benecerraf, P. and H. Putnam (eds.), *Philosophy of Mathematics: Selected Readings* , Cambridge, Cambridge University Press, 1983. pp.470-486.

[R12.32] Gorsky, D.R., *Definition,* Moscow, Progress Publishers, 1974.

[R12.33] Gray, William and Nicholas D. Rizzo(eds.), *Unity Through Diversity.* New York, Gordon and Breach, 1973.

[R12.34] Hart, W. D. (ed.), *The Philosophy of Mathematics*, Oxford, Oxford University Press, 1996.

[R12.35] Hartkamper, a and H.- Schmidt, Structure and Approximation in Physical Theories, New York, Plenum Press, 1981

[R12.36] Hausman, David, M., *The Exact and Separate Science of Economics,* Cambridge, Cambridge University Press, 1992.

[R12.37] Helmer, Olaf and Nicholar Rescher, *On the Epistemology of the Inexact Sciences,* P-1513, Santa Monica, CA, Rand Corporation, October 13, 1958.

[R12.38] Hempel, C. G., "Studies in the Logic of Confirmation," *Mind,* Vol.54, Part I, 1945, pp 1-26.

[R12.39] Hempel, Carl G., "The Theoretician's Dilemma," in Herbert Feigl and M. Scriven (eds.) *Minnesota Studies in the Philosophy of Science*, Vol.II, 1958, pp. 37-98.

[R12.40] Hempel, C. G. and P. Oppenheim, "Studies in the Logic of Explanation," *Philosophy of Science*, Vol. 15, 1948,pp. 135-175. [also in Brody, Baruch A. (ed.) *Reading in the Philosophy of Science,* Englewood Cliffs, NJ., Prentice-Hall,1970, pp. 8-27.

[R12. 41] Heyting, A., *Intuitionism: An Introduction,* Amsterdam: North-Holland, 1971.

[R12.42] Hintikka, Jackko (ed.), *The Philosophy of Mathematics*, London, Oxford University Press, 1969.

[R12.43] Hockney D. et al. (eds.), *Contemporary Research in Philosophical Logic and Linguistic Semantics*, Dordrecht-Holland, Reidel Pub., Co. 1975.

[R12.44] Hoyninggen-Huene, Paul and F. M. Wuketits, (eds.), *Reductionism and Systems Theory in the Life Science: Some Problems and Perspectives*, Dordrencht, Kluwer Academic Pub. 1989.

[R12.45] Ilyenkov, E.V., *Dialectical Logic: Essays on Its History and Theory*, Moscow, Progress Publishers, 1977.

[R12.46] Kedrov, B. M., "Toward the Methodological Analysis of Scientific Discovery,"*Soviet Studies in Philosophy*, Vol. 11962, pp45 – 65.

[R12.47] Kemeny, John G, and P Oppenheim, "On Reduction," in Brody, Baruch A.(ed.) *Reading in the Philosophy of Science*, Englewood Cliffs, NJ., Prentice-Hall,1970, 307-318.

[R12.48] Klappholz, K., "Value Judgments of Economics," *British Jour. of Philosophy*, Vol. 15, 1964, pp. 97-114.

[R12.49] Kleene, S.C.,"On the Interpretation of Intuitionistic Number Theory," Journal of Symbolic Logic, Vol 10, 1945, pp. 109-124.

[R12.50] Kmita, Jerzy, "The Methodology of Science as a Theoretical Discipline," *Soviet Studies in Philosophy*, Spring, 1974, pp. 38 –49

[R12.51] Krupp, Sherman R.,(ed.), *The Structure of Economic Science*, Englewood Cliff, N. J., Prentice-Hall, 1966.

[R12.52] Kuhn, T., *The Structure of Scientific Revolution*, Chicago, University of Chicago Press, 1970.

[R12.53] Kuhn, Thomas, "The Function of Dogma in Scientific Research," in Brody, Baruch A. (ed.) *Reading in the Philosophy of Science*, Englewood Cliffs, NJ., Prentice-Hall,1970 pp.356-374.

[R12.54] Kuhn, Thomas, *The Essential Tension: Selected Studies in Scientific Tradition and Change*, Chicago, University of Chicago Press, 1979.

[R12.55] Lakatos, I. (ed.), *The Problem of Inductive Logic*, Amsterdam, North Holland, 1968.

[R12.56] Lakatos, I., *Proofs and Refutations: The Logic of Mathematical Discovery*, Cambridge, Cambridge University Press, 1976.

[R12.57] Lakatos, I., *Mathematics, Science and Epistemology: Philosophical Papers Vol.2*, edited by J. Worrall and G. Currie, Cambridge, Cambridge Univ. Press, 1978.

[R12.58] Lakatos, I., *The Methodology of Scientific Research Programmes*, Vol 1, New York, Cambridge University Press, 1978.

[R12.59] Lakatos, Imre and A. Musgrave (eds.)., *Criticism and the Growth of Knowledge*, New York, Cambridge University Press, 1979. Holland, 1979, pp. 153 – 164.

[R12.60] Lawson, Tony, *Economics and Reality*, New York, Routledge, 1977.

[R12.61] Lenzen, Victor, "Procedures of Empirical Science," in Neurath, Otto et al.(eds.), *International Encyclopedia of Unified Science, Vol. 1 – 10,* Chicago, University of Chicago Press, 1955, pp. 280-338.

[R12.62] Levi, Isaac, "Must the Scientist make Value Judgments?," in Brody, Baruch A. (Ed.) *Reading in the Philosophy of Science,* Englewood Cliffs, NJ., Prentice-Hall,1970 pp.559-570.

[R12.63] Tse-tung, Mao, On Practice and Contradiction, in Selected works of Mao Tse-tung, Piking, 1937. Also, London, Revolutions, 2008.

[R12.64] Lewis, David, *Convention: A Philosophical Study,* Cambridge, Mass., Harvard University Press, 1969.

[R12.65] Mayer, Thomas, *Truth versus Precision in Economics,* London, Edward Elgar 1993.

[R12.66] Menger, Carl, *Investigations into the Method of the Social Sciences with Special Reference to Economics,* New York, New York University Press, 1985

[R12.67] Mirowski, Philip (ed.), *The Reconstruction of Economic Theory,* Boston, Mass. Kluwer Nijhoff, 1986.

[R12.68] Mueller, Ian, *Philosophy of Mathematics and Deductive Structure in Euclid's Elements,* Cambridge, Mass., MIT Press, 1981.

[R12.69] Nagel, Ernest, "Review: Karl Niebyl, Modern Mathematics and Some Problems of Quantity, Quality, and Motion in Economic Analysis," *The Journal of Symbolic Logic,* 1940, p.74.

[R12.70] Nagel, E. et al. (ed.), *Logic, Methodology, and the Philosophy of Science,* Stanford, Stanford University Press 1962.

[R12.71] Narens, Louis, "A Theory of Belief for Scientific Refutations," *Synthese,* Vol.145, 2005, pp. 397-423.

[R12.72] Narskii, I. S., "On the Problem of Contradiction in Dialectical Logic," *Soviet Studies in Philosophy,* Vol. vi, #4 pp.3-10, 1965

[R12.73] Neurath, Otto et al. (eds.), *International Encyclopedia of Unified Science, Vol. 1 – 10,* Chicago, University of Chicago Press, 1955.

[R12.74] Neurath Otto, "Unified Science as Encyclopedic," in Neurath, Otto et al. (eds.),*International Encyclopedia of Unified Science, Vol. 1 – 10,* Chicago, University of Chicago Press, 1955, pp.1-27.

[R12.75] Planck, Max, *Scientific Autobiography and Other Papers,* Westport, Conn.Greenwood 1971.

[R12.76] Planck, Max, "The Meaning and Limits of Exact Science," in Max Planck, *Scientific Autobiography and Other Papers,* Westport, Conn. Greenwood, 1971, pp. 80-120.

[R12.77] Polanyi, Michael, "Genius in Science," in Robert S. Cohen, and Marx W. Wartofsky (eds.), *Methodological and Historical Essays in the Natural and Social Sciences*, Dordrecht, D. Reidel Publishing Co. 1974, pp.57-71.

[R12.78] Popper, Karl, *The Nature of Scientific Discovery*, New York, Harper and Row, 1968.

[R12.79] Putnam, Hilary., "Models and Reality," in Benecerraf, P. and H. Putnam (eds.), *Philosophy of Mathematics: Selected Readings* , Cambridge, Cambridge University Press, 1983. pp. 421-444.

[R12.80] Reise, S., *The Universe of Meaning*, New York, The Philosophical Library, 1953.

[R12.81] Robinson, R., *Definition*, Oxford, clarendon Press, 1950

[R12.82] Rudner, Richard, "The Scientist qua Scientist Makes Value Judgments," *Philosophy of Science*, Vol. 20, 1953, pp 1-6.

[R12.83] Russell, B., *Our Knowledge of the External World*, New York, Norton, 1929.

[R12.84] Russell, B., *Human Knowledge, Its Scope and Limits*, London, Allen and Unwin, 1948.

[R12.85] Russell, B., *Logic and Knowledge: Essays 1901-1950*,New York, Capricorn Books, 1971.

[R12.86] Russell, B., *An Inquiry into Meaning and Truth*, New York, Norton, 1940.

[R12.87] Russell, Bertrand, *Introduction to Mathematical Philosophy*, London, George Allen and Unwin, 1919.

[R12.88] Russell, Bertrand, *The Problems of Philosophy*, Oxford, Oxford University Press, 1978.

[R12.89] Rutkevih, M. N., "Evolution, Progress, and the Law of Dialectics," *Soviet Studies in Philosophy*, Vol. IV, #3, PP. 34-43, 1965.

[R12.90] Ruzavin, G. I., "On the Problem of the Interrelations of Modern Formal Logic and Mathematical Logic," *Soviet Studies in Philosophy*, Vol. 3, #1, 1964, pp.34- 44.

[R12.91] Scriven, Michael, "Explanations, Predictions, and Laws," in Brody, Baruch A. (ed.) *Reading in the Philosophy of Science, Englewood Cliffs*, NJ., Prentice- Hall, 1970, pp.88-104.

[R12.92] Sellars, Wilfrid, The Language of Theories," in Brody, Baruch A. (ed.) *Reading in the Philosophy of Science*, Englewood Cliffs, NJ., Prentice-Hall,1970 pp.343-353.

[R12.93] Sterman, John, "The Growth of Knowledge: Testing a Theory of Scientific Revolutions with a Formal Model," *Technological Forecasting and Social Change*, Vol. 28, 1995, pp. 93-122.

[R12.94] Tsereteli, S. B. "On the Concept of Dialectical Logic,", *Soviet Studies in Philosophy*, Vol. V, #2, pp. 15-21, 1966.

[R12.95] Tullock, Gordon, *The Organization of Inquiry*, Indianapolis, Indiana, Liberty Fund Inc. 1966.

[R12.96] Van Fraassen, B., *Introduction to Philosophy of Space and Time*, New York, Random House, 1970.

[R12.97] Veldman, W., "A Survey of Intuitionistic Descriptive Set Theory," in P.P. Petkov (ed.), Mathematical Logic: Proceedings of the Heyting Conference, New York, Plenum Press, 1990, pp. 155-174.

[R12.98] Vetrov, A. A., "Mathematical Logic and Modern Formal Logic," *Soviet Studies in Philosophy*, Vol. 3, #1, 1964 pp. 24 – 33.

[R12.99] von Mises, Ludwig, *Epistemological Problems in Economics*, New York, New York University Press, 1981.

[R12.100] Wang, Hao, *Reflections on Kurt* Gödel, Cambridge, Mass. MIT Press,1987

[R12.101] Watkins, J. W. N., "The Paradoxes of Confirmation"," in Brody, Baruch A. (ed.) *Reading in the Philosophy of Science*, Englewood Cliffs, NJ., Prentice-hall, 1970 pp. 433-438.

[R12.102] Whitehead, Alfred North, *Process and Reality*, New York, The Free Press, 1978.

[R12.103] Wittgenstein, Ludwig, *Ttactatus Logico-philosophicus*, Atlantic Highlands, N.J., The Humanities Press Inc.1974.

[R12.104] Woodger, J. H., *The Axiomatic Method in Biology*, Cambridge, Cambridge University Press, 1937.

[R12.105] Zeman, Jiři', "Information, Knowledge and Time," in Kubát , Libor and J. Zeman (eds.), *Entropy and Information*, Amsterdam, 1975.

R13 THEORY OF PLANNING, THE PRESCRIPTIVE SCIENCE AND COST-BENEFIT ANALYSIS IN TRASFORMATIONS

[R13.1] Alexander Ernest R., *Approaches to Planning*, Philadelphia, Pa. Gordon and Breach, 1992.

[R13.2] Bailey, J., *Social Theory for Planning*, London, Routledge and Kegan Paul, 1975.

[R13.3] Burchell R.W. and G. Sternlieb (eds.), *Planning Theory in the 1980's: A Search for Future Directions*, New Brunswick, N. J., Rutgers University Center for Urban and Policy Research, 1978.

[R13.4] Camhis, Marios, *Planning Theory and Philosophy*, London, Tavistock Publicationa, 1979.

[R13.5] Chadwick, G., *A Systems View of Planning*, Oxford, Pergamon, 1971.

[R13.6] Cooke, P., *Theories of Planning and Special Development*, London, Hutchinson, 1983.

[R13.7] Dompere, Kofi K., and Taresa Lawrence, "Planning," in Syed B Hussain, *Encyclopedia of Capitalism*, Vol. II, New York, Facts On File, Inc., 2004, pp.649-653.

[R13.8] Dompere, Kofi K., *Social Goal-Objective Formation, Democracy and National Interest: A Theory of Political Economy under Fuzzy Rationality*, (Studies in Systems, Decision and Control, Vol. 4), New York, Springer, 2014

[R13.9] Dompere, Kofi K., *Fuzziness, Democracy Control and Collective Decision-Choice System: A Theory on Political Economy of Rent-Seeking and Profit-Harvesting*,(Studies in Systems, Decision and Control, Vol. 5), New York, Springer, 2014.

[R13.10a] Dompere, Kofi K., *The Theory of Aggregate Investment in Closed Economic Systems*, Westport, CT, Greenwood Press, 1999.

[R13.10b] Dompere, Kofi K., *The Theory of Aggregate Investment and Output Dynamics in Open Economic Systems*, Westport, CT, Greenwood Press, 1999.

[R13.11a] Faludi, A., *Planning Theory*, Oxford, Pergamon, 1973.

[R13.11b] Faludi, A.(ed.), *A Reader in Planning Theory*, Oxford, Pergamon, 1973.

[R13.12] Harwood, E.C. (ed.), Reconstruction of Economics, American Institute For Economic Research, Great Barrington, Mass, 1955., Also in John Dewey and Arthur Bently, 'Knowing and the known', Boston, Beacon Press, 1949, p. 269.

[R13.13] Kickert, W.J.M., *Organization of Decision-Making A Systems-Theoretic Approach*, New York, North-Holland, 1980.

[R13.14] Knight, Frank H. *Risk, Uncertainty and Profit,* Chicago, University of Chicago Press, 1971.

[R13.15] Knight, Frank H. *On History and Method of Economics,* Chicago, University of Chicago Press, 1971.

R14. SOCIAL SCIENCES, MATHEMATICS AND THE PROBLEMS OF EXACT AND INEXACT METHODS OF THOUGHT

[R14.1] Ackoff, R.L., *Scientific Methods: Optimizing Applied Research Decisions*, New York, John Wiley, 1962.

[R14.2] Angyal, A. "The Structure of Wholes," *Philosophy of Sciences*, Vol.6, #1, 1939, pp 23-37.

[R14.3] Bahm, A.J., "Organicism: The Philosophy of Interdependence" *International Philosophical Quarterly*, Vol. VII # 2, 1967.

[R14.4] Bealer, George, *Quality and Concept*, Oxford, Clarendon Press, 1982.

[R14.5] Black, Max, *Critical Thinking*, Englewood Cliffs, N.J., Prentice-Hall, 1952.

[R14.6] Brewer, Marilynn B., and Barry E Collins (eds.) *Scientific Inquiry and Social Sciences,* San Francisco, Ca, Jossey-Bass Pub., 1981.

[R14.7] Campbell, D.T., "On the Conflicts Between Biological and Social Evolution and Between Psychology and Moral Tradition", *American Psychologist*, Vol. 30, 1975, pp1103-1126.

[R14.8] Churchman, C. W. and P. Ratoosh (eds.) *Measurement: Definitions and Theories*, New York, John Wiley, 1959.

[R14.9] Foley, Duncan, "Problems versus Conflicts Economic Theory and Ideology" American Economic Association Papers and Proceedings, Vol. 65, May 1975, pp. 231-237.

[R14.10] Garfinkel, Alan, *Forms of Explanation:Structures of Inquiry in Social Science*, New Haven, Conn., Yale University Press, 1981.

[R14.11] Georgescu-Roegen, Nicholas, *Analytical Economics*, Cambridge, Harvard University Press, 1967.

[R14.12] Gilolispie, C., *The Edge of Objectivity*, Princeton, Princeton University press, 1960.

[R14.13] Hayek, F.A., *The Counter-Revolution of Science*, New York, Free Press of Glencoe Inc, 1952.

[R14.14] Laudan, L., *Progress and Its Problems: Towards a Theory of Scientific Growth*, Berkeley, CA, University of California Press, 1961.

[R14.15] Marx, Karl, *The Poverty of Philosophy*, New York, International Pub. 1971.

[R14.16] Phillips, Denis C., *Holistic Thought in Social Sciences*, Stanford, CA, Stanford University Press, 1976.

[R14.17] Popper, K., *Objective Knowledge*, Oxford, Oxford University Press, 1972.

[R14.18] Rashevsky, N. "Organismic Sets: Outline of a General Theory of Biological and Social Organism," *General Systems*, Vol XII, 1967, pp. 21-28.

[R14.19] Roberts, Blaine, and Bob Holdren, *Theory of Social Process*, Ames, Iowa University Press, 1972.

[R14.20] Rudner, Richard S., *Philosophy of Social Sciences*, Englewood Cliff, N.J., Prentice Hall, 1966.

[R14.21] Simon, H. A., "The Structure of Ill-Structured Problems," *Artificial Intelligence*, Vol. 4, 1973, pp. 181-201.

[R14.22] Toulmin, S., *Foresight and understanding: An Enquiry into the Aims of Science*, New York, Harper and Row, 1961.

[R14.23] Winch, Peter, *The Idea of a Social Science*, New York, Humanities Press, 1958.

R15. TRANFORMATION, POLARITY, DIALECTICS AND CATEGORIAL CONVERSION

[R15.1] Anovsky, 0mely M.E., *Linin and Modern Natural Science, Moscow*, Progress Pub. 1978

[R15.2] Arrow, Kenneth J., "Limited Knowledge and Economic Analysis", American Economic Review, Vol. 64, 1974, pp. 1-10.

[R15.3] Berkeley, George, *Treatise Concerning the Principles of Human Knowledge*, Works, Vol. I (edited by A. Fraser), Oxford, Oxford University Press, 1871-1814.

[R15.4] Berkeley, George, "Material Things are Experiences of Men or God" in [R1.5], 1967, pp. 658-668.

[R15.5] Brody, Baruch A. (ed.), *Readings in the Philosophy of Science*, Englewood Cliffs, NJ., Prentice-Hall Inc., 1970.

[R15.6] Brouwer, L.E.J., "Consciousness, Philosophy, and Mathematics," in Benecerraf, P. and H. Putnam (eds.), *Philosophy of Mathematics: Selected Readings*, Cambridge, Cambridge University Press, 1983. pp90-96

[R15.7] Brown, B. and J Woods (eds.), *Logical Consequence; Rival Approaches and New Studies in exact Philosophy: Logic, Mathematics and Science*, Vol. II Oxford, Hermes, 2000.

[R15.8] Cornforth, Maurice, *Dialectical Materialism and Science*, New York, International Pub.1960

[R15.9] Cornforth, Maurice, *Materialism and Dialectical Method*, New York, International Pub.1953

[R15.10] Cornforth, Maurice, *Science and Idealism: an Examination of "Pure Empiricism"*, New York International Pub.1947

[R15.11] Cornforth, Maurice, *The Open Philosophy and the Open Society: A Reply to Dr. Karl Popper's Refutations of Marxism* New York, International Pub.1968.

[R15.12] Cornforth, Maurice, *The Theory of Knowledge*, New York, International Pub.1960

[R15.13] Dompere, Kofi K., "On Epistemology and Decision-Choice Rationality" in R. Trapple (ed.), *Cybernetics and System Research*, New York, North Holland, 1982, pp. 219-228.

[R15.14] Dompere, Kofi K. and M. Ejaz, *Epistemics of Development Economics: Toward a Methodological Critique and Unity*, Westport, CT, Greenwood Press, 1995.

[R15.15] Dompere, Kofi K., *The Theory of Categorial Conversion: Rational Foundations of Nkrumaism*, Working Monograph on Mathematics, Philosophy, Economic and Decision Theories, Washington, D.C., Howard University, 2013.

[R15.16] Dompere, Kofi K., Nkrumaism: Socio-Political Philosophy and Ideology, Working Monograph on Mathematics, Philosophy, Economic and Decision Theories, Washington, D.C., Howard University, 2013.

[R15.17] Dompere, Kofi K., *Polyrhythmicity: Foundations of African Philosophy*, London, Adonis and Abbey Pub, 2006.

[R15.18] Dompere, Kofi K., "On Epistemology and Decision-Choice Rationality, in R. Trappl (ed.), *Cybernetics and System Research,* New York, North-Holland, 1982, pp219-228.

[R15.19] Engels, Frederick, *Dialectics of Nature*, New York, International Pub., 1971

[R15.20] Engels, Frederick, *Origin of the Family, Private Property and State*, New York, International Pub., 1971.

[R15.21] Ewing, A.C., "A Reaffirmation of Dualism" in [R1.5], pp. 454-461.

[R15.22] Fedoseyer, P.N. *et al., Philosophy in USSR: Problems of Dialectical Materialism*, Moscow, Progress Pub., 1977

[R15.23] Kedrov, B. M., "On the Dialectics of Scientific Discovery," *Soviet Studies in Philosophy*, Vol. 6 1967, pp16 – 27.

[R15.24] Lenin, V. I. *Materialism and Empirio-Criticism: Critical Comments on Reactionary Philosophy*, New York, International Pub., 1970.

[R15.25] Lenin, V. I. *Collected Works Vol. 38: Philosophical Notebooks*, New York, International Pub., 1978

[R15.26] Lenin, V. I., *On the National Liberation Movement*, Peking, Foreign Language Press, 1960

[R15.27] Hegel, George, *Collected Works*, Berlin, Duncher und Humblot, 1832 – 1845 [also *Science of Logic*, translated by W. H. Johnston and L. G. Struther , London, 1951].

[R15.28] Hempel, Carl G. and P. Oppenheim, "Studies in the Logic of Explanation," in [R15.5], pp. 8 – 27.

[R15.29] Ilyenkov, E.V., *Dialectical Logic: Essays on its History and Theory*, Moscow, Progress Pub. 1977.

[R15.30] Keirstead, B.S., "The Conditions of Survival," American Economic Review,Vol. 40, #2, pp.435- 445.

[R15.31] Kühne, Karl, *Economics and Marxism, Vol.I: The Renaissance of the Marxian System*, New York, St Martin's Press, 1979.

[R15.32] Kühne, Karl, *Economics and Marxism, Vol.II: The Dynamics of the Marxian System*, New York, St Martin's Press, 1979.

[R15.33] March, J. C., "Bounded Rationality, Ambiguity and Engineering of Choice," *The Bell Journal of Economics*, Vol. 9 (2), 1978

[R15.34] Marx, Karl, *Contribution to the Critique of Political Economy*, Chicago, Charles H. Kerr and Co. 1904.

[R15.35] Marx, Karl, *Economic and Philosophic Manuscripts of 1884*, Moscow, Progress Pub., 1967.

[R15.36] Marx, Karl, *The Poverty of Philosophy*, New York, International Publishers, 1963.

[R15.37] Marx, Karl, Economic and Philosophic Manuscripts of 1844, Moscow, Progress Pub, 1967

[R15.38] Niebyl, Karl, H., "Modern Mathematics and Some Problems of Quantity, Quality and Motion in Economic Analysis," *Philosophy of Science*, Vol 7, # 1, January, 1940, pp. 103 – 120.

[R15.39] Price, H. H., *Thinking and Experience*, London, Hutchinson, 1953.

[R15.40] Putman, H., *Reason, Truth and History*, Cambridge, Cambridge University Press, 1981.

[R15.41] Putman. H., *Realism and Reason*, Cambridge, Cambridge University Press, 1983.

[R15.42] Robinson, Joan, *Economic Philosophy*. New York, Anchor Books, 1962.

[R15.43] Robinson, Joan, *Freedom and Necessity: An Introduction to the Study of Society*, New York, Vintage Books, 1971.

[R15.44] Robinson, Joan, *Economic Heresies: Some Old-Fashioned Questions in Economic Theory*, New York, Basic Books, 1973.

[R15.45] Rostow, W. W., *The Stages of Economic Growth*, Cambridge, Cambridge University Press, 1960

[R15.46] Rostow, W. W., *Theory of Economic Growth from David Hume to the Present*, New York Oxford University Press, 1990

[R15.47] Schumpeter, Joseph A., *The Theory of Economic Development*, Cambridge, Mass. Harvard University Press, 1934.

[R15.48] Schumpeter, Joseph A., *Capitalism, Socialism and Democracy*, New York, Harper & Row, 1950.

[R15.49]] Schumpeter, Joseph A., "March to Socialism," *American Economic Review*, Vol. 40 May 1950, pp. 446 456.

[R15.50] Schumpeter, Joseph A., "Theoretical Problems of Economic Growth" *Journal of Economic History* Vol. 8, Supplement 1947, pp. 1-9.

[R15.51] Schumpeter, Joseph A., "The Analysis of Economic Change," *Review of Economic Statistics*, Vol. 17, 1935, pp 2-10.

R16. VAGUENESS, APPROXIMATION AND REASONING IN THE KNOWLEDGE DEVELOPMENT AND CATEGORIAL CONVERSION

[R16.1] Adams, E. w., and H. F. Levine, "On the Uncertainties Transmitted from Premises to Conclusions in deductive Inferences," *Synthese* Vol. 30, 1975, pp. 429 – 460.

[R16.2] Arbib, M. A., *The Metaphorical Brain*, New York, McGraw-Hill, 1971.

[R16.3] Bečvář, Jiří , " Notes on Vagueness and Mathematics," in Skala, Heinz J. et al., (eds.), *Aspects of Vagueness*, Dordrecht, D. Reidel Co. 1984, pp.1-11.

[R16.4] Black, M, "Vagueness: An Exercise in Logical Analysis," *Philosophy of Science*, Vol. 17, 1970, pp141-164.

[R16.5] Black, M. "Reasoning with Loose Concepts," *Dialogue*, Vol. 2, 1973, pp. 1-12.

[R16.6] Black, Max, *Language and Philosophy*, Ithaca, N.Y.: Cornell University Press. 1949

[R16.7] Black, Max, *The Analysis of Rules*, in Black, Max [] *Models and Metaphors: Studies in Language and Philosophy*, 1962 pp. 95-139.

[R16.8] Black, *Max Models and Metaphors: Studies in Language and Philosophy*, Ithaca, NewYork: Cornell University Press. 1962.

[R16.9] Black, Max Margins *of Precision*, Ithaca: Cornell University Press, 1970.

[R16.10] Boolos, G. S. and R. C. Jeffrey, *Computability and Logic,* New York, Combridge University Press, 1989.

[R16.11] Cohen, P. R., *Heuristic Reasoning about uncertainty: An Artificial Intelligent Approach*, Boston, Pitman, 1985.

[R16.12] Darmstadter, H., "Better Theories," *Philosophy of Science*, Vol. 42, 1972, pp.20 – 27.

[R16.13] Davis, M., *Computability and Unsolvability*, New York, McGraw-Hill, 1958.

[R16.14] Dummett, M., "Wang's Paradox," *Synthese,* Vol. 30, 1975, pp301 – 324.

[R16.15] Dummett, M., *Truth and Other Enigmas*, Cambridge, Mass. Harvard University Press, 1978.

[R16.16] Endicott, Timothy, *Vagueness in the Law*, Oxford, Oxford University Press, 2000.

[R16.17] Evans, Gareth, "Can there be Vague Objects?," *Analysis,* Vol. 38, 1978, p. 208.

[R16.18] Fine, Kit, "Vagueness, Truth and Logic," *Synthese,* Vol.54, 1975, pp. 235-259.

[R16.19] Gale, S., "Inexactness, Fuzzy Sets and the Foundation of Behavioral Geography," *Geographical Analysis,* Vol.4, #4, 1972, pp.337-349.

[R16.20] Ginsberg, M. L. (ed.), *Readings in Non-monotonic Reason,* Los Altos, Ca., Morgan Kaufman, 1987.

[R16.21] Goguen, J. A., "The Logic of Inexact Concepts," *Synthese,* Vol. 19, 1968/69, pp. 325 – 373.

[R16.22] Grafe, W., "Differences in Individuation and Vagueness," in A. Hartkamper and H. –J. Schmidt, *Structure and Approximation in Physical Theories,* New York, Plenum Press, 1981. Pp.113– 122.

[R16.23] Goguen, J. A, "The Logic of Inexact Concepts" *Synthese,* Vol.19, 1968-1969.

[R16.24] Graff, Delia and Timothy (eds.), *Vagueness,* Aldershot, Ashgate Publishing, 2002.

[R16.25] A. Hartkämper and H.J. Schmidt (eds.), *Structure and Approximation in Physical Theories,* New York, Plenum Press, 1981.

[R16.26] Hersh, H.M. et. al., "A Fuzzy Set Approach to Modifiers and Vagueness in Natural Language," *J. Experimental,* Vol. 105, 1976, pp. 254-276.

[R16.27] Hilpinen, R., "Approximate Truth and Truthlikeness," in M. Pprelecki et al. eds.) *Formal Methods in the Methodology of Empirical Sciences,* Wroclaw, Reidel, Dordrecht and Ossolineum, 1976 pp. 19 – 42.

[R16.28] Hockney D. et al. (eds.), *Contemporary Research in Philosophical Logic and Linguistic Semantics,* Dordrecht-Holland, Reidel Pub. Co. 1975.

[R16.29] Höhle Ulrich et al (eds.), *Non-Clasical Logics and their Applications to Fuzzy Subsets: A Handbook of the Mathematical Foundations of Fuzzy Set Theory,* Boston, Mass. Kluwer, 1995.

[R16.30] Katz, M., "Inexact Geometry," *Notre-Dame Journal of Formal Logic,* Vol.21, 1980, pp. 521-535.

[R16.31] Katz, M., "Measures of Proximity and Dominance," *Proceedings of the Second World Conference on Mathematics at the Service of Man*, Universidad Politecnica de Las Palmas, 1982, pp. 370 – 377.

[R16.32] Katz, M., "The Logic of Approximation in Quantum Theory," *Journal of Philosophical Logic*, Vol. 11, 1982, pp. 215 – 228.

[R16.33] Keefe, Rosanna, *Theories of Vagueness*, Cambridge, Cambridge University Press, 2000.

[R16.34] Keefe, Rosanna and Peter Smith (eds.) *Vagueness: A Reader*, Cambridge, MIT Press, 1996.

[R16.35] Kling, R., "Fuzzy Planner: Reasoning with Inexact Concepts in a Procedural Problem-solving Language," *Jour. Cybernetics*, Vol. 3, 1973, pp. 1-16.

[R16.36] Kruse, R.E. et. al., *Uncertainty and Vagueness in Knowledge Based Systems: Numerical Methods*, New York, Springer-Verlag 1991.

[R16.37] Ludwig, G., "Imprecision in Physics," in A. Hartkämper and H.J. Schmidt (eds.), *Structure and Approximation in Physical Theories*, New York, Plenum Press, 1981, pp. 7 – 19.

[R16.38] Kullback, S. and R. A. Leibler, "Information and Sufficiency," *Annals of Math.Statistics*, Vol. 22, 1951, pp. 79 – 86.

[R16.39] Lakoff, George, "Hedges: A Study in Meaning Criteria and Logic of Fuzzy Concepts," in, Hockney D. et al. (eds.), *Contemporary Research in Philosophical Logic and Linguistic Semantics*, Dordrecht-Holland, Reidel Pub. Co. 1975, pp.221-271.

[R16.40] Lakoff, G., "Hedges: A Study in Meaning Criteria and the Logic of Fuzzy Concepts," *Jour. Philos. Logic*, Vol. 2, 1973, pp. 458-508.

[R16.41] Levi, I., *The Enterprise of Knowledge*, Cambridge, Mass. MIT Press 1980.

[R16.42] Łucasiewicz, J., *Selected Works: Studies in the Logical Foundations of Mathematics*, Amsterdam, North-Holland, 1970.

[R16.43] Machina, K.F., "Truth, Belief and Vagueness," *Jour. Philos. Logic*, Vol. 5, 1976, pp. 47-77.

[R16.44] Menges, G., et. al., "On the Problem of Vagueness in the Social Sciences," in Menges, G. (ed.), *Information, Inference and Decision*, D. Reidel Pub., Dordrecht, Holland, 1974, pp. 51-61.

[R16.45] Merricks, Trenton, "Varieties of Vagueness," *Philosophy and Phenomenological Research*, Vol.53, 2001, pp. 145-157.

[R16.46] Mycielski, J., "On the Axiom of Determinateness," *Fund. Mathematics*, Vol. 53, 1964, pp. 205 – 224.

[R16.47] Mycielski, J., "On the Axiom of Determinateness II," *Fund. Mathematics*, Vol. 59, 1966, pp. 203 – 212.

[R16.48] Naess, A., "Towards a Theory of Interpretation and Preciseness," in L. Linsky (ed.) *Semantics and the Philosophy of Language*, Urbana, Ill. Univ. of Illinois Press, 1951.

[R16.49] Narens, Louis, "The Theory of Belief," *Journal of Mathematical Psychology*, Vol. 49, 2003, pp1-31.

[R16.50] Narens, Louis, "A Theory of Belief for Scientific Refutations," *Synthese*, Vol.145, 2005, pp397-423.

[R16.51] Netto, A. B., "Fuzzy Classes," *Notices, Amar, Math. Society*, Vol. 68T- H28, 1968, pp.945.

[R16.52] Neurath, Otto et al. (eds.), *International Encyclopedia of Unified Science, Vol. 1 – 10,* Chicago, University of Chicago Press, 1955.

[R16.53] Niebyl, Karl, H., "Modern Mathematics and Some Problems of Quantity, Quality and Motion in Economic Analysis," *Science*, Vol 7, # 1, January, 1940, pp. 103 – 120.

[R16.54] Orlowska, E., "Representation of Vague Information," *Information Systems*, Vol. 13, #2, 1988, pp. 167-174.

[R16.55] Parrat, L. G., *Probability and Experimental Errors in Science*, New York, John Wiley and Sons, 1961.

[R16.56] Raffman. D., "Vagueness and Context-sensitivity," *Philosophical Studies*, Vol. 81, 1996, pp.175-192.

[R16.57] Reiss, S., *The Universe of Meaning*, New York, The Philosophical Library, 1953.

[R16.58] Russell, B., "Vagueness," *Australian Journal of Philosophy*, Vol.1, 1923 pp. 84-92.

[R16.59] Russell, B., *An Inquiry into Meaning and Truth*, New York, Norton, 1940.

[R16.60] Shapiro, Stewart, *Vagueness in Context*, Oxford, Oxford University Press, 2006.

[R16.61] Skala, H. J. "Modelling Vagueness," in M. M. Gupta and E. Sanchez, *Fuzzy Information and Decision Processes*, Amsterdam North-Holland, 1982, pp 101–109.

[R16.62] Skala, Heinz J. et al., (eds.), *Aspects of Vagueness*, Dordrecht, D. Reidel Co. 1984.

[R16.63] Sorensen, Roy, *Vagueness and Contradiction*, Oxford, Oxford University Press, 2001.

[R16.64] Tamburrini, G. and S. Termini, "Some Foundational Problems in Formalization of Vagueness," in M. M. Gupta et al (eds.), *Fuzzy Information and Decision Processes*, Amsterdam, North Holland, 1982, pp. 161-166.

[R16.65] Termini, S. "Aspects of Vagueness and Some Epistemological Problems Related to their Formalization," in Skala, Heinz J. et al., (eds.), *Aspects of Vagueness*, Dordrecht, D. Reidel Co. 1984, pp.205 – 230.

[R16.66] Tikhonov, Andrey N. and Vasily Y. Arsenin., *Solutions of Ill-Posed Problems*, New York, John Wiley and Sons, 1977.

[R16.67] Tversky, A. and D. Kahneman, "Judgments under Uncertainty: Heuristics and Biases," *Science*, Vil 185 September 1974, pp. 1124-1131.

[R16.68] Ursul, A. D., "The Problem of the Objectivity of Information," in Kuba't, Libor and J. Zeman (eds.), *Entropy and Information*, Amsterdam, Elsevier, 1975.pp. 187-230.

[R16.69] Vardi, M. (ed.), *Proceedings of Second Conference on Theoretical Aspects of Reasoning about Knowledge*, Asiloman, Ca, Los Altos, Ca, Morgan Kaufman, 1988.

[R16.70] Verma, R.R., "Vagueness and the Principle of the Excluded Middle," *Mind*, Vol. 79, 1970, pp. 66-77.

[R16.71] Vetrov, A. A., "Mathematical Logic and Modern Formal Logic," *Soviet Studies in Philosophy*, Vol. 3, #1, 1964 pp. 24 – 33.

[R16.72] von Mises, Richard, *Probability, Statistics and Truth*, New York, Dover Pub. 1981.

[R16.73] Williamson, Timothy, *Vagueness*, London, Routledge, 1994.

[R16.74] Wiredu, J.E., "Truth as a Logical Constant With an Application to the Principle of the Excluded Millde," *Philos. Quart.*, Vol. 25, 1975, pp. 305-317.

[R16.75] Wright, C., "On Coherence of Vague Predicates," *Synthese*, Vol. 30, 1975. pp. 325 – 365.

[R16.76] Wright, Crispin, "The Epistemic Conception of Vagueness," *Southern Journal of Philosophy*, Vol. 33, Supplement, 1995, pp. 133-159.

[R16.77] Zadeh, L. A., A Theory of Commonsense Knowledge," in Skala, Heinz J. et al., (eds.), *Aspects of Vagueness*, Dordrecht, D. Reidel Co. 1984., pp 257 – 295.

[R16.78] Zadeh, L. A., "The Concept of Linguistic Variable and its Application to Approximate reasoning," *Information Science*, Vol. 8, 1975, pp. 199 – 249 (Also in Vol. 9, pp. 40 – 80.

R17. VAGUENESS AND FUZZY GAME THEORY IN CATEGOTIAL CONVERSION AND PHILOSOPHICAL CONSCIENCISM

[R17.1] Aubin, J.P. " Cooperative Fuzzy Games", Mathematics of Operations Research, Vol.6, 1981, pp1 – 13.

[R17.2] Aubin, J.P. Mathematical Methods of Game and Economics Theory, New York, North Holland.

[R17.3] Butnaria, D., "Fuzzy Games: A description pf the concepts," Fuzzy sets and systems Vol.1, 1978, pp. 181 – 192.

[R17.4] Butnaria, D., "Stability and shapely value for a n – persons Fuzzy Games," Fuzzy sets and systems, Vol. 4, #1, 1980, pp 63 – 72.

[R17.5] Nurmi, H.., "A Fuzzy Solution to a Majority Voting Game, "Fuzzy sets and systems, Vol.5, 1981 pp187-198.

[R17.6] Regade., R. K., " Fuzzy Games in the Analysis of Options," jour. Of Cybernetics, Vol. 6, 1976, pp 213 – 221.

[R17.7] Spillman, B. et al., "Coalition Analysis with Fuzzy Sets," Kybernetes, Vol. 8, 1979, pp. 203-211.

[R17.8] Wernerfelt, B., "Semifuzzy Games" Fuzzy sets and systems, Vol. 19, 1986, pp 21 – 28.

Index

A

455